AF581730

Modern Applications in Optics and Photonics

Modern Applications in Optics and Photonics: From Sensing and Analytics to Communication

Editors

Lourdes S. M. Alwis
Kort Bremer
Bernhard Wilhelm Roth

MDPI • Basel • Beijing • Wuhan • Barcelona • Belgrade • Manchester • Tokyo • Cluj • Tianjin

Editors
Lourdes S. M. Alwis
Edinburgh Napier University
UK

Kort Bremer
Gottfried Wilhelm Leibniz
Universität
Germany

Bernhard Wilhelm Roth
Gottfried Wilhelm Leibniz
Universität
Germany

Editorial Office
MDPI
St. Alban-Anlage 66
4052 Basel, Switzerland

This is a reprint of articles from the Special Issue published online in the open access journal *Applied Sciences* (ISSN 2076-3417) (available at: https://www.mdpi.com/journal/applsci/special_issues/Appl_Opt_Photonics).

For citation purposes, cite each article independently as indicated on the article page online and as indicated below:

LastName, A.A.; LastName, B.B.; LastName, C.C. Article Title. *Journal Name* **Year**, *Volume Number*, Page Range.

ISBN 978-3-0365-0866-5 (Hbk)
ISBN 978-3-0365-0867-2 (PDF)

Contents

About the Editors

Lourdes S. M. Alwis obtained a First Class (Hons) degree in Electrical, Electronic and Information Engineering from City University London, U.K., in 2005. She completed her Ph.D. thesis in the field of grating-based fiber optic sensors in 2013 at the same establishment, while working for R&D at Alcatel-Lucent Ltd., a company specializing in design, implementation, building, and testing of optical fiber telecommunications products. Completing her PhD and industrial experience in 2013, she joined Edinburgh Napier University, U.K., as a Lecturer in Electronic Engineering. Her current research focuses on avenues where optical fiber sensors can be utilized, such as civil infrastructure monitoring, chemical and biomedical sensing, and wearable technology.

Kort Bremer is a scientific staff member at the Hannover Centre for Optical Technologies (HOT) at Leibniz University Hannover, Germany. He graduated from the Hochschule Wismar, Germany, in Electrical Engineering, and received a Ph.D. from the University of Limerick, Ireland. His research interests include optical sensors and optical communication; for instance, fiber optic pressure and temperature sensors, fiber optic gas sensors, reinforcement structures which are functionalized with fiber optic sensors for structural heath monitoring applications, mode-selective fiber couplers for spatial division multiplexing, as well as optical biosensors.

Bernhard Wilhelm Roth received a Ph.D. degree in Atomic and Particle Physics from the University of Bielefeld, Germany, in 2001, and a State Doctorate (Habilitation) degree in Experimental Quantum Optics from the University of Duesseldorf, Germany, in 2007. From 2002 to 2007, he was a Research Group Leader, and from 2007 to 2010, he was an Associate Professor with the University of Duesseldorf. From 2011 to 2012, he was Center Manager at the Centre for Innovation Competence innoFSPEC Potsdam of the Leibniz Institute for Astrophysics Potsdam (AIP) and the University Potsdam. Since 2012, he has been the Scientific and Managing Director with the Hanover Centre for Optical Technologies, and a Professor of Physics at Leibniz University Hannover, Germany, since 2014. His scientific activities include applied and fundamental research in laser development and spectroscopy, polymer and fiber optical sensing, micro- and nano-optics fabrication, and optical technologies for illumination, information technology, the life sciences, and medicine.

Editorial

Editorial on Special Issue "Modern Applications in Optics and Photonics: From Sensing and Analytics to Communication"

Lourdes S. M. Alwis [1], Kort Bremer [2] and Bernhard Roth [2,*]

1 School of Engineering and the Build Environment, Edinburgh Napier University, Edinburgh EH10 5DT, UK; L.Alwis@napier.ac.uk
2 Hannover Centre for Optical Technologies (HOT), Gottfried Wilhelm Leibniz Universität, 30167 Hannover, Germany; Kort.Bremer@hot.uni-hannover.de
* Correspondence: Bernhard.Roth@hot.uni-hannover.de; Tel.: +49-511-7621-7907

Keywords: fiber optical sensing; biosensing; optofluidics; integrated optics and photoncis; optical analytics; medical imaging and diagnostics; optical communication technology; distributed sensing

Optics and photonics are among the key technologies of the 21st century and offer the potential for novel applications in areas as diverse as sensing and spectroscopy, analytics, monitoring, biomedical imaging and diagnostics, as well as optical communication technology, among others. The high degree of control over optical fields that is possible today, for example, by using micro- and nano-optics together with the tremendous capabilities of modern processing and integration technology, enables new optical measurement systems with enhanced functionality and unprecedented sensitivity. Such systems are thus attractive for a wide range of applications that have been previously inaccessible and may ultimately lead to the democratization of optics and photonics. This Special Issue aims to provide an overview on some of the most advanced application areas in optics and photonics and indicate the broad potential for the future. The Special Issue contains 15 papers which all underwent substantial peer-review under the guidelines of the MDPI *Applied Sciences* journal. The fifteen papers are divided into five categories: Fiber Optic Sensing, Optical Communication, Distributed Sensing, Optical Imaging and Laser Technology.

Citation: Alwis, L.S.M.; Bremer, K.; Roth, B. Editorial on Special Issue "Modern Applications in Optics and Photonics: From Sensing and Analytics to Communication". *Appl. Sci.* **2021**, *11*, 1589. https://doi.org/10.3390/app11041589

Received: 28 January 2021
Accepted: 29 January 2021
Published: 10 February 2021

1. Fiber Optical Sensing

In the area of fiber optic sensing, five papers are presented in this Special Issue. For example, Bremer et al. [1] reported an investigation on the feasibility of utilizing Mode Division Multiplexing (MDM) for simultaneous measurement of Surrounding Refractive Index (SRI) and temperature using a single sensor element based on an etched OM4 Graded Index Multi Mode Fiber (GI-MMF) with an integrated fiber Bragg Grating (BG). In another research article by Bremer et al. [2], the durability of functionalized carbon structures (FCS) that are equipped with fiber optic sensors in a highly alkaline concrete environment are investigated. In their investigation, the suitability of optical fibers with different coatings as well as different integration techniques for the FCS were analyzed. In terms of fiber optic strain sensing, a new concept is presented by Mądry et al. [3], where, an intensity-modulated Sagnac loop sensor based on Polarization-Maintaining Photonic Crystal Fiber (PM-PCF) in a setup with a Dense Wavelength Division Multiplexer (DWDM) for strain measurement, is presented. The proposed setup uses an optical power measurement scheme, i.e., as opposed to measurement of wavelength, and thus would be relatively cheaper when compared to the utilization of complex optical spectrum analyzers for the target application. Moreover, two additional research papers presented by Xie et al. [4] and Xue et al. [5] have also been categorized under the rubric "fiber optic sensing". For instance, Xie et al. [4] reported the application of an integrated optical electric-field sensor on the measurements of transient voltages in AC high-voltage power grids. They developed an integrated optical electric-field sensor based on the Pockels effect to measure the transient voltages of high-voltage conductors and achieved a response speed faster than 6 ns and a wide bandwidth ranging from 5 Hz to 100 MHz [4]. In contrast, Xue et al. [5] present an electro-optic dual-comb Doppler velocimeter for high-accuracy velocity measurement by generating two optical combs using electro-optic phase modulators and tracing their repetition frequencies to a rubidium clock and demonstrating experimentally a high accuracy in the range of 100–300 mm/s with a maximum deviation of 0.44 mm/s.

2. Optical Communications

The category of optical communications is represented by a research article from Kirrbach et al. [6] and a review article from Lallas [7]. The research article from Kirrbach et al. [6] investigates the use of Optical Wireless Communications (OWC) for on-axis rotary communication scenarios by discussing different realization approaches for bi-directional full-duplex links as well as designing a monolithic hybrid transmitter–receiver lens and studying its performance using ray trancing simulations. The article by Lallas [7] reviews current and future state-of-the-art plasmonic system implementations for THz communications.

3. Distributed Sensing

In terms of distributed sensing, one paper is presented by Wiesmeyr et al. [8]. In their research article, the real-time train tracking from distributed acoustic sensing data is reported by presenting an algorithm that extracts the positions of moving trains for a given point in time from Distributed Acoustic Sensing (DAS) signals.

4. Optical Imaging

In the area of optical imaging, five papers are included in this Special Issue. For example, Kerrouche et al. [9] report the development of rapid and low-cost pathogen detection systems using microfluidic technology and optical image processing by employing a cost-effective microscopic camera and computational algorithms and detecting small size microbeads (1–5 µm) from a measured water sample. In contrast, Dong et al. [10] propose a method based on dependence analysis to identify and then eliminate the measurement configurations with redundant information in optical scatterometry for fast nanostructure reconstruction. In terms of Optical Coherence Tomography (OCT), Yi et al. [11] report a mesh-based Monte Carlo model in order to study OCT signals reflecting the structural and functional activities of brain tissue as well as to improve the quantitative accuracy of chromophores in tissue. Furthermore, Wang et al. [12] evaluate the performance of different closed path determination methods in order to measure the topological charge (TC) of an optical vortex (OV) beam and Fricke et al. [13] present a non-contact dermatoscope with ultra-bright light source and liquid lens-based autofocus function. Moreover, Fricke et al. [13] could demonstrate, i.e., with their prototype, feature resolution of up to 30 µm and feature size scaling fulfilling the requirements to apply the device in regular skin cancer screening.

5. Laser Technology

Furthermore, two research papers of this Special Issue can be categorized as relating to laser technology. For instance, Čehovski et al. [14] report on the integration of organic thin-film lasers directly into polymeric single-mode ridge waveguides forming a monolithic laser device and obtaining single-mode characteristics even with high pump energy densities and thus demonstrating its suitability for lab-on-a-chip (LoC) applications. In another research paper by Huang et al. [15], a sub-nanosecond Nd:YVO4 laser system at 1 kHz repetition rate without Stimulated Raman Scattering (SRS) with high peak power and high beam quality is reported with a maximum output energy of 65.4 mJ and a pulse duration of 600 ps which corresponds to a pulse peak power of 109 MW. Huang et al. could also achieve 532 nm green light with an average power of 40.5 W and a power stability of 0.28% by frequency doubling with an LBO crystal.

Acknowledgments: This Special Issue benefited from the outstanding coordination efforts by Wing Wang from MDPI.

Conflicts of Interest: The authors declare no conflict of interest.

References

1. Bremer, K.; Alwis, L.S.M.; Zheng, Y.; Roth, B.W. Towards Mode-Multiplexed Fiber Sensors: An Investigation on the Spectral Response of Etched Graded Index OM4 Multi-Mode Fiber with Bragg grating for Refractive Index and Temperature Measurement. *Appl. Sci.* **2020**, *10*, 337. [CrossRef]
2. Bremer, K.; Alwis, L.S.M.; Zheng, Y.; Weigand, F.; Kuhne, M.; Helbig, R.; Roth, B. Durability of Functionalized Carbon Structures with Optical Fiber Sensors in a Highly Alkaline Concrete Environment. *Appl. Sci.* **2019**, *9*, 2476. [CrossRef]
3. Mądry, M.; Alwis, L.; Bereś-Pawlik, E. Intensity-Modulated PM-PCF Sagnac Loop in a DWDM Setup for Strain Measurement. *Appl. Sci.* **2019**, *9*, 2374. [CrossRef]
4. Xie, S.; Zhang, Y.; Yang, H.; Yu, H.; Mu, Z.; Zhang, C.; Cao, S.; Chang, X.; Hua, R. Application of Integrated Optical Electric-Field Sensor on the Measurements of Transient Voltages in AC High-Voltage Power Grids. *Appl. Sci.* **2019**, *9*, 1951. [CrossRef]
5. Xue, B.; Zhang, H.; Zhao, T.; Jing, H. A Traceable High-Accuracy Velocity Measurement by Electro-Optic Dual-Comb Interferometry. *Appl. Sci.* **2019**, *9*, 4118. [CrossRef]
6. Kirrbach, R.; Faulwaßer, M.; Schneider, T.; Meißner, P.; Noack, A.; Deicke, F. Monolitic Hybrid Transmitter-Receiver Lens for Rotary On-Axis Communications. *Appl. Sci.* **2020**, *10*, 1540. [CrossRef]
7. Lallas, E. Key Roles of Plasmonics in Wireless THz Nanocommunications—A Survey. *Appl. Sci.* **2019**, *9*, 5488. [CrossRef]
8. Wiesmeyr, C.; Litzenberger, M.; Waser, M.; Papp, A.; Garn, H.; Neunteufel, G.; Döller, H. Real-Time Train Tracking from Distributed Acoustic Sensing Data. *Appl. Sci.* **2020**, *10*, 448. [CrossRef]
9. Kerrouche, A.; Lithgow, J.; Muhammad, I.; Romdhani, I. Towards the Development of Rapid and Low-Cost Pathogen Detection Systems Using Microfluidic Technology and Optical Image Processing. *Appl. Sci.* **2020**, *10*, 2527. [CrossRef]
10. Dong, Z.; Chen, X.; Wang, X.; Shi, Y.; Jiang, H.; Liu, S. Dependence-Analysis-Based Data-Refinement in Optical Scatterometry for Fast Nanostructure Reconstruction. *Appl. Sci.* **2019**, *9*, 4091. [CrossRef]
11. Yi, L.; Sun, L.; Zou, M.; Hou, B. A Mesh-Based Monte Carlo Study for Investigating Structural and Functional Imaging of Brain Tissue Using Optical Coherence Tomography. *Appl. Sci.* **2019**, *9*, 4008. [CrossRef]
12. Wang, D.; Huang, H.; Toyoda, H.; Liu, H. Topological Charge Detection Using Generalized Contour-Sum Method from Distorted Donut-Shaped Optical Vortex Beams: Experimental Comparison of Closed Path Determination Methods. *Appl. Sci.* **2019**, *9*, 3956. [CrossRef]
13. Fricke, D.; Denker, E.; Heratizadeh, A.; Werfel, T.; Wollweber, M.; Roth, B. Non-Contact Dermatoscope with Ultra-Bright Light Source and Liquid Lens-Based Autofocus Function. *Appl. Sci.* **2019**, *9*, 2177. [CrossRef]
14. Čehovski, M.; Becker, J.; Charfi, O.; Johannes, H.-H.; Müller, C.; Kowalsky, W. Single-Mode Polymer Ridge Waveguide Integration of Organic Thin-Film Laser. *Appl. Sci.* **2020**, *10*, 2805. [CrossRef]
15. Huang, Y.; Zhang, H.; Yan, X.; Kang, Z.; Lian, F.; Fan, Z. A High Peak Power and High Beam Quality Sub-Nanosecond Nd:YVO_4 Laser System at 1 kHz Repetition Rate without SRS Process. *Appl. Sci.* **2019**, *9*, 5247. [CrossRef]

Article

Towards Mode-Multiplexed Fiber Sensors: An Investigation on the Spectral Response of Etched Graded Index OM4 Multi-Mode Fiber with Bragg grating for Refractive Index and Temperature Measurement

Kort Bremer [1,*], Lourdes Shanika Malindi Alwis [2], Yulong Zheng [1] and Bernhard Wilhelm Roth [1,3]

1 Hannover Centre for Optical Technologies (HOT), Leibniz University Hannover, 30167 Hannover, Germany; Zheng.Yulong@slm-solutions.com (Y.Z.); bernhard.roth@hot.uni-hannover.de (B.W.R.)

2 School of Engineering and the Built Environment, Edinburgh Napier University, Edinburgh EH10 5DT, UK; l.alwis@napier.ac.uk

3 Cluster of Excellence PhoenixD, (Photonics, Optics, and Engineering–Innovation Across Disciplines), 30167 Hannover, Germany

* Correspondence: Kort.Bremer@hot.uni-hannover.de; Tel.: +49-511-762-17905

Received: 18 November 2019; Accepted: 28 December 2019; Published: 2 January 2020

Abstract: An investigation on the feasibility of utilizing Mode Division Multiplexing (MDM) for simultaneous measurement of Surrounding Refractive Index (SRI) and temperature using a single sensor element based on an etched OM4 Graded Index Multi Mode Fiber (GI-MMF) with an integrated fiber Bragg Grating (BG), is presented. The proposed work is focused on the concept of principle mode groups (PMGs) generated by the OM4 GI-MMF whose response to SRI and temperature would be different and thus discrimination of the said two parameters can be achieved simultaneously via a single sensor element. Results indicate that the response of all PMGs to temperature to be equal, i.e., 11.4 pm/°C, while the response to SRI depends on each PMG. Thus, it is evident that temperature "de-coupled" SRI measurement can be achieved by deducing the temperature effects experienced by the sensor element. Sensitivity of the PMGs to applied SRI varied from 3.04 nm/RIU to of 0.22 nm/RIU from the highest to lowest PMG, respectively. The results verify that it is feasible to obtain dual measurement of SRI and temperature simultaneously using the same, i.e., single, sensing element.

Keywords: Mode Division Multiplexing (MDM); Few-Mode Fiber (FMF); principle mode groups (PMG); Bragg grating (BG); multi-mode fiber bragg grating; fiber optic sensor (FOS); multi-parameter sensing

1. Introduction

Despite the progress that had been already achieved on Mode Division Multiplexing (MDM) in the fiber optic communications sector, Fiber Optic Sensor (FOS) applications utilizing MDM is relatively unexplored. In this publication, the authors investigate the application of MDM that would allow the development of new sensor concepts. For instance, one FOS approach based on MDM could be Multi-Mode Fibers (MMF) incorporated with Bragg grating (BG) structures. Since every mode of the MMF has a different propagation constant, the inscription of a BG structure into the MMF would cause individual Bragg reflections for every mode. When combining a mode-selective excitation, such as grating assisted mode multiplexer [1,2] or Spatial Light Modulators (SLMs) [3], and a spectral interrogator, several BG structures could be wavelength multiplexed along the same optical fiber and the additional mode-multiplexing would allow the discrimination between the different Bragg reflection of each BG structure. This, for instance, could allow the simultaneous measurement of the Surrounding Refractive Index (SRI) and temperature using only one BG structure, since different modes depends differently on the SRI.

Recent work by Yang et al. [4] demonstrated the capability of this concept by showing the simultaneous measurement of refractive index (RI) and temperature using an etched Few-Mode Fiber (FMF) with BG, termed FMFBG. The main mechanism of FMF follows that although BGs are insensitive to the SRI, when the cladding is sufficiently reduced, the evanescent field could interact with the surrounding media and the effective RI of the core would depend on the RI of the surrounding media. This has a direct impact on the resonance wavelength of the grating and thus the sensor is then sensitive to the SRI. Due to the different mode field distribution in the core and cladding region (and thus the effective refractive indices, n_{eff}, of the modes depend differently on the SRI), the Bragg wavelength of the *LP01* and *LP11* modes depend differently on applied SRI and temperature (the temperature sensitivity of the modes depends on thermally induced refractive index change and the thermal expansion coefficient [5]) and this feature allows simultaneous detection of these two parameters using only a single sensor element.

Compared to step-index MMF, Graded-Index (GI)-MMF have the unique property that some of the modes propagating inside the fiber have almost identical propagation constants, which causes the formation of principle mode groups (PMGs) [3]. This unique feature leads to the grouping of all modes propagating inside the GI-MMF into only a few PMGs, where each PMG is characterized by a single mode propagation constant. Consequently, when a BG is inscribed in a GI-MMF the corresponding transmission and reflection spectrum consists of only several dips and peaks, respectively [6]. In addition, lower order PMGs contain modes which are confined more in the center of the core whereas higher order PMGs contain modes that are confined at the core/cladding interface [7]. Therefore, when the cladding material of the GI-MMF is removed, higher order PMGs are thus more sensitive to RI changes of the surrounding compared to lower order PMGs [7] and thus this kind of fibers could form the basis for mode-multiplexed FOS concepts.

The work presented herein investigates whether GI-MMFs can be applied for mode-multiplexing in FOS applications and the simultaneous measurement of SRI and temperature. A schematic of the proposed fiber optic sensor-head is shown in Figure 1. By etching the fiber optic sensor-head the higher order PMG become sensitive to applied SRI variation and by inscribing the BG the higher order and lower order PMG can be discriminated at the optical detector by observing the individual Bragg reflection peaks. The advantage of this approach is that only a single sensor element would be required for the simultaneous detection of these two parameters and due to the relatively large core diameter of 50 µm (in case of the OM4 GI-MMF) the sensor elements remains stable even after etching compared to FMF (core etched down to 17 µm [8]) and SMF (core etched down to 3.4 µm [9]) approaches and thus more applicable to distributed FOSs in the field.

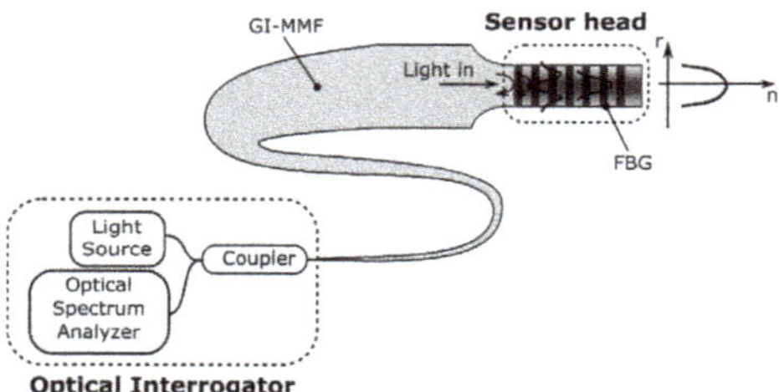

Figure 1. Schematic of the experimental setup for the etched OM4 Graded-Index Multi Mode Fiber (GI-MMF) with Bragg Grating (BG) for simultaneous measurement of Surrounding Refractive Index (SRI) and temperature using Principle Mode Group (PMG) multiplexing. The inset figure visualizes the general refractive index (n) distribution of a GI-MMF as a function of the fiber core radius (r).

2. Materials and Methods

2.1. Theoretical Background

When inscribing a BG into the core of GI-MMF the PMG formation causes that the Bragg condition is satisfied for modes of the same PMG or modes of neighboring PMGs [6]. The Bragg condition is given by:

$$\beta_1 - \beta_2 = 2\pi/\Lambda \tag{1}$$

where β_1 and β_2 are the propagation constants of the forward and backward propagating modes to be coupled and Λ is the grating period. In case of the coupling between modes of the same PMG, the propagation constants are equal $\beta = \beta_1 = -\beta_2 = 2{\cdot}\pi{\cdot}n_{eff}/\lambda$, where n_{eff} is the effective RI of the modes to be coupled, and from this the reflected Bragg wavelength for the coupled modes is calculated as [6]:

$$\lambda_B = 2n_{eff}\Lambda. \tag{2}$$

However, in case of the mode coupling between modes of neighboring PMGs the propagation constants are unequal with $\beta_1 = 2{\cdot}\pi{\cdot}n_{eff1}/\lambda$ and $\beta_2 = 2{\cdot}\pi{\cdot}n_{eff2}/\lambda$ and thus the corresponding reflected Bragg wavelength can be determined as follows [6]:

$$\lambda_B = \left(n_{eff1} + n_{eff2}\right)\Lambda \tag{3}$$

where n_{eff1} and n_{eff2} are the effective refractive indices of the coupled modes of the neighboring PMGs. From this it follows that the spectrum of a GI-MMF comprising a BG contains several peaks in reflection mode and several dips in transmission mode, where each peak and dip represents the mode coupling within the PMG of the same order and neighboring PMGs, respectively [6]. Consequently, lower order and higher order PMGs can be distinguished relatively easily by monitoring the reflection and/or transmission spectrum of the GI-MMF with BG, respectively, and observing the corresponding Bragg wavelength [6].

Furthermore, in Equations (2) and (3) the effective refractive indices depend on the field distribution of the modes to be coupled and thus this parameter can be sensitive to SRI, depending on whether a fraction of the electrical field of the mode is propagating inside the cladding or in the surrounding region.

2.2. Fabrication of GI-MMF FBGs

The GI-MMF BGs were fabricated on common OM4 GI-MMF from Corning©, using a KrF excimer laser (ATL Laser, UV inscription at 248 nm) and the phase mask technique. The OM4 GI-MMF was acrylate coated and had core and cladding diameters of 50 μm and 125 μm, respectively. To enable the inscription of BGs in the Germanium (Ge)-doped silica core, the OM4 GI-MMF was hydrogen loaded at a pressure of 200 bar for two weeks at room temperature (23 °C). Immediately after the hydrogen loading, the acrylate fiber coating was removed and 300 mm long fiber samples were exposed to the UV light from the KrF excimer laser with a repetition rate of 15 Hz, a pulse energy of 5 mJ and an inscription time of 2 min. The BG inscription length was 7 mm and the period of the applied phase mask was 1070 nm. The transmission spectrum of the inscribed BG was carefully monitored using a broadband light source (Thorlabs SLS201L/M) and an optical spectrum analyzer (Ando AQ6317B). An example of a transmission spectrum of a BG inscribed in the hydrogen loaded OM4 GI-MMF is shown in Figure 2. In total, 21 absorption dips of the fiber BG transmission spectrum were measured, where each absorption dip represents the Bragg reflection of an individual mode group and adjacent mode groups at the grating structure inside the GI-MMF [6] and the wavelength of the measured Bragg reflection peak decreases with increasing order of the mode groups, i.e., Bragg reflections peaks of the higher order mode groups are in the blue wavelength region.

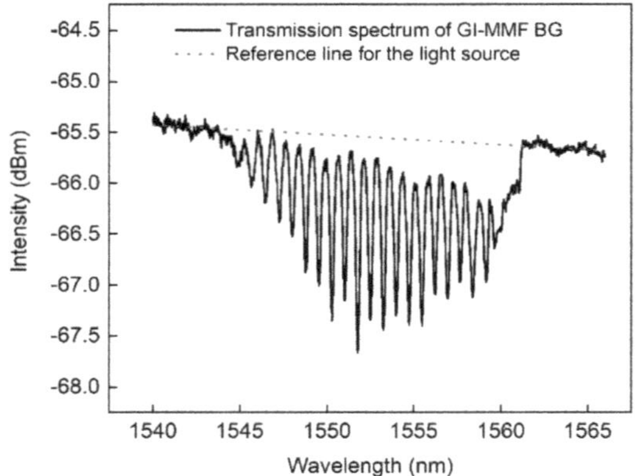

Figure 2. Measured transmission spectrum at a temperature of 20 °C of a BG written in OM4 GI-MMF using the phase mask technique. The red dashed line shows the reference intensity level of the applied light source.

Following the inscription of the BG, the cladding of the modified GI-MMF was removed over a length of 55 mm using hydrogen fluoride (HF) acid using the etching technique described elsewhere [2]. The final diameter of the etched fiber part was approximately 43 µm. In addition, after the etching process was complete, one end of the fiber containing the BG was spliced to the OM4 GI-MMF pigtail and the other end was cut close to the spatial position of the BG in order to obtain a single ended sensor probe, i.e., for RI and temperature measurement.

2.3. Experimental Set-Up

The experimental set-up shown in Figure 1 was developed to evaluate the response of the modified GI-MMF with BG to SRI and temperature variations. The sensor element was interrogated using a broadband light source (Opto-Link C-Band ASE) and an optical spectrum analyzer (Ando AQ6317B). In order to receive the reflection spectrum from the GI-MMF BG, a MMF 1 × 2 (all4fiber) coupler was applied. Since the output of the broadband light source is fiber coupled (SM, 9/125 µm core/cladding diameter), it was connected to a 2000 m OM4 GI-MMF coil where at the beginning of the coil the OM4 fiber was periodically bent with a bending radius of 5 mm and 30 turns in order to obtain equal excitation of the modes propagating in the GI-MMF (equilibrium mode distribution). A fiber length of 2000 m and a bend radius of 5 mm were chosen in order to prevent modal interferences as well as to obtain a relatively high fiber bending and enabling modal cross-talk without breaking the fiber.

In order to alter the SRI, different RI liquids (series A and AA) from Cargille Labs (n = 1.40, n = 1.42, n = 1.43, n = 1.45 and n = 1.47) and deionized water (n = 1.33) were applied. Cargille Labs specifies the refractive index of the liquids for an optical wavelength of 589.3 nm and a temperature of 25 °C. After each RI measurement, the sensor element was cleaned subsequently with acetone, ethanol and deionized water. The temperature response was measured by inserting the fiber sensor into a rectangular aluminum bar (100 × 20 × 20 mm^3) with an 8 mm diameter and 70 mm long borehole. The temperature of the aluminum bar was controlled using electrical resistors which were used as heating elements that were connected to a temperature controller (Quick-Cool QC-PC-CC-12). The reflection spectrum of each measurement was observed and saved using Labview and the measured sensor response was analyzed using Matlab. The wavelength shifts of the individual Bragg reflection peaks were obtained by determining the "center of mass" of each peak. Moreover, the Bragg reflection peaks in the wavelength region from 1549–1558 nm were analyzed and presented since reflection peaks in this wavelength range were consistent for all measurements.

3. Results

3.1. Response to Refractive Index Variations

The reflection spectrum of the etched GI-MMF BG at 20 °C for different SRI values is illustrated in Figure 3.

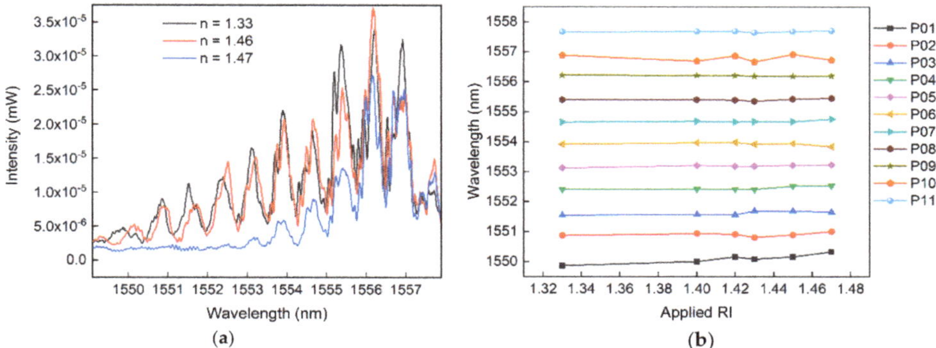

Figure 3. Reflection spectrum of etched GI-MMF with BG for different SRI values (**a**) and the response of the individual Bragg reflection peaks to different applied SRI values (**b**). *P01–P11* indicate the Bragg reflection peaks in the reflection spectrum from left to right, i.e., the Bragg reflection peak of the highest PMG is marked with *P01*, whereas the Bragg reflection peak of the lowest PMG is marked with *P11*.

Since the diameter of the GI-MMF containing the BG was only 43 µm after the etching procedure, the measured reflection spectrum illustrated in Figure 3a contains a lower number of reflection peaks in the left-hand side (blue wavelength range) compared to the measured transmission spectrum shown in Figure 2, i.e., no Bragg reflection peaks below 1549 nm were measured after etching the optical fiber. In addition, at an SRI of $n = 1.47$ the reflection amplitudes of the higher order PMGs decrease, as depicted in Figure 3a. This can only be explained by modes that are confined at the core/cladding interface, i.e., when the etched fiber sensor-head is immersed into a solution with a higher RI compared to the fiber core, these modes of the higher order PMGs are coupled out of the optical glass fiber which thus results in power loss. Therefore, only modes of higher order PMGs are able to interact with the environment and thus are capable to shift the Bragg reflection wavelength depending on the RI of the surrounding. Moreover, as depicted in Figure 3a the envelope of the measured reflection spectra of the sensor varies with different SRI and thus makes the exact determination of the Bragg reflection peak wavelength for each SRI value difficult (discussed in Section 4). In Figure 3b also the response of the fabricated sensor element to RI variations is illustrated. In total of 10 measurements were performed for each refractive index solution and only the obtained mean values are depicted in Figure 3b. The corresponding sensitivity of the individual Bragg reflection peaks are summarized in Table 1.

Table 1. Sensitivities of the different Bragg reflection peaks to applied SRI.

	P01	*P02*	*P03*	*P04*	*P05*	*P06*	*P07*	*P08*	*P09*	*P10*	*P11*
nm/RIU	3.04	0.47	0.98	0.81	0.71	−0.35	0.47	0.24	−0.21	−0.61	0.22
R^2	0.88	0.12	0.56	0.49	0.81	0.11	0.38	0.12	0.80	0.07	0.19

From the obtained response, it follows that for increasing order of the PMG, a trend toward higher sensitivity to SRI can be seen, which agrees well with the literature [7]. The value R^2 in Table 1 specifies the coefficient of linear regression of the sensitivity for each PMG. From the obtained R^2 values it follows that for a decreasing PMG order, a trend towards negligible or no sensitivity to applied RI was measured. Moreover, the PMGs *P06*, *P09* and *P10* show a negative sensitivity to applied RI. However, this error in measurement is attributed to the optical interrogation system (see Section 4). In Figure 4 the relative wavelength shifts of the Bragg reflection peaks *P01* and *P11* for different SRI are shown for the 10 measurements performed for each refractive index solution. A standard deviation of 0.064 nm has been obtained for the refractive index measurements using the Bragg reflection peak *P01*.

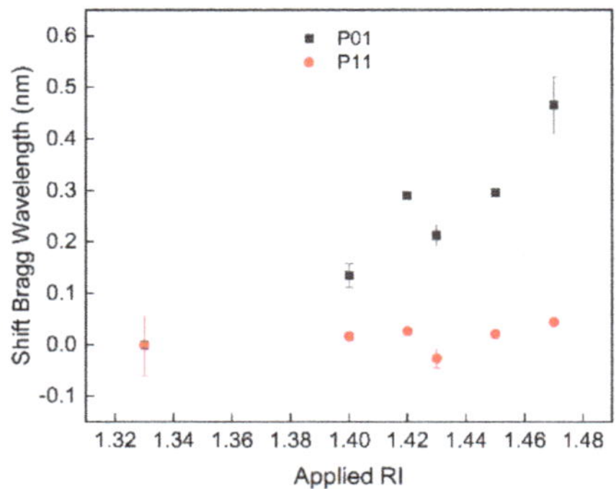

Figure 4. Bragg wavelength shift for the highest (*P01*) and lowest (*P11*) PMG for different SRI solutions.

3.2. Response to Temperature Variations

In Figure 5 as well as Table 2, the response of the individual Bragg reflection peaks to applied temperature is illustrated for a constant RI (air). Ten measurements were performed for every temperature step and the obtained mean values for the measured Bragg reflection peaks per temperature step are depicted in Figure 5.

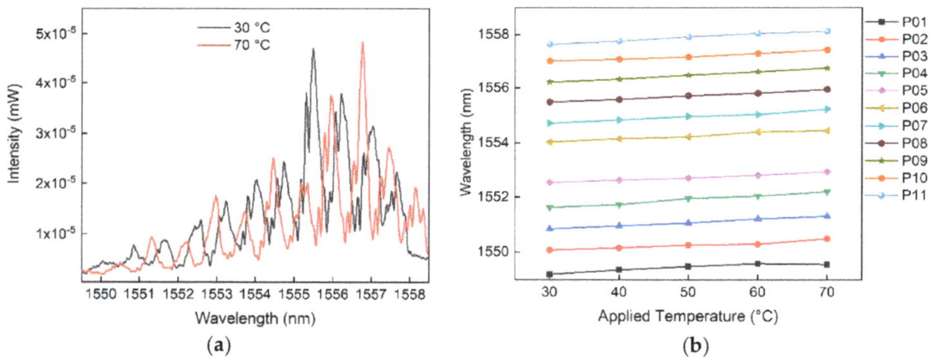

Figure 5. Reflection spectrum of etched GI-MMF with BG for the measurement of different surrounding temperatures (**a**) and the response of the individual Bragg reflection peaks to different applied temperature values (**b**) *P01–P11* indicate the Bragg reflection peaks in the reflection spectrum from the left to the right, i.e., the Bragg reflection peak of the highest PMG is marked with *P01*, whereas the Bragg reflection peak of the lowest PMG is marked with *P11*.

Table 2. Sensitivities of the different Bragg reflection peaks to applied temperature.

	P01	*P02*	*P03*	*P04*	*P05*	*P06*	*P07*	*P08*	*P09*	*P10*	*P11*
pm/°C	9.6	9.7	11.6	14.3	9.8	10.9	11.8	11.6	13.4	10.5	12.6
R^2	0.88	0.95	0.99	0.98	0.99	0.97	0.98	0.99	0.99	0.98	0.99

According to the results obtained, all individual Bragg reflection peaks show linear behavior (R^2 value ≥ 0.88) as well as an almost identical sensitivity to applied temperature (on average 11.4 pm/°C). This result is consistent with the literature (11.5 pm/°C) for GI-MMF [6]. Since all modes have almost the same thermo-optic coefficient all Bragg reflection peaks show almost identical temperature sensitivity [6]. A standard deviation of 0.014 nm has been obtained for the temperature measurements using the Bragg reflection peak *P11* (since *P11* exhibit negligible or no sensitivity to applied RI, it is more suitable for the determination of applied temperature), for instance.

4. Discussion

The obtained results verify that higher order PMGs are more sensitive to SRI variations compared to lower order PMGs. In addition, results obtained are consistent with reported results for GI-MMF [6] (in case of the response to temperature) and FMF [4] (in case of the response to SRI) in the literature. Consequently, it can be shown that by applying the single sensor element and PMG multiplexing, the simultaneous measurement of RI and temperature is feasible. However, the applied interrogation system which is based on equilibrium mode excitation and the applied Bragg wavelength peak detection ("center of mass" determination) is not sufficient to prove the simultaneous measurement of both parameters at the moment as the envelope of the measured Bragg reflection peaks of the etched GI-MMF with BG varies for different SRI values (see Figure 3a). The change of the envelope for every SRI value can be explained by the different sensitivities to SRI of the modes (the amount of the evanescent field spread in the surrounding varies depending on the mode) of each PMG and thus the corresponding Bragg wavelength change of each mode varies differently with different RI values. This causes a broadening of the individual Bragg reflection peaks. Another reason for the change of the envelope of the measured spectrum for different refractive indices might be the handling of the sensor element and immersion into the different RI solutions. Since the fiber was moved and held in another position for each RI solution, the power of the different modes might vary slightly (through change in the equilibrium mode distribution) and thus the envelope of the reflected FBG spectrum might fluctuate.

Since the ultimate goal of achieving a mode-multiplexed sensor system requires the excitation of individual modes, the interrogation system shown in Figure 1 is currently being modified by adding a SLM to the optical path. Therefore, in the next configuration, the new interrogation system will allow the individual excitation of certain modes of each PMG (modes that propagate mostly at the core/cladding interface or only within the core of the sensor element) and thus the uncertainty of the peak detection due to envelope changes will be eliminated. Compared to other BG based fiber optic RI sensors, which had reported, i.e., a sensitivity of 71.2 nm/RIU [10], the obtained RI sensitivity of the proposed sensor system is relatively low. Therefore, in future research different techniques to increase the RI sensitivity will also be explored, such as coating the fiber optic sensor with thin polymer overlays [11].

5. Conclusions

In this work a single sensor element based on an etched OM4 GI-MMF with integrated BG was proposed and investigated for simultaneous determination of applied temperature and SRI. Experimental results verify that different PMGs respond differently to applied SRI but equally to applied temperature. Therefore, by applying PMG multiplexing and a single sensor element consisting of an etched GI-MMF with BG, the simultaneous measurement of SRI and temperature is feasible. Experiments show that the sensitivity to SRI increases with increasing order of the PMG. Sensitivities of 3.04 nm/RIU and of 0.22 nm/RIU to applied SRI in the range from 1.33 and 1.47 for the highest and lowest PMG, respectively, were obtained. The temperature sensitivity of all PMGs was almost equal. The obtained standard deviation of the SRI and temperature measurement were 0.064 nm (for *P1*) and 0.014 nm (for *P11*). Further work is being carried out presently on a new interrogation system which would allow the excitation of individual modes of the etched GI-MMF. Thus, the interrogation system will then compensate for the envelope fluctuations of the measured reflection spectrum which limits the detection of the wavelength shift of the Bragg reflection peaks due to applied SRI. Additionally, the enhancement of the sensitivity to SRI will be investigated by using thin polymer overlays [11] as well as whether the new sensor element can be applied for simultaneous measurement of humidity [12] and temperature or strain and temperature by using different sensor coatings. In future experiments, the temperature range of the proposed sensor element will also be investigated.

Author Contributions: Conceptualization: K.B.; software: K.B.; validation: K.B., L.S.M.A. and Y.Z.; investigation: K.B., Y.Z., L.S.M.A.; writing—original draft preparation: K.B. and L.S.M.A.; writing—review and editing: B.W.R.; supervision: B.W.R. All authors have read and agreed to the published version of the manuscript.

Funding: Bernhard Roth acknowledges funding from the Deutsche Forschungsgemeinschaft (DFG, German Research Foundation) under Germany's Excellence Strategy within the Cluster of Excellence PhoenixD (EXC 2122, Project ID 390833453).

Acknowledgments: The authors would like to acknowledge the work from Jörg Neumann, Michael Steinke and Sebastian Böhm from the Laser Zentrum Hannover for etching the GI-MMF.

Conflicts of Interest: The authors declare no conflict of interest.

References

1. Bremer, K.; Lochmann, S.; Roth, B. Grating assisted optical waveguide coupler to excite individual modes of a multi-mode waveguide. *Opt. Commun.* **2015**, *356*, 560–564. [CrossRef]
2. Schlangen, S.; Bremer, K.; Böhm, S.; Wellmann, F.; Steinke, M.; Neumann, J.; Roth, B.; Overmeyer, L. Grating assisted glass fiber coupler for mode selective co-directional coupling. *Opt. Lett.* **2019**, *44*, 2342–2345. [CrossRef] [PubMed]
3. Franz, B.; Bulow, H. Experimental Evaluation of Principal Mode Groups as High-Speed Transmission Channels in Spatial Multiplex Systems. *IEEE Photonics Technol. Lett.* **2012**, *24*, 1363–1365. [CrossRef]
4. Yang, H.Z.; Ali, M.M.; Islam, M.R.; Lim, K.; Gunawardena, D.S.; Ahmad, H. Cladless few mode fiber grating sensor for simultaneous refractive index and temperature measurement. *Sens. Actuators A* **2015**, *228*, 62–68. [CrossRef]
5. Kashyap, R. *Fiber Bragg Gratings*, 2nd ed.; Academic Press: New York, NY, USA, 2009.
6. Mizunami, T.; Djambova, T.V.; Niiho, T.; Gupta, S. Bragg gratings in multimode and few-mode optical fibers. *J. Lightwave Technol.* **2000**, *18*, 230–235. [CrossRef]
7. Kezmah, M.; Donlagic, D. Multimode all-fiber quasi-distributed refractometer sensor array and cross-talk mitigation. *Appl. Opt.* **2007**, *46*, 4081–4091. [CrossRef] [PubMed]
8. Fernandes, D.; Winter, R.; Da Silva, J.C.C.; Kamikawachi, R.C. Influence of temperature on the refractive index sensitivities of fiber Bragg gratings refractometers. *J. Microw. Optoelectron. Electromagn. Appl.* **2017**, *16*, 385–392. [CrossRef]
9. Chryssis, A.N.; Lee, S.M.; Lee, S.B.; Saini, S.S.; Dagenais, M. High sensitivity evanescent field fiber Bragg grating sensor. *IEEE Photonics Technol. Lett.* **2005**, *17*, 1253–1255. [CrossRef]
10. Wei, L.; Huang, Y.; Xu, Y.; Lee, R.K.; Yariv, A. Highly sensitive fiber Bragg grating refractive index sensors. *Appl. Phys. Lett.* **2005**, *86*, 151122. [CrossRef]
11. Alwis, L.; Bremer, K.; Sun, T.; Grattan, K.T.V. Analysis of the Characteristics of PVA-Coated LPG-Based Sensors to Coating Thickness and Changes in the External Refractive Index. *IEEE Sens. J.* **2013**, *13*, 1117–1124. [CrossRef]
12. Alwis, L.; Sun, T.; Grattan, K.T.V. Analysis of Polyimide-Coated Optical Fiber Long-Period Grating-Based Relative Humidity Sensor. *IEEE Sens. J.* **2012**, *13*, 767–771. [CrossRef]

Article

Durability of Functionalized Carbon Structures with Optical Fiber Sensors in a Highly Alkaline Concrete Environment

Kort Bremer [1,*], Lourdes S. M. Alwis [2], Yulong Zheng [1], Frank Weigand [3], Michael Kuhne [4], Reinhard Helbig [3] and Bernhard Roth [1,5]

[1] Hannover Centre for Optical Technologies (HOT), Leibniz University Hannover, 30167 Hannover, Germany; Zheng.Yulong@slm-solutions.com (Y.Z.); Bernhard.Roth@hot.uni-hannover.de (B.R.)

[2] School of Engineering and the Built Environment, Edinburgh Napier University, Edinburgh EH10 5DT, UK; L.Alwis@napier.ac.uk

[3] Saxon Textile Research Institute (STFI), 09125 Chemnitz, Germany; Frank.Weigand@stfi.de (F.W.); Reinhard.Helbig@stfi.de (R.H.)

[4] Materialforschungs-und-Prüfanstalt an der Bauhaus-Universität Weimar (MFPA), 99423 Weimar, Germany; michael.kuhne@mfpa.de

[5] Cluster of Excellence PhoenixD, Leibniz University Hannover, Welfengarten 1, 30167 Hannover, Germany

* Correspondence: Kort.Bremer@hot.uni-hannover.de; Tel.: +49-511-762-17905

Received: 4 May 2019; Accepted: 12 June 2019; Published: 18 June 2019

Abstract: The paper presents an investigation into the durability of functionalized carbon structures (FCS) in a highly alkaline concrete environment. First, the suitability of optical fibers with different coatings—i.e., acrylate, polyimide, or carbon—for the FCS was investigated by subjecting fibers with different coatings to micro/macro bending and a 5% sodium hydroxide (NaOH) (pH 14) solution. Then, the complete FCS was also subjected to a 5% NaOH solution. Finally, the effects of spatial variation of the fiber embedded in the FCS and the bonding strength between the fiber and FCS was evaluated using different configurations —i.e., fiber integrated into FCS in a straight line and/or with offsets. All three coatings passed the micro/macro bending tests and show degradation after alkaline exposure, with the carbon coating showing least degradation. The FCS showed relative stability after exposure to 5% NaOH. The optimum bonding length between the optical fiber and the carbon filament was found to be ≥150 mm for adequate sensitivity.

Keywords: structural health monitoring (SHM); functionalized carbon structure (FCS); carbon reinforced concrete (CRC); fiber optic sensor (FOS); optical glass fiber

1. Introduction

The replacement of conventional techniques used for structural health monitoring (SHM) based on electrical means (i.e., strain gauges for strain measurement) [1,2] with fiber optic sensors (FOSs) has seen increased popularity over the last decade due to the number of advantages they possess over conventional schemes. The glass (silica) construction of fiber optics renders them robust and capable of withstanding harsh and corrosive environments [3–7]. The low attenuation of optical glass fiber enables them to be interrogated over a long length (i.e., several 100 s of kilometers). The fact that a multitude of sensors can be multiplexed within a single strand of fiber together with the possibility of utilizing a wider bandwidth makes FOSs most suitable for applications that require extraction of data from a vast amount of sensing elements such as those required for SHM of large structures [8–12]. Moreover, the sensors, although multiplexed along one fiber, have the capability to monitor several parameters simultaneously and thus provide the opportunity to assess not only, for example, the strain levels but also other chemical parameters, such as humidity and pH. Thus, it is evident from the current literature that the utilization of FOSs for SHM has not yet reached its full potential and requires further investigations paired with advances in chemical and materials engineering. In fact, the recent merging between textile and sensor engineering has seen rapid developments in wearable technology, especially in the biomedical engineering field [13–15], which provides much motivation for adopting advances in chemical/textile engineering in other areas of sensing, i.e., SHM.

Conventional FOSs used for SHM have been mainly grating-based and require adequate attachment of the sensor region to the structure. This had been mainly achieved using binding agents, such as epoxy [16,17]. The installation of such a sensing scheme not only requires specialist trained personnel in fiber optics to carry out the installation, but also, the accuracy of the sensing mechanism relies heavily on the efficiency of the intermediate transfer agent, i.e., epoxy. In addition, the hassle of having to somehow attach the fiber on the structure makes it unattractive to civil engineers, who are not used to dealing with cables that are of micro-meter-diameter-level dimension [18]. An ideal solution would be to embed the sensors on to the strengthening element of the structure itself, which saves time and costs, increasing the efficiency of the overall scheme.

Projects such as the Carbon Concrete Composite (C3) initiative in Germany aim to replace traditional steel reinforced concrete (SRC) as a building material with carbon reinforced concrete (CRC) [19]. The inherent characteristics of CRC possess a considerable number of advantages over conventional SRC, i.e., light weight, lifespan, thermal and electrical conductivity properties, flexibility of fabrication, efficiency and immunity to risk of corrosion. Further, CRC can be constructed in thin layers with high tensile strength, which makes it most suitable for intelligent building construction. The direct integration of FOSs into CRC forms an advanced field of research, which needs careful investigation into the durability of the said integration and thus its practicality in long-term use. To this end, it is of vital importance to evaluate whether the physical construction of the optical glass fiber (preferably single-mode fiber with acrylate, polyimide or carbon coating) is able to withstand its host environment, i.e., concrete.

The combination of the high alkaline environment in concrete and the concrete mechanical stress adversely affects not only the integrity of the optical fiber but the strength of the bonding between the carbon fiber reinforcement and the optical fiber. In a worst-case scenario, the optical fiber can be damaged to a level of non-operation. The sensing region must not detach from the structure/reinforcement as a result of mechanical stress or be decomposed due to the high alkaline surrounding. These considerations emphasize the level of attention needed to evaluate the mechanical properties of the optical fiber embedded into concrete.

Over the past few decades, much work has been focused on the survival of glass in high alkaline concrete environments [20], leading to embedding FOSs in concrete to evaluate its durability [21,22]. Research has shown that polymeric coatings increase the durability of optical fibers in concrete [21]. In particular, optical silica glass fibers with high alkaline resistance were obtained by coating them with polyetherimide [23], fluorine-polymer [24], and carbon coating [25].

Previous work by the authors presented a Functionalized Carbon Structure (FCS), where fiber optic sensors were "woven" into carbon fiber reinforcement polymer (CFRP) strands in order to evaluate the bonding strength between the sensor element and concrete [26]. This was focused on FOSs embedded in FCSs and textile net structures (TNSs) based on alkaline resistant glass. This work was further extended to embed FCSs in concrete blocks to evaluate the performance of the FOSs and the detailed analysis on the performance and the results can be found elsewhere [27]. However, when FCSs are embedded into concrete structures, the highly alkaline environment would degrade the FCS, thus limiting its life-span and operation. The strength of resistance to these highly alkaline environments would depend on the type of fiber used and the material of the protective-coating surrounding the fiber, i.e., acrylate, polyimide, or carbon. To this end, the paper presented here consists of a detailed analysis on the durability of FCS embedded with optical glass fiber sensors in a highly alkaline concrete environment. In addition, the spatial variation of the optical fiber inside the FCS was also analyzed to evaluate its effects on the bonding between the optical glass fiber and FCS.

2. Materials and Methods

2.1. Fabrication of one Dimensional (1D) and Two Dimensional (2D)-FCS

Figure 1 depicts the experimental concept of embedding the FCS in concrete for its SHM, as evaluated in the work presented herewith.

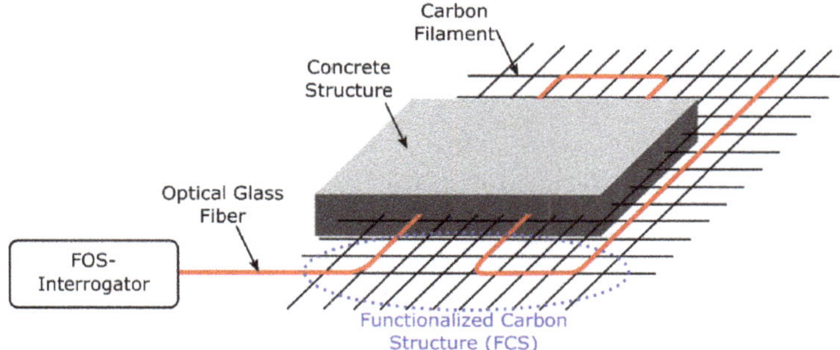

Figure 1. Schematic of the application of functionalized carbon structures (FCS) for the reinforcement and structural health monitoring (SHM) of concrete structures.

The fabrication of the FCS has already been reported [26,27]. The fabrication technique was developed at the Saxon Textile Research Institute (STFI) in Chemnitz, Germany and is based on embroidering carbon fibers and optical glass fibers simultaneously in a grid-like format on a polyvinyl alcohol (PVA) nonwoven substrate. PVA was chosen for this purpose since it can be easily dissolved in hot water as well as it protects the FCS. The advantage of this fabrication technique is that depending on the application, tailored FCS can be produced with various forms of lattice structures, multiple layers of carbon filaments as well as with different configuration of the optical glass fibers inside the FCS. The latter is, in particular, interesting to optimize the bond between the textile carbon structure and the optical glass fiber and thus to optimize the sensor response.

In order to explore the sensor response and the resistance to highly alkaline concrete environment of the FCS, one dimensional (1D) and two dimensional (2D) FCSs have been fabricated. The 1D-FCSs have been fabricated, in particular, to investigate the sensitivity of the FCS to the highly alkaline concrete environment. To do this, a single strand of carbon filaments of 400, 800 or 1600 tex have been simultaneously embroidered with SM fiber Corning ClearCurve (CC) on the PVA nonwoven substrate. After dissolving the PVA, the length of each 1D-FCS was measured to be 300 mm.

The 2D-FCSs have been fabricated in order to evaluate whether the shape of the optical glass fiber inside the FCS has an impact on the sensor response. To explore this, FCSs have been fabricated with different configurations of the optical glass fiber, as shown in Figure 2. In total, the fiber is integrated into the FCS in three different configurations. In the first case, (a) the fiber is straight; in the second case, (b) the fiber is placed with a slight offset; and in the third case, (c) the fiber meanders along the material. The purpose of the second and third configurations is to explore spatial variation of the optical fiber inside the FCS and whether this results in a stronger bond between the optical glass fiber and the FCS and thus facilitates an optimized sensor response. The FCS fabricated for the investigation had a dimension of 500×110 mm^2 and a grid size of 10 mm $\times$ 10 mm. These were fabricated using 1600 tex carbon filaments and optical glass fibers of type Corning CC with acrylate coating. For the second configuration, the offset of the optical glass fiber was introduced at half length of the FCS with an offset of one grid element. In case of the third configuration, the size of each meander was 2×3 grid elements. In case of the offset and meander configuration, no significant attenuation of the light propagating the optical fiber of the FCS could be measured.

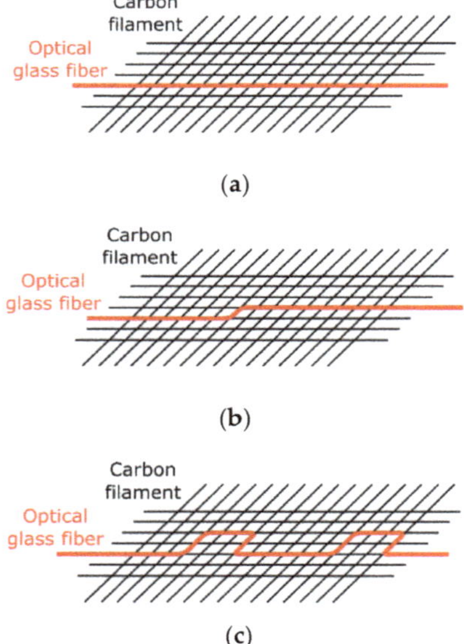

Figure 2. 2D-FCSs with different optical fiber sensor configurations. Straight (**a**), with an offset (**b**) and meander (**c**) fiber configurations were applied to investigate the bond between optical fiber and the carbon filament.

2.2. Optical Glass Fibers

Since FCSs are designed for reinforced concrete structures, the optical fibers inside FCSs have to withstand the highly alkaline concrete environment and still be fully capable of expected operations in terms of sensing and light guidance. Usually a fiber coating is applied in order to protect the optical silica glass fiber against impacts and thus to avoid mechanical degradation. However, when, for instance, the fiber is subjected to harsh environments, such as highly alkaline concrete environments, the fiber coating might be destroyed and thus the optical glass fiber might lose its mechanical stability. Therefore, commercially available optical fibers with different fiber coatings have been investigated in terms of their suitability to be applied for FCS applications. For the investigation, the Corning CC with acrylate coating (ClearCurve ZBL), the OFS Fitel Clearlite with polyimide (PI) (F21976) coating as well as the OFS GEOSIL with carbon/polyimide (C/PI) (BF06159) coating were chosen.

2.3. Evaluating Sensitivity to Micro- and Macro-Bending of Optical Fibers

In addition, an advantage of optical fibers applied for FCSs, discussed in Section 2.2, is that they can be processed during the embroidery fabrication process, i.e., any breaks while embroidering on the PVA nonwoven substrate can be avoided as well as the introduced light attenuation inside the optical glass fiber due to micro- and macro-bending can be neglected. Hence, in order to investigate their sensitivity to micro- and macro-bending when embroidered on the substrate, an experimental set-up was designed that consisted of a 3D printed mold and a power meter (dB-meter from FiboTec). The 3D printed mold enables the periodic and reproducible bending of the glass fiber with a bending radius of five millimeters and the power meter is used to monitor the introduced light attenuation inside the optical glass fiber. The minimum bending radius of 5 mm was determined from the fabrication process of the FCS. In Figure 3, a picture of the experimental setup consisting of the 3D printed mold and the power meter is illustrated.

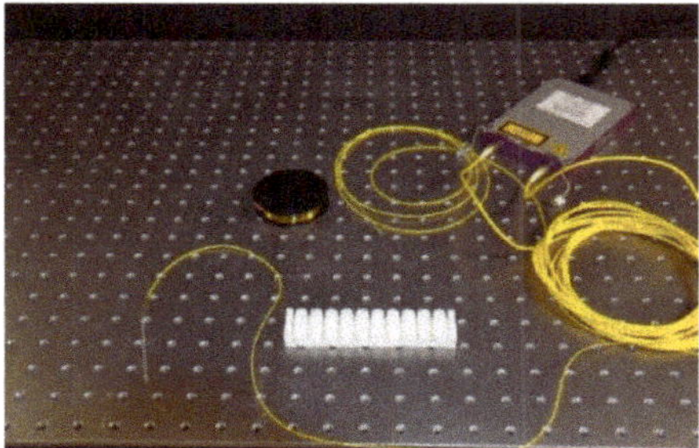

Figure 3. Experimental setup to evaluate the sensitivity to micro- and macro-bending effects on the applied optical glass fibers.

2.4. Simulated Highly Alkaline Concrete Environment

Moreover, the impact of the highly alkaline concrete on the optical fiber was simulated using a 5% NaOH solution (pH 14) [24]. The failure stress of the optical fibers as well as the sensor response to the FCSs was determined using the tensile testing machine (MFC Sensortechnik T3000) after being exposed to the 5% NaOH solution.

The 5% NaOH solution was used to simulate the influence of concrete pore water on the glass fiber and on the fiber coating, respectively. In case of determining the failure stress of the optical glass fibers for each exposure time, the mean value from five fiber samples were calculated. Moreover, the length of the fiber samples under test was 385 mm for the Corning CC and 280 mm for the OFS Clearlite and OFS GEOSIL. The length of the acrylate coated fiber was different due to the fixation of this fiber to the tensile testing machine. However, the length of all fibers-under-test was kept equal. Prior to the tensile tests, all fiber samples were strained up to 1 N and to additionally visualize the fiber surface after exposure to the highly alkaline 5% NaOH solution, images of the fibers were taken using a scanning electron microscope (SEM) at the Saxon textile research institute (STFI). To determine the sensor response of the FCSs after being exposed to simulated concrete pore water, the 300 mm long 1D-FCS was immersed in the 5% NaOH solution for three months. After the exposure, the 1D-FCSs was dried for one day and then subsequently mounted on the tensile testing machine. The sensor response was measured using the experimental setup described in Section 2.5.

2.5. Evaluating Force Transfer of FCS

In order to investigate the sensor response of the FCS a fiber optic Mach-Zehnder (MZ) interferometer was applied [25]. The previously developed fiber optic MZ interferometer enables the investigation of the force transfer between the FCS and the optical glass fiber (fiber optic sensors) and thus the characterization of the bond between these two elements of the FCS. As illustrated in Figure 4, the applied fiber optic MZ interferometer consists of a broadband light source (BBS) (Opto-Link C-Band ASE), two fiber optic 3 dB couplers, two fiber arms, and a spectrometer (OSA) (Ando AQ6317B). The fiber optic MZ interferometer was set-up with only SM fiber components and all fiber components were spliced together. Moreover, the FCS was spliced to one arm of the developed interferometer in order to determine the strain transfer between the optical fiber and the FCS. When force is applied to the FCS using the tensile testing machine, the related length change of the optical fiber of the FCS causes a phase difference ($\Delta\varphi$) between the light (of wavelength λ) travelling in both fiber arms. This in turn results in a change of the interference pattern (displayed in the subset of Figure 4) from which the length change ΔL of the optical fiber inside the FCS can be determined as follows:

$$L = \frac{\lambda_c^2}{n_{core} \cdot \Delta\lambda} \quad (1)$$

In Equation (1), n_{core}, λ_c, and $\Delta\lambda$ are the refractive index of the fiber core, the central wavelength of the light source, and spectral difference between two adjacent maximums of the measured transmission spectrum of the Mach-Zehnder interferometer, respectively. The refractive index value for the fiber core was taken as 1.45.

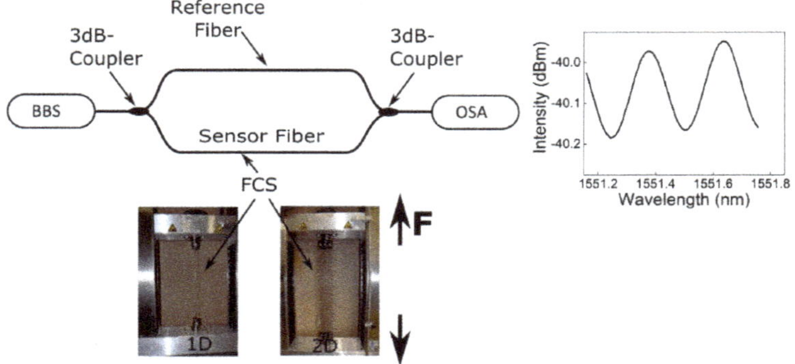

Figure 4. Fiber optic Mach-Zehnder (MZ) interferometer to characterize the sensor response from the FCS under test.

3. Results

3.1. Response of Optical Fibers to Macro- and Micro-Bending

The results of the micro- and macro-bending tests of the optical glass fibers from Section 2.3 are summarized in Table 1. All three optical fibers show relatively less sensitivity to the applied bending. However, the best results were obtained for the two optical glass fibers from OFS (NA = 0.17). Therefore, due to less sensitivity to bending and thus the relatively low light attenuation, all three fiber types are suitable for the application in FCS.

Table 1. Results of the micro- and macro-bending test of the optical glass fibers with different protective coatings. CC: ClearCurve; PI: Polyimide; C/PI: Carbon/Polyimide.

Fiber-Type	10 turns [dB]	20 turns [dB]	30 turns [dB]	40 turns [dB]
Corning CC	−0.06	−0.11	−0.2	−0.28
OFS PI	0	0	0	0
OFS C/PI	0	0	−0.1	0

3.2. Response of Optical Fibers to Simulated Alkaline Pore Water

The impact of the highly alkaline environment of the concrete on the optical glass fibers is summarized in Table 2 and Figure 5. The failure strain tests of the polyimide coated fiber had to be stopped after 14 days. This is due to the detachment of the polyimide coating from the glass fiber (Figure 5b). In addition, for the test on the polyimide coated fiber after been exposed to alkaline 5% NaOH solution, only three fiber samples could be measured, since all other fiber samples were already broken during mounting to the tensile testing machine. In case of the optical glass fiber with carbon coating, two different fabrication batches were evaluated. Fiber samples of the carbon coated optical glass fiber from the first fabrication batch were already consumed after 28 days of exposure to alkaline pore water. Therefore, a second batch of this fiber was ordered to at least conduct the resistant experiments to alkaline pore water over a period of one year. As can be seen in Table 2, the carbon coated optical glass fiber from the first fabrication batch did not show a clear trend towards a lower failure strain for increasing alkaline attack. However, in the case of the optical fibers with carbon coating from the second fabrication batch, a continuous degradation of the failure strain for increasing alkaline attack was observed. It is assumed that the defects in the coating from the second fabrication batch of the carbon coated optical glass fiber had allowed simulated pore water to attack the silica glass matrix of the fiber cladding. It can also be seen in Table 2 that the optical glass fiber with an acrylate coating shows continuous mechanical degradation with increasing alkaline attack.

Table 2. Change of the mean failure strain of the optical glass fibers with different fiber coating for different exposure times to highly alkaline 5% NaOH solution. N specifies the number of samples per test.

	Acrylate	Polyimide	Carbon/Polyimide (Batch #1)	Carbon/Polyimide (Batch #2)
Before exposure	1.27 GPa N = 5	3.29 GPa N = 5	2.29 GPa N = 5	2.10 GPa N = 5
After exposure	0.90 GP (14 days) N = 5	0.59 GPa (14 days) N = 3	2.20 GPa (14 days) N = 5	1.94 GPa (14 days) N = 5
	0.83 GPa (3 months) N = 5		2.23 GPa (28 days) N = 5	0.91 GPa (30 days) N = 5
				0.34 GPa (1 year) N = 5

Figure 5 shows different fiber coatings after being subjected to simulated alkaline attack. As can be clearly seen in Figure 5b, the polyimide coating is detached from the fiber after only 14 days of alkaline attack. No defects on the fiber coating could be detected for the optical glass fibers with acrylate (Figure 5a) and carbon (Figure 5c) coatings. The results obtained are consistent with [24,25], where acrylate and carbon coated fibers showed resistance against the highly alkaline environment.

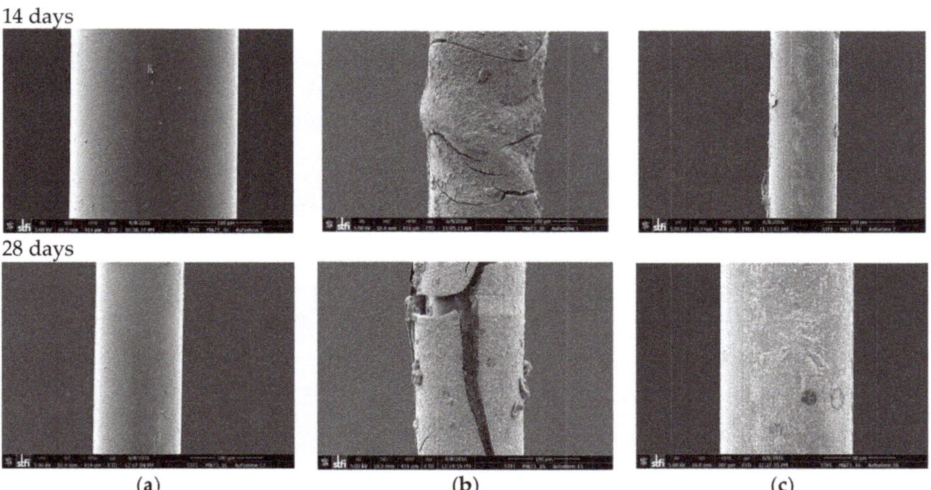

Figure 5. Scanning electron microscope (SEM) images of the optical glass fiber with acrylate (**a**), polyimide (**b**), and carbon/polyimide (**c**) coatings after 14- and 28-days of alkaline attack, respectively.

3.3. Response of 1D-FCS to Simulated Alkaline Pore Water

The response of the FCS (equipped with Corning CC) to alkaline pore water was investigated using 400, 800, and 1600 tex 1D-FCSs. In Table 3, the sensitivity of 300 mm long FCSs to applied force are illustrated before and after exposure to the simulated alkaline pore water solution over a period of three months. For each tex number, three 1D-FCS samples were measured for increasing and decreasing force before and after the alkaline attack. In addition, before the measurements commenced, the 1D-FCS samples were strained prior to the measurement up to 5 N and the maximum applied strain was measured to be 6.7 mε during the tests. From Table 3, it follows that in general the FCSs are relatively stable against alkaline pore water attack. One reason for this might be the PVA that is used to stabilize the FCS and known to be relatively inert against chemicals. Furthermore, FCSs with higher tex numbers show less sensitivity to applied force (due to the increased cross-section) but in turn also less variation of the sensor response before and after the alkaline attack.

Table 3. Change of the sensor response from 1D-FCS due to exposure times to highly alkaline 5% NaOH solution. σ specifies the standard deviation of the measurement.

	400 Tex Filament	800 Tex Filament	1600 Tex Filament
Before exposure	0.0093 mm/N ($\sigma = 3.5 \times 10^{-4}$)	0.0065 mm/N ($\sigma = 2.1 \times 10^{-4}$)	0.0018 mm/N ($\sigma = 1.1 \times 10^{-4}$)
After exposure	0.0174 mm/N ($\sigma = 7.3 \times 10^{-4}$)	0.0077 mm/N ($\sigma = 4.1 \times 10^{-4}$)	0.0023 mm/N ($\sigma = 1.7 \times 10^{-4}$)

3.4. Sensor Response of FCS with Different Optical Fibre Configurations

In Figure 6, the sensor response of the FCSs with different optical fiber configurations are illustrated (for all three different fiber configurations the Corning CC was applied). Before the measurements were performed, the FCS samples were strained to a force of 40 N and the maximum applied strain was 2.4 mε during the tests. In case of configuration one (straight arrangement, Figure 6a) and two (offset arrangement, Figure 6b), a linear response to an applied force of 6.4×10^{-4} mm/N ($R^2 = 0.96$) and 6.7×10^{-4} mm/N ($R^2 = 0.99$), respectively, with a relatively low hysteresis of 4.4×10^{-5} mm/N and 1.9×10^{-5} mm/N, respectively, were obtained. In case of the third configuration (meander arrangement, Figure 6c), no correlation between the applied force and the length change of the optical fiber of the FCS could be observed ($R^2 = 0.25$). The obtained result for configuration three suggests that due to the meander shape and thus due to the periodic spatial variation of the optical fiber inside the FCS, the bond between the optical fiber and the carbon filament is less strong (less bonding length) and thus the FCS is less sensitive to applied force. Therefore, the obtained results (from the three different fiber configurations) suggest that for sufficient sensor sensitivity, the bonding length between the optical fiber and the carbon filament to be at least ≥150 mm long (compare configuration two).

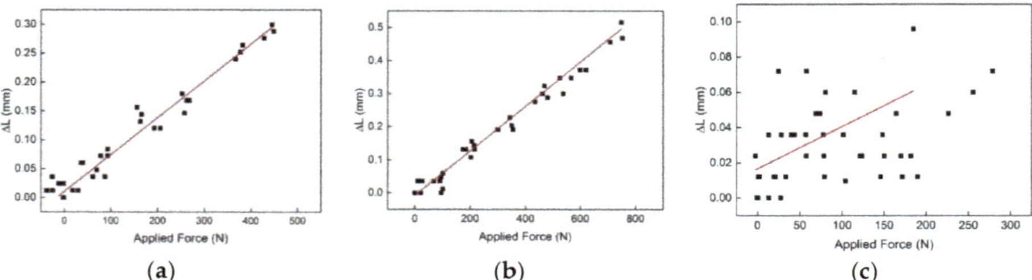

Figure 6. Sensor response for the 2D-FCSs with straight (bonding length 300 mm) (**a**), offset (bonding length 150 mm) (**b**), and meander (bonding length 60 mm) (**c**) optical fiber configurations.

4. Conclusions

The results verified that all three different optical glass fibers can be applied for FCS applications. Following this, the resistance of the optical glass fibers to highly alkaline environment was evaluated where all three fiber coating materials showed degradation and thus not provide full protection against highly alkaline attack. However, optical glass fibers with carbon coating showed the most promising results. Carbon coated optical glass fibers of two different fabrication batches were evaluated. One fabrication batch showed almost no degradation due to alkaline attack while the failure strain of fibers of the second fabrication batch decreased with increasing exposure time to highly alkaline solution. Therefore, it is assumed that micro defects of the fiber coating of the second fabrication batch caused the degradation of the failure strain. The evaluation of the sensor response of the 1D-FCS before and after exposure to the NaOH solution indicated that the FCSs are relatively stable against alkaline pore water attack. One reason for this might be that the PVA that is used to stabilize the FCS is known to be relatively inert against chemicals. In addition, the response of the fiber optic sensors inside the FCS were investigated for three different fiber configurations. The purpose of the second and third configurations was to explore the spatial variation of the optical fiber inside the FCS and whether this results in a stronger bond between the optical glass fiber and FCS. The obtained results suggest that the bonding length between the optical fiber and the carbon filament should be at least ≥150 mm long in order to provide sufficient sensor sensitivity. Currently, work is ongoing to develop a theoretical model that will be used to simulate the work presented herewith, paving way to evaluate the optimum conditions for improved durability.

Author Contributions: Conceptualization, K.B., F.W., and B.R.; software, K.B. and M.K.; validation, K.B. and Y.Z., F.W. and M.K.; formal analysis, K.B., M.K., and F.W.; investigation, K.B., Y.Z., F.W., L.S.M.A., and M.K.; writing—original draft preparation, K.B. and L.S.M.A.; writing—review and editing, B.R.; supervision, B.R. and R.H.; project administration, B.R. and R.H.

Funding: This research was funded by the Bundesministerium fuer Bildung und Forschung (BMBF) within Grant Number 03ZZ0345 (Carbon Concrete Composite (C3)). B.R. acknowledges support by the Deutsche Forschungsgemeinschaft (DFG, German Research Foundation) under Germany's Excellence Strategy within the Cluster of Excellence PhoenixD (EXC 2122). The publication of this article was funded by the Open Access fund of Leibniz Universität Hannover.

Conflicts of Interest: The authors declare no conflict of interest.

References

1. Kouroussis, G.; Caucheteur, C.; Kinet, D.; Alexandrou, G.; Verlinden, O.; Moeyaert, V. Review of trackside monitoring solutions: From strain gages to optical fiber sensors. *Sensors* **2015**, *15*, 20115–20139. [CrossRef] [PubMed]
2. Vardanega, P.J.; Webb, G.T.; Fidler, P.R.A.; Middleton, C.R. *Innovative Bridge Design Handbook: Construction, Rehabilitation and Maintenance*; Elsevier Inc.: Amsterdam, The Netherlands, 2015; pp. 759–775.
3. Mihailov, S.J. Fiber Bragg grating sensors for harsh environments. *Sensors* **2012**, *12*, 1898–1918. [CrossRef] [PubMed]
4. Pal, A.; Dhar, A.; Ghosh, A.; Sen, R.; Hooda, B.; Rastogi, V.; Ams, M.; Fabian, M.; Sun, T.; Grattan, K.T.V. Sensors for Harsh Environment: Radiation Resistant FBG Sensor System. *J. Lightwave Technol.* **2016**, *35*, 3393–3398. [CrossRef]
5. Berruti, G.; Consales, M.; Giordano, M.; Sansone, L.; Petagna, P.; Buontempo, S.; Breglio, G.; Cusano, A. Radiation hard humidity sensors for high energy physics applications using polyimide-coated fiber Bragg gratings sensors. *Sens. Actuators B Chem.* **2013**, *177*, 94–102. [CrossRef]
6. Hassan, M.R.A.; Bakar, M.H.A.; Dambul, K.; Adikan, F.R.M. Optical-based sensors for monitoring corrosion of reinforcement rebar via an etched cladding Bragg grating. *Sensors* **2012**, *12*, 15820–15826. [CrossRef]
7. Alwis, L.S.M.; Bustamante, H.; Bremer, K.; Roth, B.; Sun, T.; Grattan, K.T.V. Evaluation of the durability and performance of FBG-based sensors for monitoring moisture in an aggressive gaseous waste sewer environment. *J. Lightwave Technol.* **2017**, *35*, 3380–3386. [CrossRef]

8. Bremer, K.; Meinhardt-Wollweber, M.; Thiel, T.; Werner, G.; Sun, T.; Grattan, K.T.V.; Roth, B. Sewerage tunnel leakage detection using a fiber optic moisture-detecting sensor system. *Sens. Actuators A Phys.* **2014**, *220*, 62–68. [CrossRef]
9. Bremer, K.; Weigand, F.; Kuhne, M.; Wollweber, M.; Helbig, R.; Rahlves, M.; Roth, B. Fiber optic sensor systems for the structural health monitoring of building structures. *Procedia Technol.* **2016**, *26*, 524–529. [CrossRef]
10. Thevenaz, L.; Facchini, M.; Fellay, A.; Robert, P.; Inaudi, D.; Dardel, B. Monitoring of large structures using distributed Brillouin fiber sensing. *SPIE Proc.* **1999**, *3746*, 345–348.
11. Measures, R.M.; Alavie, A.T.; Maaskant, R.; Ohn, M.M.; Karr, S.E.; Huang, S.Y. Bragg grating structural sensing system for bridge monitoring. *SPIE Proc.* **1994**, *2294*, 53–60.
12. Ansari, F. Practical Implementation of Optical Fiber Sensors in Civil Structural Health Monitoring. *J. Intell. Mater. Syst. Struct.* **2007**, *18*, 879–889. [CrossRef]
13. Khan, S.; Ali, S.; Bermak, A. Recent Developments in Printing Flexible and Wearable Sensing Electronics for Healthcare Applications. *Sensors* **2019**, *19*, 1230. [CrossRef] [PubMed]
14. Han, T.; Kundu, S.; Nag, A.; Xu, Y. 3D Printed Sensors for Biomedical Applications: A Review. *Sensors* **2019**, *19*, 1706. [CrossRef] [PubMed]
15. Chen, X.; An, J.; Cai, G.; Zhang, J.; Chen, W.; Dong, X.; Zhu, L.; Tang, B.; Wang, J.; Wang, X. Environmentally Friendly Flexible Strain Sensor from Waste Cotton Fabrics and Natural Rubber Latex. *Sensors* **2019**, *11*, 404. [CrossRef]
16. Wu, L.; Maheshwari, M.; Yang, Y.; Xiao, W. Selection and Characterization of Packaged FBG Sensors for Offshore Applications. *Sensors* **2018**, *18*, 3963. [CrossRef] [PubMed]
17. Luyckx, G.; Voet, E.; Lammens, N.; Degrieck, J. Strain measurements of composite laminates with embedded fiber Bragg gratings: Criticism and opportunities for research. *Sensors* **2011**, *11*, 384–408. [CrossRef] [PubMed]
18. Barrias, A.; Casas, J.R.; Villalba, S. A review of distributed optical fiber sensors for civil engineering applications. *Sensors* **2016**, *16*, 748. [CrossRef] [PubMed]
19. Scheerer, S.; Schütze, E.; Curbach, M. Strengthening and Repair with Carbon Concrete Composites—The First General Building Approval in Germany. In Proceedings of the International Conference on Strain-Hardening Cement-Based Composites (SHCC 2017), Dresden, Germany, 18–20 September 2018.
20. Wiederhorn, S.M.; Johnson, H. Effect of electrolyte pH on crack propagation in glass. *J. Am. Ceram. Soc.* **1973**, *56*, 192–197. [CrossRef]
21. Mendez, A.; Morse, T.F.; Mendez, F. Applications of Embedded Optical Fiber Sensors in Reinforced Concrete Buildings and Structures. *SPIE Proc.* **1990**, *1170*, 60–70.
22. Ansari, F. State-of-the-art in the Applications of Fiber-optic Sensors to Cementitious Composites. *Cem. Concr. Compos.* **1997**, *19*, 3–19. [CrossRef]
23. Kuwabara, T.; Katsuyama, Y. Polyetherimide Coated Optical Silica Glass Fiber with High Resistance to Alkaline Solution. In *Optical Fiber Communication Conference*; Vol. 5 of 1989 OSA Technical Digest Series; Optical Society of America: Washington, DC, USA, 1989; p. WA4.
24. Habel, W.R.; Hopcke, M.; Basedau, R.; Polster, H. The influence of concrete and alkaline solutions on different surfaces of optical fibers for sensors. In Proceedings of the 2nd European Conference on Smart Structures and Materials, Glasgow, UK, 12–14 October 1994; pp. 168–171.
25. Habel, W.R.; Hofmann, D.; Hillemeier, B. Deformation measurements of mortars at early ages and of large concrete components on site by means of embedded fiber-optic microstrain sensors. *Cem. Concr. Compos.* **1997**, *19*, 81–102. [CrossRef]
26. Bremer, K.; Weigand, F.; Zheng, Y.; Alwis, L.S.; Helbig, R.; Roth, B. Structural Health Monitoring Using Textile Reinforcement Structures with Integrated Optical Fiber Sensors. *Sensors* **2017**, *17*, 345. [CrossRef] [PubMed]
27. Bremer, K.; Alwis, L.S.; Weigand, F.; Kuhne, M.; Zheng, Y.; Krüger, M.; Helbig, R.; Roth, B. Evaluating the Performance of Functionalized Carbon Structures with Integrated Optical Fiber Sensors under Practical Conditions. *Sensors* **2018**, *18*, 3923. [CrossRef] [PubMed]

Article

Intensity-Modulated PM-PCF Sagnac Loop in a DWDM Setup for Strain Measurement

Mateusz Mądry [1,*], Lourdes Alwis [2] and Elżbieta Bereś-Pawlik [1]

1 Telecommunications and Teleinformatics Department, Faculty of Electronics, Wroclaw University of Science and Technology, Wyb. Wyspiańskiego 27, 50-370 Wrocław, Poland; elzbieta.pawlik@pwr.edu.pl

2 School of Engineering and The Built Environment, Edinburgh Napier University, 10 Colinton Road, Edinburgh EH10 5DT, UK; l.alwis@napier.ac.uk

* Correspondence: mateusz.madry@pwr.edu.pl; Tel.: +48-7-1340-7637

Received: 9 April 2019; Accepted: 1 June 2019; Published: 11 June 2019

Featured Application: Potential application of the presented setup for strain measurement in a DWDM configuration.

Abstract: A novel intensity-modulated Sagnac loop sensor based on polarization-maintaining photonic crystal fiber (PM-PCF) in a setup with a dense wavelength division multiplexer (DWDM) for strain measurement is presented. The sensor head is made of PM-PCF spliced to single-mode fibers. The interferometer spectrum shifts in response to the longitudinal strain experienced by the PM-PCF. After passing the Sagnac loop, light is transmitted by a selected DWDM channel, resulting in a change in the output optical power due to the elongation of PM-PCF. Hence, appropriate adjustment of spectral characteristics of the DWDM channel and the PM-PCF Sagnac interferometer is required. However, the proposed setup utilizes an optical power measurement scheme, simultaneously omitting expensive and complex optical spectrum analyzers. An additional feature is the possibility of multiplexing of the PM-PCF Sagnac loop in order to create a fiber optic sensor network.

Keywords: fiber optic sensor; Sagnac loop; intensity-modulated; DWDM; strain sensor

---

1. Introduction

Optical fiber sensors have been widely explored due to their potential advantages, i.e., immunity to electromagnetic interferences, compact size, lightweight and high sensitivity [1]. They could be applied as sensors for temperature [2,3], refractive index [4,5], humidity [6,7] or strain [8–14]. In fact, strain determination is one of the most important factors in structural health monitoring (SHM) [8]. So far, different fiber strain sensors have been presented, for example Mach–Zehnder interferometers [9,10], Fabry–Pérot interferometers [11], fiber Bragg gratings [12] or long period gratings [13,14]. In addition, Sagnac interferometers with highly birefringent fibers are also used for strain measurement. For instance, polarization-maintaining fibers (PMF) were proposed to be sensor heads for strain detection [15]. However, conventional PMFs, i.e., PANDA or bow-tie, are susceptible not only to strain, but also to ambient temperature. To overcome the temperature cross-sensitivity of PMF, Dong et al. [16] and Han [17] demonstrated the use of polarization-maintaining photonic crystal fiber (PM-PCF) in order to achieve strain sensing, which is inherently not sensitive to ambient temperature. Fu et al. presented a pressure sensor by applying PM-PCF within a Sagnac loop achieving a sensitivity of 3.42 nm/MPa [18]. The PM-PCF Sagnac loop was also incorporated within a fiber ring laser (FRL) to evaluate its environmental stability [19]. However, current setups based on spectral analysis in order to convert shifts in the transmission spectrum of the interferometer in response to elongation or pressure are not convenient and portable. Therefore, intensity-modulated optical fiber sensor setups are of interest in order to avoid expensive and advanced spectral analysis instruments [10,20]. Hence a novel, cost-effective, highly sensitive PM-PCF Sagnac loop strain sensor connected to a DWDM (Dense Wavelength Division Multiplexer) based on optical power measurement is herewith proposed. The incident light is modulated by the interferometer and output optical power is measured after passing through the DWDM. The elongation of PM-PCF causes a shift in the interferometer spectrum, resulting in different output spectra. Dong et al. showed a possibility of ~30 mε elongation of PM-PCF, which indicates good sensitivity, durability and efficient application in the field of strain sensing [16]. In the meantime, PM-PCF has low temperature sensitivity due to its structure, which was previously shown to be ~0.3 pm/1 °C [16,21].

The idea of the presented setup relies on a shift in the interferometer spectrum, which has an influence on the transmitted light modulated by DWDM. Therefore, investigation of axial strain is performed only by optical power meters. Approximately 11 dB of output optical power change was experimentally measured for elongation in a range of 0–2000 με, which proves good sensing capabilities. The highest achieved sensitivity is approximately 0.01 dB/με with maximal resolution of 1 με. By applying reference measurement, incident light fluctuation could be eliminated. Additionally, the proposed setup may be easily multiplexed to create a sensor network consisting of PM-PCF Sagnac interferometers.

The work presented here considers one possible sensor configuration utilizing PM-PCF. It sheds light on a new approach with the use of PM-PCF Sagnac loops and optical power meters in order to determine the elongation of the fiber itself. One great advantage of the proposed system is that it is cost-effective in comparison to the current literature that utilizes spectral analysis measurement, where an interrogation unit is also quite expensive [16,17,22]. Currently, the PM-PCF fabrication technology is well developed as evidenced by the wide commercial availability of these fibers. Only a small length of PM-PCF is needed to complete a single sensor unit. Another feature of the proposed setup is its fairly high resolution compared to other wavelength-based setups employing this type of photonic crystal fiber [16,17,22]. A disadvantage of the demonstrated setup is the need for careful adjustment of PM-PCF Sagnac interferometer spectrum with the DWDM channel. However, preparation of proper PM-PCF length to correlate the spectrum should not to be a challenging problem. Another issue, which needs to be taken into consideration regarding practical implementation, is the protection of single-mode fibers in order to avoid any bending and elongation of the non-sensing length of the fiber.

This paper, for the first time to the best of the authors' knowledge, presents a PM-PCF Sagnac loop sensor setup connected to a DWDM for sensing applications. The paper presents the theoretical background followed by the proposal for a sensor network as well as experimental results.

2. Theory

The Sagnac interferometer relies on the phase difference between two counter propagating light beams. Introducing highly birefringent fibers inside a loop provide different light paths, which results in a specific interferometer pattern at the output. The phase difference can be formulated as follows [16]:

$$\varphi = \frac{2\pi BL}{\lambda} \tag{1}$$

where B is the birefringence of the PM-PCF known as the difference between effective refractive indices of the fast and slow axis, respectively, L is the length of the fiber and λ refers to the light wavelength. Due to the determination of the phase difference (φ), the transmission spectrum of the interferometer could be presented according to the following equation [16]:

$$T = 1 - \frac{\cos\varphi}{2} \tag{2}$$

Indeed, the transmission spectrum is a period function depending on the phase difference (φ). The wavelength spacing between two adjacent interferometer fringes could be approximated by the following function [16]:

$$\Delta\lambda = \frac{\lambda^2}{BL} \tag{3}$$

According to Equation (3), the wavelength spacing of interferometer fringes directly depends on parameters of the fiber, i.e., birefringence and length. Elongation of the fiber leads to a change of phase difference (Δφ) between counter propagating light waves, which could be expressed by the following formula [16]:

$$\Delta\varphi = \frac{2\pi}{\lambda}(\Delta LB + L\Delta B) \tag{4}$$

As a consequence, this change of phase difference causes the interferometer spectrum to red-shift due to an increase in longitudinal strain. The temperature effect is negligible because of the PCF structure. In the proposed setup, the output power could be estimated as an integral over the common spectrum of the Sagnac interferometer (T_{INT}) and filter function of a given DWDM channel (T_{DWDM}) with respect to incident broadband light emission (T_{BLS}):

$$P_{out} \approx \int T_{BLS}(\lambda)T_{DWDM}(\lambda)T_{INT}(\lambda)d\lambda \tag{5}$$

The strain induces change in phase difference, which results in a shift in the interferometer spectrum. Thus, the change of optical power could be approximated as follows:

$$\Delta P \approx \int T_{BLS}(\lambda)T_{DWDM}(\lambda)\left(\frac{1-\cos\Delta\varphi}{2}\right)d\lambda \tag{6}$$

Change of phase difference caused by elongation has an influence on the transmitted power. Both the sensitivity and the measurement range are related to the edge slope of the interferometer and the spectral characteristics of DWDM. According to Equation (3), the wavelength spacing of interferometer fringes depends on both incident light wavelength and parameters of the fiber, i.e., birefringence and length. Thus, by the adjustment of the length of the PM-PCF, the measurement range could be modified.

3. Proposed Sensor Network

The sensor network consists of a broadband light source, which is split into *N* number of Sensor Units. A small fraction of light is coupled out to the optical power reference measurement in order to eliminate power fluctuations. One sensor unit refers simply to the PM-PCF Sagnac loop. The main part of light propagates through the fiber coupler, demultplexer (DEMUX) and multiplexer (MUX). At the end, each DWDM channel is assigned to a given detector, which corresponds to the PM-PCF Sagnac loop. The whole scheme of the sensor network proposed is shown in the Figure 1.

In Figure 1 the OPM (REF) refers to the optical power meter used for reference measurement. OPM corresponds to the optical power meter used for measurement of optical power at the output of the n-th sensor unit, MUX is a multiplexer and DEMUX is a demultiplexer. The experimental setup was performed with one PM-PCF Sagnac loop (sensor unit) in order to prove the concept of the sensor network. Multiplexing of sensor units could lead to the obtaining of the proposed sensor network by selecting different wavelengths as it operates on a wavelength division multiplexing scheme. The presented sensor network proposal could then be implemented practically. The operating wavelength range of DWDM is consistent with the International Telecommunication Union (ITU) recommendations.

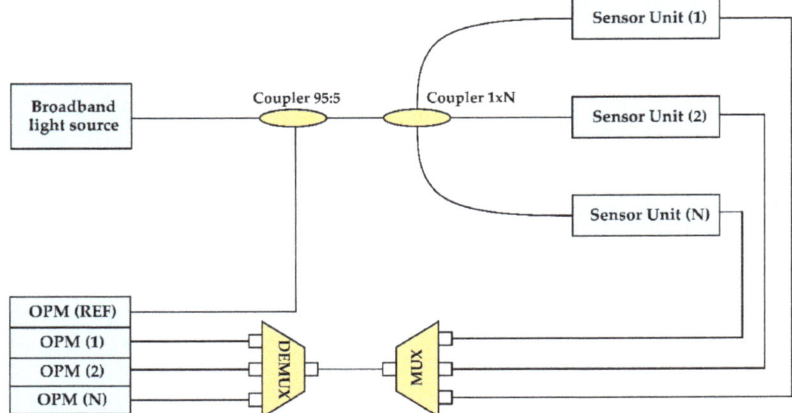

Figure 1. The proposed sensor network based on DWDM (Dense Wavelength Division Multiplexer) PM-PCF (Polarization-Maintaining Photonic Crystal Fiber) Sagnac loops.

4. Experimental Setup

The experimental sensor setup is presented in Figure 2 and consists of a broadband light source (superluminescent light-emitting diode, λ_{peak} = 1544.4 nm, FWHM = 45.5 nm, Thorlabs), two fiber couplers 1 × 2 (split ratio—95:5), a 3-port circulator, a fiber coupler 1 × 2 (split ratio—50:50), a polarization controller, polarization-maintaining photonic crystal fiber (PM-PCF), a dense wavelength division multiplexer (DWDM, 100 GHz, 8 channels, Fiberon) and two hand-held optical power meters (OPM, Detector: InGaAs, Resolution 0.01 dB, Grandway).

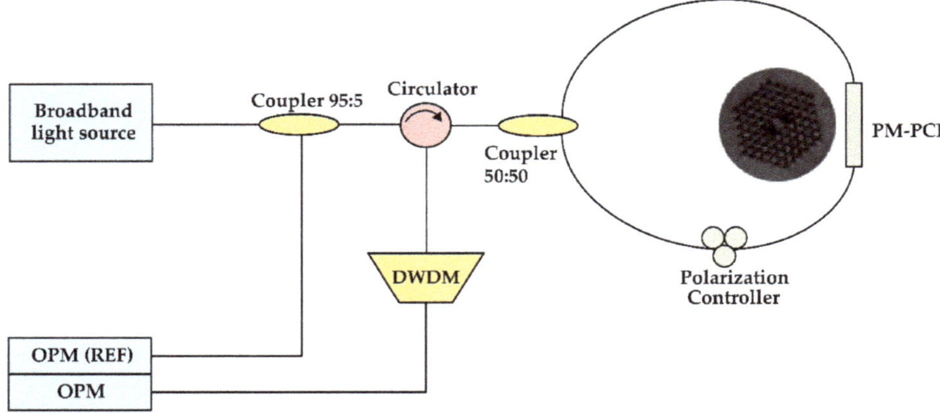

Figure 2. The scheme for the intensity-modulated PM-PCF based Sagnac loop strain sensor. Inset: The image of PM-PCF cross-section.

The used PM-PCF is commercially available and its properties are herein presented: core diameter 6.3/4.4 ± 0.5 µm, outer cladding 125 ± 5 µm, coating diameter 240 ± 10 µm, length—40 cm, birefringence: $\geq 4 \times 10^{-4}$, attenuation ≤3 dB/km at 1550 nm. The image of the cross-section of fiber is fully presented as an inset in Figure 2.

The first fiber coupler 1 × 2 (95:5) was used to eliminate fluctuations from the light source. The fluctuation of any light source output power could be observed in a function of time. In order to prevent fluctuations, the fiber coupler is proposed to be implemented in the setup. Otherwise, the sensor output power value could be disturbed by light source power variation, thereby affecting strain determination accuracy. A circulator was placed within the sensor setup so as to provide the light propagation in the proper direction and to prevent any light reflections from affecting the superluminescent diode. In order to prepare the Sagnac interferometer, the splicing process of PM-PCF to SMF (single-mode fiber) was taken into consideration to enhance repeatability and minimize losses. Following the parameters presented by Xiao et al. [23], the PM-PCF was spliced to SMF using a commercial fusion splicer (FSU975, Ericsson). Additionally, appropriate adjustment of the polarization state within the Sagnac loop was made to provide adequate interferometer fringe visibility. A DWDM was incorporated to modulate the intensity of light by adjusting its spectrum with the edge slope of the PM-PCF Sagnac interferometer. Both spectrum of the DWDM channel and the Sagnac loop interferometer are shown in Figure 3.

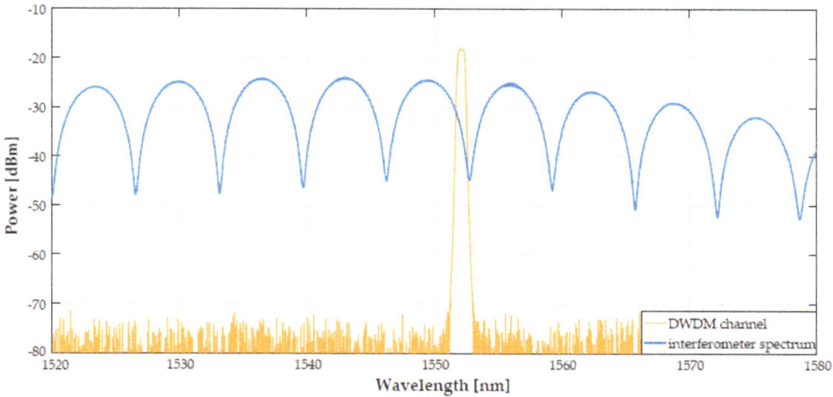

Figure 3. The spectrum of selected DWDM channel and PM-PCF Sagnac loop interferometer.

Spectral analysis from Figure 3 reveals that the interferometer fringe spacing is approximately 6.5 nm (near λ = 1550 nm) with a fringe visibility of ~20 dB. The selected DWDM channel spectrum overlaps the slope of the interferometer spectrum, which ensures adequate operation of the proposed sensor. A shift in the interferometer spectrum provides different values of output optical power. The slope of the interferometer spectrum is also crucial for sensor setup capabilities as it is directly related to the fringe spacing and the measurement range of the sensor. Hence, the length of the PM-PCF needs to be controlled in order to meet specific requirements.

In summary, Table 1 presents all components used within the experimental sensor setup with specified parameters.

Table 1. The parameters of used components.

Components of Sensor Setup	Parameters
Broadband light source (Thorlabs)	Superluminescent light emitting diode, λpeak = 1544.4 nm, FWHM = 45.5 nm, Noise: <0.1%
Optical power meter (Grandway)	InGaAs detector, resolution: 0.01 dB, measuring range: −70 to +10 dBm (1550 nm)
Polarization maintaining photonic crystal fiber (PM-1550-01, Thorlabs)	Core diameter 6.3/4.4 ± 0.5 µm, outer cladding 125 ± 5 µm, coating diameter 240 ± 10 µm, length—40 cm, attenuation ≤3 dB/km (1550 nm), birefringence: ≥4·× 10−4.
Fiber coupler (Cellco)	Single mode, 1 × 2, split ratio: 50:50
Fiber coupler (Cellco)	Single mode, 1 × 2, split ratio: 95:5
Fiber circulator (Cellco)	3-Port circulator, operating wavelength range: 1520–1580 nm
Dense wavelength division multiplexer (DWDM) (Fiberon)	Channel spacing: 100 GHz, transmission insertion loss: typical—1.2 dB, transmission isolation: 28 dB

5. Results and Discussion

5.1. Strain Response of the PM-PCF Sagnac Interferometer

Firstly, the strain response of the PM-PCF Sagnac interferometer was investigated over the range of 0–1500 με in steps of 250 με through the use of translation stages in order to prove the sensing idea. The spectra of the interferometers are presented in Figure 4 as well as the wavelength shift as a function of elongation.

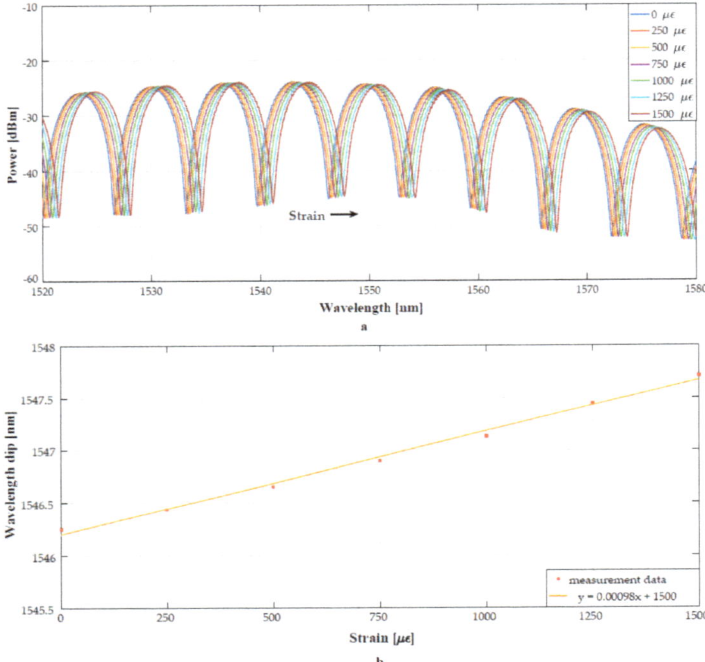

Figure 4. (**a**) The interferometer spectrum shift due to strain, (**b**) the strain response of a selected wavelength dip from the interferometer.

An interferometer fringe at 1546.3 nm was selected to analyze shifts due to elongation of the PM-PCF. A linear response to strain is observed, which is in agreement with the literature. The sensitivity of the PM-PCF Sagnac loop is determined to be approximately 0.98 pm/με.

5.2. Strain Response of the Intensity-Modulated DWDM PM-PCF Sagnac Loop Sensor

The proposed sensor setup as depicted in Figure 2 was investigated by monitoring the output light after passing the Sagnac loop and the DWDM due to elongation of PM-PCF. The fiber was stretched over a range of 0–2000 με in steps of 250 με. The output transmission spectra are shown in Figure 5.

It could be observed from Figure 5 that the intensity of the output light is different due to the applied strain, i.e., the elongation of PM-PCF influences the spectrum of PM-PCF, which shifts towards longer wavelengths and thus the light coupling out from the Sagnac loop is accordingly modulated by the DWDM. The integral over the output spectrum corresponds to measured optical power, which determines the elongation of the fiber. An increase in strain causes an increase in output light. The intensity levels of the output spectrum are different due to the influence of edge slope spectrum of the interferometer. Thus, an investigation into the optical power change was conducted using the reference (P_{ref}) and output power (P_{out}), which eliminates the fluctuation of incident light. Multiple measurements were performed in order to examine the proposed experimental setup. The relationship between the change of optical power and the axial strain is presented in Figure 6.

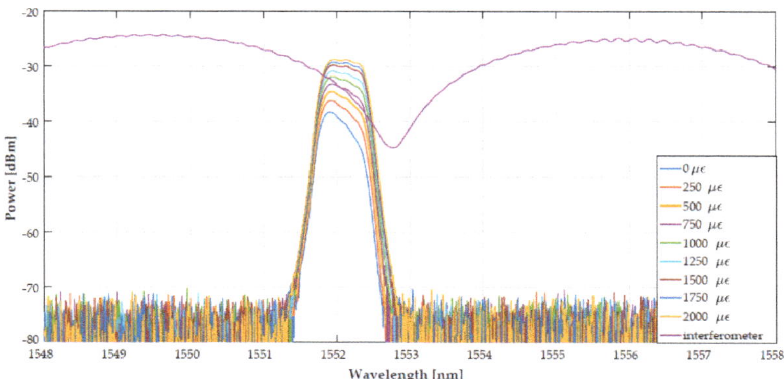

Figure 5. The output light spectra for different stretched PM-PCF.

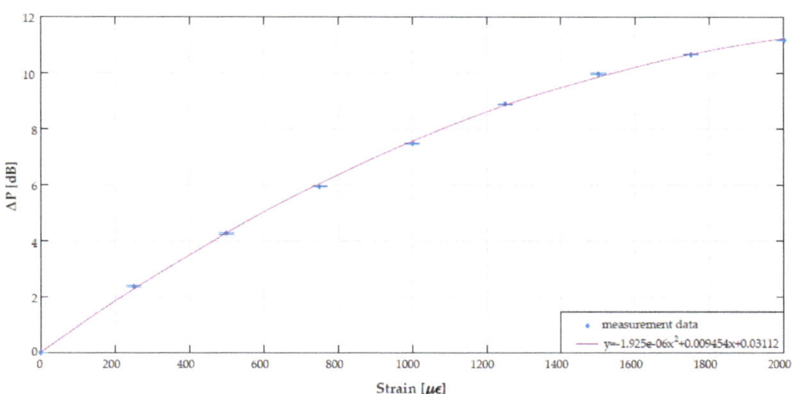

Figure 6. The optical power change as a function of axial strain.

The analysis of measurement data shows that the change in optical power exceeds 11 dB (~11.2 dB) within 2000 με with a negligible deviation between the performed measurements, maximally ± 0.03 dB. A nonlinear response to strain is observed, which could be a result of the spectral correlation between the interferometer and DWDM. The quadratic function was selected as a fitting function (R^2 = 0.999) determined by the following equation:

$$\Delta P = -1.925 \cdot 10^{-6} \cdot S^2 + 0.009454 \cdot S + 0.03112 \tag{7}$$

In the equation above S refers to the strain value applied on PM-PCF (με) and ΔP corresponds to the change in output power (dB). To determine sensitivity at given strain values, the first order derivative of the fitting function was calculated as follows:

$$\frac{\partial(\Delta \mathrm{P})}{\partial \mathrm{S}} = -3.85 \cdot 10^{-6} \cdot \mathrm{S} + 0.009454 \tag{8}$$

Thus, the sensitivity is approximately 0.01 dB/με at the initial value (0 με) and 0.002 dB/με at 2 mε, respectively. Assuming the resolution of the standard optical power meter used in the experiment

(0.01 dB), the resolution of the proposed setup changes within the range of 1 με to 5 με was the result of the measurement.

Due to the use of an optical spectrum analyzer with a resolution of 0.01 nm and assuming the use of the experimental PM-PCF Sagnac interferometer (sensitivity of 0.98 pm/με according to the Figure 4), the possible resolution could be ~10 με, which is lower than expected from the proposed intensity-modulated sensor setup.

The setup presented by Dong et al. had a sensitivity of 0.23 pm/με and a resolution of ~43 με [16]. A similar setup demonstrated by Han achieved a sensitivity of ~1.3 pm/με [17]. Another comparable sensitivity was achieved by Orlando et al. [22], i.e., 1.21 pm/με or 1.11 pm/με depending on whether PM-PCF is uncoated or coated (acrylate). Thus, compared to wavelength-based sensor setups employing this type of PCF [16,17], the presented sensor exhibits a higher resolution (~1 to 5 με) and reduces system cost through the replacement of OSA (Optical Spectrum Analyzer) by optical power meters. Moreover, an easy multiplication of PM-PCF Sagnac loops is possible in order to create a sensor network as its operation relies on DWDM.

It is also necessary to account for possible error resulting from temperature effects. This constraint had already been thoroughly investigated in the literature regarding this type of fiber, [16,21,22] where almost inherent insensitivity to ambient temperature had been found in PM-PCF (~0.3 pm/°C). Assuming a temperature variation of 50 °C (0–50 °C), the induced error could be approximately 15 με. It evidently shows that thermal effect could be negligible in the proposed setup. Nevertheless, the advantage of this setup is the utilization of cost-effective optical power measurement instead of spectral analysis. The measurement range is related to physical parameters of PM-PCF. An increase of PM-PCF length causes relatively smaller spacing of interferometer fringes, which reduces the measurement range of the sensor setup.

6. Conclusions

An intensity-modulated PM-PCF Sagnac loop for strain measurement has been presented and experimentally verified. The sensing part contains a Sagnac interferometer using a PM-PCF, which was subjected to axial strain. Due to the elongation of the PM-PCF, the interferometer spectrum shifts. By adjusting the DWDM, elongation of PM-PCF could be determined by direct output optical power measurement without a need for an OSA. Experimental results pointed out an increase in the output optical power as a function of longitudinal strain experienced by the PM-PCF. The setup has a maximal sensitivity of 0.01 dB/με and resolution of 1 με when measured using standard optical power meters. Additionally, this setup could be multiplexed in order to build a fiber sensor network by exploiting different DWDM wavelengths. It greatly reduces cost due to replacement of expensive and advanced OSA with optical power meters.

Author Contributions: Conceptualization, M.M. and E.B.-P.; investigation, M.M.; methodology, M.M.; supervision, E.B.-P.; validation, L.A.; writing—original draft, M.M.; writing—review & editing, L.A. and E.B.-P.

Acknowledgments: This research was co-financed by the Designated Subsidy for Young Scientists, no. 0402/0159/18 and statutory funds of the Telecommunications and Teleinformatics Department, Wrocław University of Science and Technology, no. 0401/0023/18.

Conflicts of Interest: The authors declare no conflict of interest.

References

1. Chin, K.; Fang, Z.; Qu, R. *Fundamentals of Optical Fiber Sensors*, 1st ed.; John Wiley & Sons, Inc.: Hoboken, NJ, USA, 2012; pp. 1–6.
2. Zhu, T.; Ke, T.; Rao, Y.; Chiang, K.S. Fabry–Perot optical fiber tip sensor for high temperature measurement. *Opt. Commun.* **2010**, *283*, 3683–3685. [CrossRef]
3. Geng, Y.; Li, X.; Tan, X.; Deng, Y.; Yu, Y. High-Sensitivity Mach–Zehnder Interferometric Temperature Fiber Sensor Based on a Waist-Enlarged Fusion Bitaper. *IEEE Sens. J.* **2011**, *11*, 2891–2894. [CrossRef]

4. Zhou, J.; Wang, Y.; Liao, C.; Sun, B.; He, J.; Yin, G.; Liu, S.; Li, Z.; Wang, G.; Zhong, X.; et al. Intensity modulated refractive index sensor based on optical fiber Michelson interferometer. *Sens. Actuators B Chem.* **2015**, *208*, 315–319. [CrossRef]
5. Wo, J.; Wang, G.; Cui, Y.; Sun, Q.; Liang, R.; Shum, P.P.; Liu, D. Refractive index sensor using microfiber-based Mach–Zehnder interferometer. *Opt. Lett.* **2012**, *37*, 67–69. [CrossRef] [PubMed]
6. Alwis, L.; Sun, T.; Grattan, K.T.V. Fibre optic long period grating-based humidity sensor probe using a Michelson interferometric arrangement. *Sens. Actuators B Chem.* **2013**, *178*, 694–699. [CrossRef]
7. Chen, L.H.; Li, T.; Chan, C.C.; Menon, R.; Balamurali, P.; Shaillender, M.; Neu, B.; Ang, X.M.; Zu, P.; Wong, W.C.; et al. Chitosan based fiber-optic Fabry–Perot humidity sensor. *Sens. Actuators B Chem.* **2012**, *169*, 167–172. [CrossRef]
8. López-Higuera, J.M.; Cobo, L.R.; Incera, A.Q.; Cobo, A. Fiber optic sensors in structural health monitoring. *J. Lightwave Technol.* **2011**, *29*, 587–608. [CrossRef]
9. Hu, L.M.; Chan, C.C.; Dong, X.Y.; Wang, Y.P.; Zu, P.; Wong, W.C.; Qian, W.W.; Li, T. Photonic Crystal Fiber Strain Sensor Based on Modified Mach–Zehnder Interferometer. *IEEE Photonics J.* **2012**, *4*, 114–118. [CrossRef]
10. Zhou, J.; Wang, Y.; Liao, C.; Yin, G.; Xu, X.; Yang, K.; Zhong, X.; Wang, Q.; Li, Z. Intensity-Modulated Strain Sensor Based on Fiber In-Line Mach–Zehnder Interferometer. *IEEE Photonics Technol. Lett.* **2014**, *26*, 508–511. [CrossRef]
11. Liu, S.; Wang, Y.; Liao, C.; Wang, G.; Li, Z.; Wang, Q.; Zhou, J.; Yang, K.; Zhong, X.; Zhao, J.; et al. High-sensitivity strain sensor based on in-fiber improved Fabry-Perot interferometer. *Opt. Lett.* **2014**, *39*, 2121–2124. [CrossRef]
12. Lima, H.F.; Antunes, P.F.; de Lemos Pinto, J.; Nogueira, R.N. Simultaneous Measurement of Strain and Temperature with a Single Fiber Bragg Grating Written in a Tapered Optical Fiber. *IEEE Sens. J.* **2010**, *10*, 269–273. [CrossRef]
13. Wang, Y.-P.; Xiao, L.; Wang, D.N.; Jin, W. Highly sensitive long-period fiber-grating strain sensor with low temperature sensitivity. *Opt. Lett.* **2006**, *31*, 3414–3416. [CrossRef]
14. Zhong, X.; Wang, Y.; Qu, J.; Liao, C.; Liu, S.; Tang, J.; Wang, Q.; Zhao, J.; Yang, K.; Li, Z. High-sensitivity strain sensor based on inflated long period fiber grating. *Opt. Lett.* **2014**, *39*, 5463–5466. [CrossRef] [PubMed]
15. Frazão, O.; Baptista, J.M.T.; Santos, J.L. Recent Advances in High-Birefringence Fiber Loop Mirror Sensors. *Sensors* **2007**, *7*, 2970–2983. [CrossRef] [PubMed]
16. Dong, X.; Tam, H.Y. Temperature-insensitive strain sensor with polarization-maintaining photonic crystal fiber based Sagnac interferometer. *Appl. Phys. Lett.* **2007**, *90*, 151113. [CrossRef]
17. Han, Y.G. Temperature-insensitive strain measurement using a birefringent interferometer based on a polarization-maintaining photonic crystal fiber. *Appl. Phys. B* **2009**, *95*, 383–387. [CrossRef]
18. Fu, H.Y.; Tam, H.Y.; Shao, L.-Y.; Dong, X.; Wai, P.K.A.; Lu, C.; Khijwania, S.K. Pressure sensor realized with polarization-maintaining photonic crystal fiber-based Sagnac interferometer. *Appl. Opt.* **2008**, *47*, 2835–2839. [CrossRef] [PubMed]
19. Tang, Z.; Lou, S.; Wang, X. Experimental investigation on environmental stability of erbium-doped fiber laser based on Sagnac interferometer with a polarization-maintaining photonic crystal fiber. *Opt. Laser Technol.* **2018**, *107*, 186–191. [CrossRef]
20. Qian, W.W.; Zhao, C.L.; Dong, X.Y.; Jin, W. Intensity measurement based temperature-independent strain sensor using a highly birefringent photonic crystal fiber loop mirror. *Opt. Commun.* **2010**, *283*, 5250–5254. [CrossRef]
21. Zhao, C.-L.; Yang, X.; Lu, C.; Jin, W.; Demokan, M.S. Temperature-insensitive Interferometer using a highly birefringent photonic Crystal fiber loop mirror. *IEEE Photonics Technol. Lett.* **2004**, *16*, 2535–2537. [CrossRef]
22. Frazão, O.; Baptista, J.M.; Santos, J.L. Temperature-Independent Strain Sensor Based on a Hi-Bi Photonic Crystal Fiber Loop Mirror. *IEEE Sens. J.* **2007**, *7*, 1453–1455. [CrossRef]
23. Xiao, L.; Demokan, M.S.; Jin, W.; Wang, Y.; Zhao, C. Fusion Splicing Photonic Crystal Fibers and Conventional Single-Mode Fibers: Microhole Collapse Effect. *J. Lightwave Technol.* **2007**, *25*, 3563–3574. [CrossRef]

Article

Application of Integrated Optical Electric-Field Sensor on the Measurements of Transient Voltages in AC High-Voltage Power Grids

Shijun Xie [1,*], Yu Zhang [1], Huaiyuan Yang [2], Hao Yu [2], Zhou Mu [1], Chenmeng Zhang [1], Shupin Cao [1], Xiaoqing Chang [1] and Ruorong Hua [1]

1 Postdoctoral Workstation, State Grid Sichuan Electric Power Research Institute, Chengdu 610041, China; zyscdl@163.com (Y.Z.); mu_zhou@126.com (Z.M.); zhangchenmeng@126.com (C.Z.); caosp5330@sc.sgcc.com.cn (S.C.); changxq4374@sc.sgcc.com.cn (X.C.); huaruorong@iCloud.com (R.H.)

2 State Key Lab of Control and Simulation of Power Systems and Generation Equipment, Tsinghua University, Beijing 100084, China; yanghuai17@mails.tsinghua.edu.cn (H.Y.); yuhao19896@163.com (H.Y.)

* Correspondence: sj-xie@163.com

Received: 28 March 2019; Accepted: 7 May 2019; Published: 13 May 2019

Featured Application: Comparing with traditional measuring techniques based on electrical engineering, the novel measuring techniques for transient voltage in AC high-voltage power grid based on optical electric-field sensor is of fast response speed, wide bandwidth, small size, light weight and safety.

Abstract: Transient voltages in the power grid are the key for the fault analysis of a power grid, optimized insulation design, and the standardization of the high-voltage testing method. The traditional measuring equipment, based on electrical engineering, normally has a limited bandwidth and response speed, which are also featured by a huge size and heavy weight. In this paper, an integrated optical electric-field sensor based on the Pockels effect was developed and applied to measure the transient voltages on the high-voltage conductors in a non-contact measuring mode. The measuring system has a response speed faster than 6 ns and a wide bandwidth ranging from 5 Hz to 100 MHz. Moreover, the sensors have the dimensions of 18 mm by 18 mm by 48 mm and a light weight of dozens of grams. The measuring systems were employed to monitor the lightning transient voltages on a 220 kV overhead transmission line. The switching transient voltages were also measured by the measuring system during the commissioning of the 500 kV middle Tibet power grid. In 2017, 307 lightning transient voltages caused by induction stroke were recorded. The characteristics of these voltage waveforms are different from the standard lightning impulse voltage proposed by IEC standards. Three types of typical switching transient voltage in 500 kV AC power grid were measured, and the peak values of these overvoltages can reach 1.73 times rated voltage.

Keywords: optical; electric-field; sensor; measurement; transient voltage; AC power grid; Pockels effect

---

1. Introduction

The wide existence of the transient voltage in the power grid threatens the insulation system of high-voltage plants, which often causes the grid faults [1]. The accurate measurement of transient voltage is the key for the fault analysis, the optimized insulation design and the standardization of high-voltage testing techniques.

Various types of transient voltages can be found in a power grid. The fastest transient voltages (i.e., very fast front overvoltage, VFFO) usually appear at a gas-insulated switchgear (GIS), which is caused by the operation of the GIS disconnectors [2–4]. A VFFO commonly has high frequency components of approximately 100 MHz and a rising time of several nanoseconds [5]. For the fast-front transient voltages, the rising time is normally in the range of 0.1 μs to 20 μs [5]. In terms of a transient voltage caused by normal operation, the rising time is in the range of microseconds. It is noted that these transient voltages are usually superimposed on the power-frequency voltage, which requires a wide bandwidth and fast response speed for the voltage measuring system to be well measured. Moreover, portability and safety are also advised for developing a voltage measuring system.

Subject to the requirements of insulation and capacity, traditional electrotechnical voltage measuring techniques usually bring huge sized and heavy weighted equipment, especially for those applied in an extra-high voltage (EHV) or ultra-high voltage (UHV) power grids. It is also noted that their bandwidth and response speed are unfavorably limited. For example, the capacitance-voltage transformer (CVT) which is a commonly used voltage measuring device in the high-voltage AC power grid typically has a weight of hundreds of kilograms (for 500 kV) and a bandwidth lower than 1 kHz [6–8]. Besides, the bushing tap is applied mainly to measure the transient voltage during a substation commissioning, of which the performance is mainly affected by the manufacturing process and the bushing structure. As a consequence, the upper bandwidth is generally not higher than 5 MHz for a bushing tap technique [9–11]. In addition, the portability cannot be achieved on these techniques.

Thus, subject to the future demand of the power system, it is highly expected to develop a safe and easy-to-apply voltage measuring technique. For a quasi-static problem, since the electric field is considered to be irrotational, the space electric field around a conductor is linear to the potential of the conductor. Based on the measuring technique for the electric field, a non-contacted measuring method for transient voltage could be possible. On the other hand, as the development of optical sensing technology, several electro-optical effects have been used for measuring electric fields. The electro-chromatic effect has a response time of ~100 s, and it is suitable only for extremely low-frequency electric field measurement [12,13]. The Kerr effect is a quadratic electro-optic effect and normally has a small electro-optic coefficient. Thus, the sensors based on the Kerr effect have a low sensitivity [14,15]. The Pockels effect is known as a linear electro-optic effect, which means the output voltage is linear to the measured electric field. The sensor based on the Pockels effect normally has a fast response speed, small size, and are passive. It is particularly used for the measurement of high-amplitude transient electric field [16].

In this paper, based on the linear correspondence between the transient voltage and the transient electric-field caused, a non-contact transient voltage measuring technique was proposed. An integrated optical electric-field sensor based on the Pockels effect was introduced to implement the field measurement. The measuring system was developed and tested for its bandwidth and response speed. A series of measured lightning transient voltages and typical switching transient voltages were presented. Finally, the characteristics of these measured transient voltages and the decoupling issues were discussed.

2. Measuring Method and Measuring System

2.1. Method

For a quasi-static problem, the voltage $U(t)$ of the conductor is linear to the electric field $E(t)$ caused by it, which can be expressed as:

$$U(t) = kE(t). \tag{1}$$

Normally, the voltage waveform of a transient process includes the power-frequency component and the transient component. Since the amplitude of the power-frequency voltage is known for a power grid, by comparing the amplitudes of the power-frequency voltage and the measured power-frequency electric field, the coefficient k in (1) can be given.

In an actual AC power system, the transmission line contains three phase conductors, which unavoidably leads to the coupling effect for the non-contact measuring technique as illustrated in

Figure 1. Beneath each phase conductor, an electric-field sensor is installed denoted as Sensor A, Sensor B, and Sensor C for the corresponding phase.

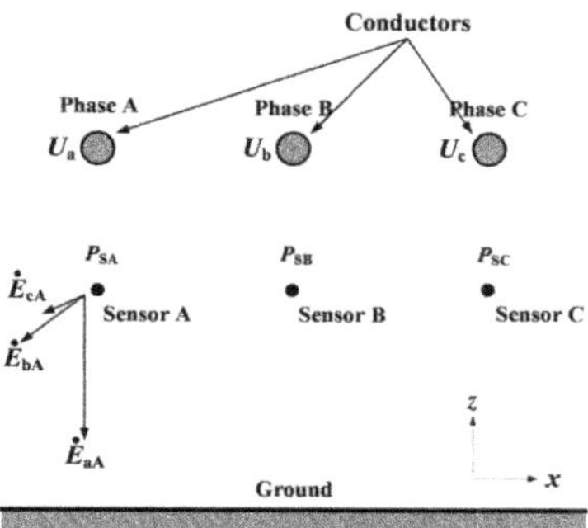

Figure 1. Schematic diagram of the measuring principle. U_a, U_b, U_c—potentials on phase conductors; P_{SA}, P_{SB}, P_{SC}—positions where electric field sensors are set; $\mathbf{E}_{aA}$, $\mathbf{E}_{bA}$, $\mathbf{E}_{cA}$—electric field produced by U_a, U_b, and U_c, respectively at position P_{SA}.

Sensors are arranged at the height level with enough air insulation clearance against corresponding conductors. As conductors are energized, the potentials of three-phase conductors produce electric fields at sensor's locations denoted by P_{SA}, P_{SB}, and P_{SC} in Figure 1. Taking P_{SA} as an example, the electric field at P_{SA} is a compound electric field contributed by $\mathbf{E}_{aA}$, $\mathbf{E}_{bA}$, and $\mathbf{E}_{cA}$ which can be decomposed into the components in the x-axis ($E_{aA\text{-}x}$, $E_{bA\text{-}x}$, $E_{cA\text{-}x}$) and z-axis ($E_{aA\text{-}z}$, $E_{bA\text{-}z}$, $E_{cA\text{-}z}$) respectively. The total electric field at P_{SA} ($\mathbf{E}_A$) can be given by:

$$\begin{aligned} \mathbf{E}_A(t) &= E_{A-x}(t) + E_{A-z}(t)i \\ &= \begin{bmatrix} k_{aA-x} & k_{bA-x} & k_{cA-x} \end{bmatrix} \begin{bmatrix} U_a(t) \\ U_b(t) \\ U_c(t) \end{bmatrix} + \begin{bmatrix} k_{aA-z} & k_{bA-z} & k_{cA-z} \end{bmatrix} \begin{bmatrix} U_a(t) \\ U_b(t) \\ U_c(t) \end{bmatrix} i, \end{aligned} \tag{2}$$

where k is a coefficient equalling to the ratio of electric field detected to the origin potential. The first letter in the subscript of k represents the phase conductor producing the aimed electric field, and the second letter represents the phase under investigation. To integrate three phases, the electric field matrix can be written as follows:

$$\begin{bmatrix} \mathbf{E}_A \\ \mathbf{E}_B \\ \mathbf{E}_C \end{bmatrix} = \begin{bmatrix} E_{A-x} \\ E_{B-x} \\ E_{C-x} \end{bmatrix} + \begin{bmatrix} E_{A-z}(t) \\ E_{B-z}(t) \\ E_{C-z}(t) \end{bmatrix} i = \begin{bmatrix} k_{aA-x} & k_{bA-x} & k_{cA-x} \\ k_{aB-x} & k_{bB-x} & k_{cB-x} \\ k_{aC-x} & k_{bC-x} & k_{cC-x} \end{bmatrix} \begin{bmatrix} U_a(t) \\ U_b(t) \\ U_c(t) \end{bmatrix} + \begin{bmatrix} k_{aA-z} & k_{bA-z} & k_{cA-z} \\ k_{aB-z} & k_{bB-z} & k_{cB-z} \\ k_{aC-z} & k_{bC-z} & k_{cC-z} \end{bmatrix} \begin{bmatrix} U_a(t) \\ U_b(t) \\ U_c(t) \end{bmatrix} i. \tag{3}$$

Equation (3) can be divided into

$$\begin{bmatrix} E_{A-x}(t) \\ E_{B-x}(t) \\ E_{C-x}(t) \end{bmatrix} = \begin{bmatrix} k_{aA-x} & k_{bA-x} & k_{cA-x} \\ k_{aB-x} & k_{bB-x} & k_{cB-x} \\ k_{aC-x} & k_{bC-x} & k_{cC-x} \end{bmatrix} \begin{bmatrix} U_a(t) \\ U_b(t) \\ U_c(t) \end{bmatrix} \tag{4}$$

and

$$\begin{bmatrix} E_{A-z}(t) \\ E_{B-z}(t) \\ E_{C-z}(t) \end{bmatrix} = \begin{bmatrix} k_{aA-z} & k_{bA-z} & k_{cA-z} \\ k_{aB-z} & k_{bB-z} & k_{cB-z} \\ k_{aC-z} & k_{bC-z} & k_{cC-z} \end{bmatrix} \begin{bmatrix} U_a(t) \\ U_b(t) \\ U_c(t) \end{bmatrix}, \tag{5}$$

Based on either (4) or (5), the voltage waveforms on a conductor can be given by measuring the electric fields in the case that the k matrix is known.

2.2. Sensor

An integrated optical electric-field sensor is developed based on the Pockels effect [17–19]. In this optical sensor, an optical waveguide is fabricated on a $LiNbO_3$ substrate by titanium diffusion. The refractive index of the optical waveguide would be changed under an external electric field (E). Thus, while a polarized light passes through the optical waveguide, there would be a phase difference between its vertical and horizontal components. The modulated polarized light was transmitted to a common path interferometer via a polarization maintaining optical fiber, while the phase difference caused by external electric-field is demodulated into light intensity. Eventually, the light intensity is transferred into an electrical signal (U_{out}) by a laser receiver. The relationship between E and U_{out} is given as [20,21]:

$$U_{out} = A \cdot \left[1 + b \cdot \cos\left(\phi_0 + \frac{E}{E_\pi}\pi\right)\right] \tag{6}$$

where A represents the photoelectric conversion coefficient and the transmission loss. b denotes the extinction ratio. ϕ_0 can be controlled near $\pi/2$ through the production of the sensor. In the case that the external electric field is much smaller than the half-wave electric field (E_π), by Taylor's expansion, (6) can be rewritten as:

$$U_{out} = A \cdot \left[1 + b \cdot \frac{E}{E_\pi}\pi\right]. \tag{7}$$

Thus, the output voltage of the optical electric-field measuring system is linear to the external electric field.

Figure 2 shows the schematic diagram of the developed optical sensor which has the dimensions of 18 mm by 18 mm by 48 mm and a weight of dozens of grams. The sensors, as well as the input and output fibres, are made of dielectric materials which could guarantee the safety under high-voltage conditions. Moreover, the sensor indicates a good directivity as it is sensitive to the electric-field component perpendicular to the front of the sensor. On the field measurement, the sensor is usually facing the measured conductor. Thus, instead of (4), (5) is normally used to compute the transient voltage.

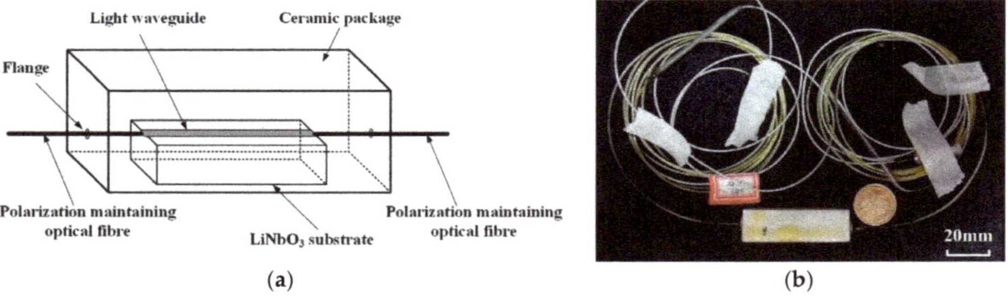

Figure 2. Integrated optical electric-field sensor. (**a**) schematic diagram; (**b**) product photo.

2.3. Measuring System

The measuring system contains an optical electric-field measuring unit and a data processing unit. The optical electric-field measuring unit includes optical sensors, a laser source, a laser receiver, and transmission fibers. The output analog signals of the laser receiver are converted into digital signals by a high-speed data acquisition card (HS-DAQ) with 4 acquisition channels and a sampling ratio of 40 MS/s. The digital signals are temporarily stored in the internal storage and further analyzed by the central processor to determine whether a transient process occurs. While the system is triggered by the transient voltage, the transient voltage waveform from 20 ms before to 80 ms after the triggering will be stored into the local storage hard disk. Simultaneously, 400 milliseconds voltage waveform after the triggering time would also be recorded with a sampling ratio of 20 kS/s. All these signals are timed by a GPS module with the error of less than 100 ns. The field measured signal will be transmitted to the central station by a 4G wireless communication module. Thus, the HS-DAQ, the central processor, the internal storage, the hard disk, the GPS module, and the communication module constitute the data processing system. Except for the sensors and optical fibers, all the other devices are integrated into a cabinet or a portable suitcase and powered by an isolated power supply. The general connection diagram of the measuring system is shown in Figure 3.

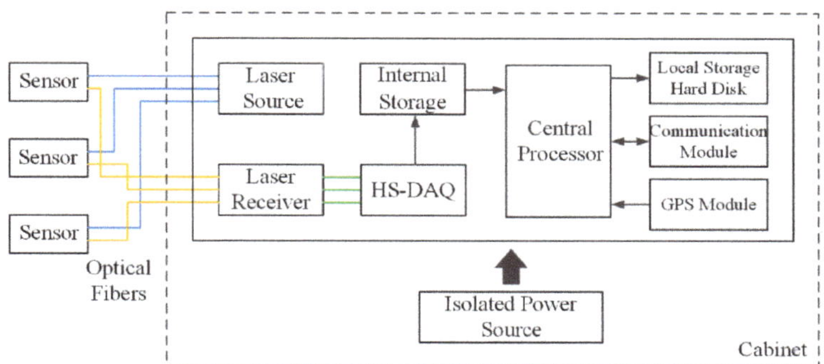

Figure 3. General connection diagram of the measuring system.

Due to the nature of randomness for the lightning strike, the sensors used for measuring lightning transient voltages are installed outdoor for a long term. To protect them from environmental effects, sensor containers are used, which are supported by a grounded metal tube to the same level of the base of adjacent insulators, as shown in Figure 4a. For the measurement of switching transient voltage, the sensors are just supported by an insulating rod temporarily, as shown in Figure 4b. On both arrangements, the clearances between the sensor and the high-voltage conductor are no shorter than the necessary distance specified in [5].

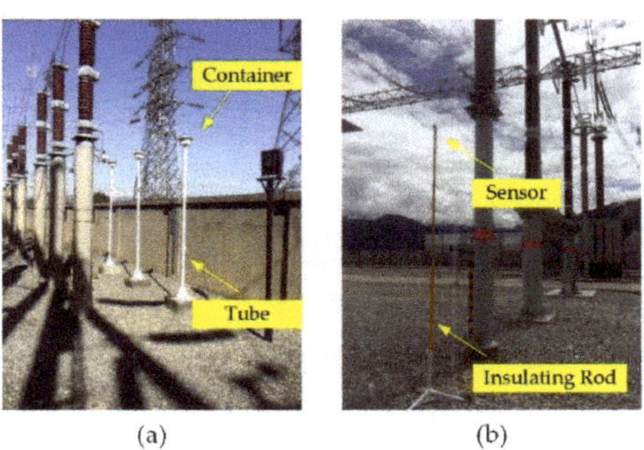

Figure 4. Field arrangement of sensors. (**a**) Field arrangement of sensors for monitoring the lightning transient voltage; (**b**) Field arrangement of sensors for measuring the switching transient voltage.

2.4. Performance Test

To ensure that the developed measuring system meets the performance requirements for the transient voltage measurement, a series of tests including the response speed and frequency response have been carried out in the laboratory.

A 5 kV 1.2/30 μs lighting impulse was imposed on a circular-shape plane-plane gap with the gap distance of 10 cm and the radius of 50 cm. A 996:1 standard impulse voltage divider was employed to offer the standard measuring result. The normalized measuring results of the developed measuring system and the standard divider are shown in Figure 5a. To determine the response time of the measuring system which is in the range of nanoseconds, a bounded wave electromagnetic pulse simulator, composed of charging transformer, impulse capacitor, sharping switch, parallel-plate transmission line, and matching resistor was employed, to generate a pulse with 6 ns for the rise time and 38 ns for the half time. A 104.2:1 standard resistive divider was used to offer standard measuring results. The normalized measuring results are shown in Figure 5b, which indicates that the response time of the measuring system is less than 6 ns.

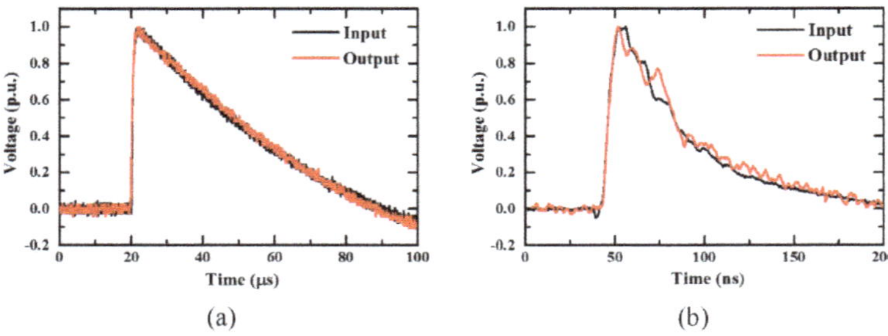

Figure 5. Test results of response speed. (**a**) Lightning impulse; (**b**) electro-magnetic pulse.

To conduct the frequency response test, a signal generator was employed to generate a standard sinusoidal voltage from 5 Hz to 250 kHz. The frequency response of a higher frequency range is determined by a sharp voltage impulse which was imposed on a transverse electric and magnetic field (TEM) cell. The amplitude-frequency response is shown in Figure 6. Except for individual resonant frequencies, the curve of the amplitude-frequency response is flat from 5 Hz to 100 MHz.

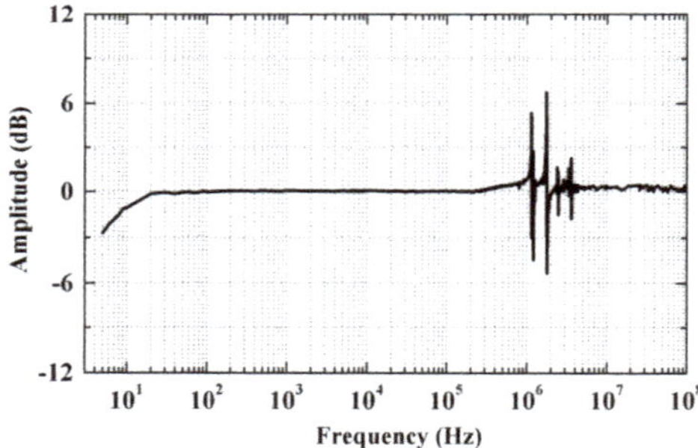

Figure 6. Amplitude-frequency response.

Based on the characteristics of transient voltages defined in [5], performances of the measuring system developed in this paper have satisfied the requirements for measuring field transient voltages in AC high-voltage power grid.

3. Results

3.1. Lightning Transient Voltages

3.1.1. Field Test Overview

Considering the randomness of lightning strikes, the measuring system was installed for a certain duration in a 220 kV Yan-Zhuang overhead transmission line connecting the Yanshenshang substation (Y-substation) and the Zhuangshang substation (Z-substation) with 78 towers and the length of 39.8 km.

The area of the transmission lines has an average ground flash density about 1.012 times/km^2 in 2015 and 2.155 times/km^2 in 2016. Measuring systems were arranged on the entrances of Y-substation and Z-substation respectively.

The causes for the transient voltage in a substation are various, including the operations, lightning strikes, electro-magnetic interferences, etc. To distinguish the actual lightning transient voltage, the measuring systems are synchronized with the lightning location system of China (LLS). The LLS includes over 940 lightning detection terminals covering most areas of China. Its average detection efficiency is higher than 90% and the location errors are smaller than 500 m. The errors of GPS clock used in the measuring system and LLS are both within one hundred nanoseconds.

A ground flash event usually contains the first stroke and several subsequent strokes. The period between two continuous strokes is usually longer than 30 ms [22]. Since the length of Yan-Zhuang line is 39.8 km, the difference of the triggering time of the measuring systems and the LLS will not be longer than 150 μs. Thus, the recorded data is considered to be a lightning transient voltage waveform caused by an involved ground flash, while the measuring system is triggered and the LLS has recorded a ground flash near the Yan-Zhuang line in 1 ms. Moreover, both the event that the measuring equipment is triggered and a ground flash strike to the point near the monitored transmission line are of small probabilities. When two small-probability events happen almost at the same time, it is believed that these two observed processes are the same event.

3.1.2. Typical Measurement Results

In 2017, a total of 832 pieces of data were recorded by the two measuring systems. 650 pieces of data were recorded by the system in Z-substation and 184 pieces of data had been recorded by the system in the Y-substation. 307 pieces of data can be synchronized with a ground flash recorded by the LLS.

Once a lightning stroke was recorded by two measuring systems, the double-end traveling wave location method (DTLM) is activated to determine the position along the Yan-Zhuang line and nearest to the lightning stroke point.

On 13:58:27, May 14th, 2017, the two measuring systems were triggered in succession and the original recorded voltage waveforms from the sensors of phase C are shown in Figure 7. Obvious voltage fluctuations can be seen in the dotted box. After filtering out the power frequency signal, the lightning transient voltage waveforms are shown in Figure 8. As shown in the figures, the arriving time of lightning transient voltage for the Y-substation and Z-substation are estimated to be 341,862.1 μs and 341,911.1 μs, respectively. The time difference is 49.0 μs. According to DTLM, the initial position (which is usually the position nearest to the lightning stroke point on the transmission line) of the lightning transient traveling wave is about 12.41 km away from the Y-substation, which is between No. 25 tower and No. 26 tower. Meanwhile, LLS indicated that there was a lightning stroke with a return stroke current of −49.5 kA near No. 23 tower. The location error could be caused by the sag of transmission lines, the distortion of transient voltage during transmitting, the nonideal traveling velocity, and the randomness of the lightning stroke.

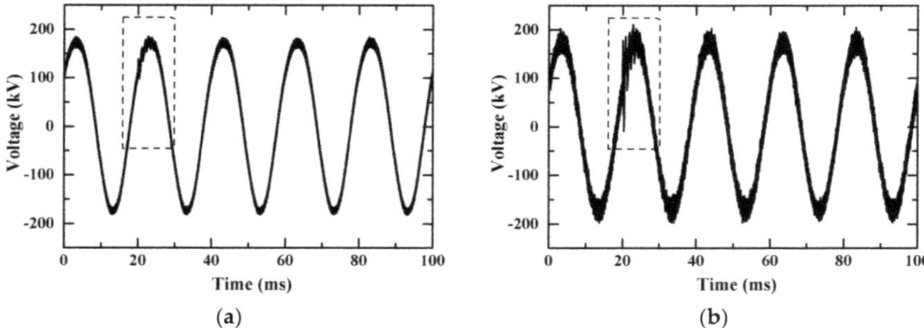

Figure 7. Voltage waveforms recorded on 13:58:27, May 14th, 2017. (**a**) Field measured results in Y-substation; (**b**) field measured results in Z-substation.

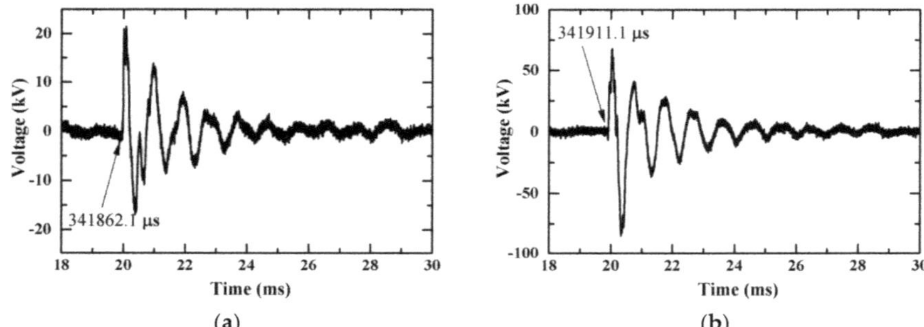

Figure 8. Lightning transient voltage waveforms. (**a**) Field measured results in Y-substation; (**b**) field measured results in Z-substation.

3.2. Switching Transient Voltages

3.2.1. Field Test Overview

Series of switching transient voltages were measured by a portable measuring system during the commissioning of 500 kV middle Tibet power grid. In this power grid, the transmission lines are mostly longer than 150 km with an average altitude beyond 3000 m. The operations for the transient voltage measurement mainly include:

(1) switching on the no-load long overhead transmission line;
(2) switching on the no-load power transformer in the substation;
(3) switching off the no-load power transformer in the substation.

For the operation (1), the sensors were arranged under the end terminal of the transmission lines and the sensors were arranged under the entrance conductors of the transformers for the operation (2) and (3).

3.2.2. Typical Measuring Results

(1) Transient voltages during switching on the Tang-Xiang transmission line

Tang-Xiang transmission line connects the 500 kV Batang substation and the 500 kV Xiangcheng substation. The length of this line is about 190 km with the common-tower double-transmission scheme. Before the energization, the line is disconnected at both sides with a suspended potential. The sensors are arranged under the conductors on the Batang substation. The energization process is accomplished by switching on the breakers on the Xiangcheng substation to connect the line to the 500 kV bus bar.

The typical measuring result after the decoupling process for this type of switching transient voltage is shown in Figure 9. The maximum overvoltage (U_{max}) of 679.6 kV appeared on the Phase C which was 1.66 times as the rated value. While the conductor of Phase C was energized, a negative edge of the voltage formed on Phase B at the Xiangcheng side which then reached the Batang side (sensor installed) at t_1 (18.226 ms). Then this edge further transmitted back to the Xiangcheng substation and then reflected back to the Batang side at t_2 (19.596 ms). During the period between t_1 and t_2, the traveling wave had passed through the transmission line twice. Thus, the actual average velocity of the traveling wave on this line was estimated to be about 2.77×10^8 m/s. Moreover, this estimated velocity from field measured data can be used in the fault determination in the future.

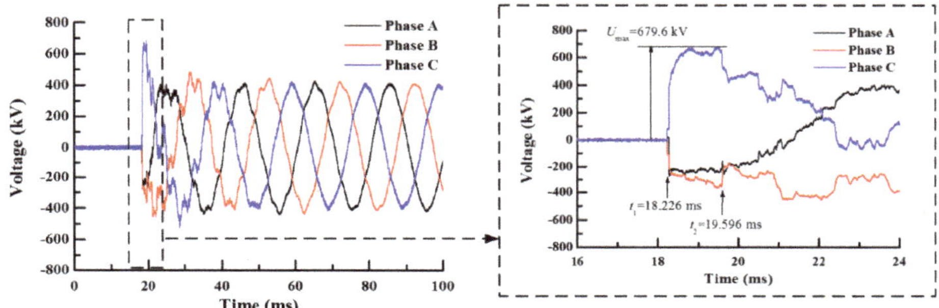

Figure 9. Transient voltage during switching on the Tang-Xiang transmission line.

(2) Transient voltages during switching on a power transformer to the bus bar

The transient voltages on switching the #3 power transformer onto the 500 kV bus bar on Batang substation were measured and the typical measured results are shown in Figure 10. To avoid involving overvoltage caused by magnetizing rush current during switching on a no-load power transformer, the closing resistors, and phase selection closing strategy were used during the operations. According to Figure 10, the breaker of Phase A were closed firstly and the voltage on the Phase A conductor reached the rated value immediately. Though the breaker for Phase B and Phase C were still open, the voltage on these two phases rose to about half of the rated value. This is because the flux linkage in the power transformer had been established by the winding of Phase A and the winding for the three phases were in the same flux linkage. About four and a quarter periods later (85.84 ms), the breakers of Phase B and Phase C were closed, and a transient overvoltage of approximately 554.2 kV (1.36 p.u.) can be found on Phase C.

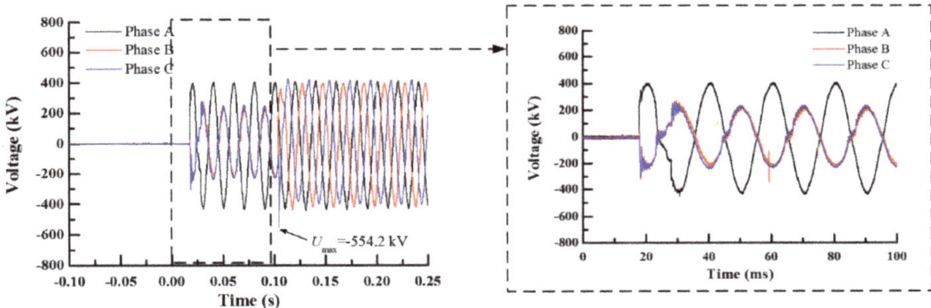

Figure 10. Transient voltage during switching on the #3 power transformer on Batang substation.

(3) Transient voltages during switching off a power transformer from the bus bar

While switching off the no-load power transformer under non-zero running current, since the current flowing through an inductance never suddenly changes, the residual running current will charge the capacitance connecting to the transformer. This energization process would produce transient overvoltage. The amplitude of the overvoltage mainly depends on the current upon switching off the power transformer. Theoretically, a larger current leads to a higher overvoltage.

The transient voltage while switching off the #3 power transformer from the 500 kV bus bar on Batang substation is shown in Figure 11. The breaker of Phase C was opened while the voltage just crossed the zero level with a low value. For a no-load power transformer, it can be treated as an inductance of which the current lags the voltage by 90°, indicating the current flowing through Phase C was near its peak value while the breaker opened. In this case, there is an overvoltage with an amplitude of −703.9 kV (1.73 p.u.) on Phase C, while the voltages on Phase A and Phase B are relatively lower.

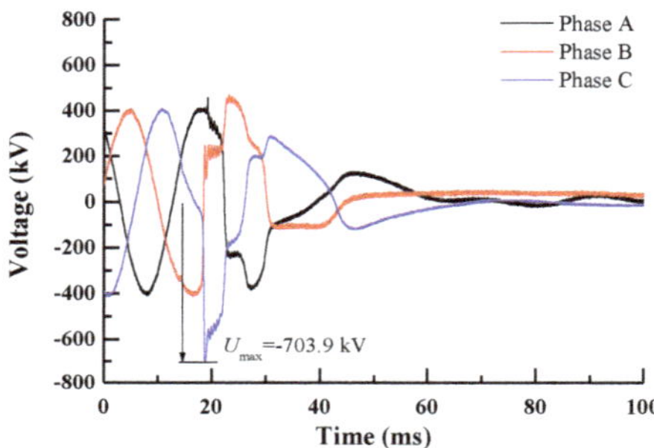

Figure 11. Transient voltage during switching off the #3 power transformer on Batang substation.

4. Discussion

4.1. Data Decoupling

According to the measuring principle, a decoupling process is necessary to recover the actual voltage waveforms on a high-voltage conductor from the original measured data, for which the coefficient matrix is the key. Under an ideal condition, the coefficient matrix can be found by static electric field analysis. However, the actual on-site situation brings about unpredictable variations to the matrix. Therefore, determining the matrix from field measured data is more effective and reasonable. In [23], an example for acquiring the coefficient matrix according to the field measured data is presented in which three capacitor voltage dividers for three phases are switched onto the bus bar by disconnectors. Since the switching processes for the three phases are not absolutely synchronized, the changes of transient voltages would appear asynchronously on the conductors. Taking Phase A as an example, if a sharp voltage change happens in Phase A, all the sensors for the three phases will record an electric field change. According to the definition of k, the first column of the coefficient matrix could be determined. Using the same method, the second and third columns could also be found by recording a sharp voltage change on Phase B and Phase C. Moreover, based on the phase difference and amplitude among three-phase power frequency voltage, several equations could also be established to solve the k coefficient in (5). Above three switching transient measurement results are all decoupled.

For the 220 kV Yan-Zhuang overhead transmission line, no asynchronous transient process has been recorded by far. Thus, it is unfortunate that no opportunity has been given to solve the decoupling coefficients. Instead, a calibration work has been planned on the next power outage of this line. In the calibration, the impulse voltage will be applied to each phase conductor by turns to obtain the decoupling coefficients. However, since the data recorded is all triggered by induction strokes, the three-phase waveforms of transient voltages are synchronized with the same variation trend.

Moreover, the coefficients in the matrix of (5) have the same polarity. With linear superposition, in (5), the variation trends of the measured electric field waveform (E) and the transient voltage waveform (U) would be the same in the time domain.

Setting the waveform on Phase A, Phase B, and Phase C are $U_A(t)$, $U_B(t)$, and $U_C(t)$, respectively. Since they have the same variation trend, they can express as:

$$U_A(\mathrm{t}) = m_a U_n(\mathrm{t}) \tag{8}$$

$$U_B(\mathrm{t}) = m_b U_n(\mathrm{t}) \tag{9}$$

$$U_C(\mathrm{t}) = m_c U_n(\mathrm{t}) \tag{10}$$

where $U_n(t)$ is a normalized waveform, and m_a, m_b, and m_c are the amplitude coefficients for $U_A(t)$, $U_B(t)$, and $U_C(t)$, respectively. Taking $E_{a\text{-}z}(\mathrm{t})$ in (5) as an example, it can be express as:

$$E_{A-z}(\mathrm{t}) = (k_{aA-z}m_a + k_{bA-z}m_b + k_{cA-z}m_c)U_n(\mathrm{t}) \tag{11}$$

It is obvious that the composite electric field has the same variation trend with the original transient voltages. In other words, they will have the same value of time parameters, such as the rising time and half wave time. However, the amplitude value needs to implement the decoupling process.

4.2. Characteristics of Lightning Transient Voltages

The general waveform of the measured lightning transient voltage is a damped oscillation curve. The duration of oscillation can be several milliseconds. The distribution of rising time (t_r) and half wave time (t_h) are shown in Figure 12. The rising time (t_r) of measured lightning transient voltage varies from several microseconds to about 200 microseconds. The distribution of t_r is mainly in the region from 10 μs to 50 μs with a middle value of approximately 34.3 μs. The half wave time (t_h) varies from several tens of microseconds to about 350 microseconds. The distribution of t_h is mainly in the region from 40 μs to 160 μs with a middle value of about 105.25 μs. These features are much different from those of the standard lightning impulse voltage [24]. Thus, it is necessary to conduct research on the insulation performances under the standard lightning impulse voltage and the actual lightning voltage they suffered.

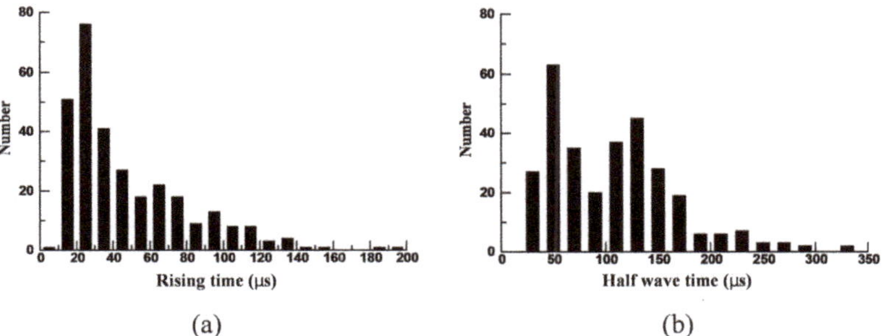

Figure 12. Distribution of rising time and half wave time. (**a**) Rising time; (**b**) half wave time.

5. Conclusions

To measure the transient voltages in an AC power grid safely, a non-contact measuring method was developed. An integrated optical electric-field sensor based on the Pockels effect was developed, as well as the measuring system. These measuring systems were used to monitor the lightning transient voltages and measure the typical switching transient voltages.

(1) The measuring system has a response time less than 6 nanoseconds and a relatively flat amplitude-frequency characteristic from 5 Hz to 100 MHz. Its performances meet the requirements for measuring transient voltages in an AC power grid.

(2) In 2017, 307 lightning transient voltage waveforms were recorded on a 220 kV overhead transmission line. All of these transients were caused by induction strokes with damped oscillation waveforms illustrated. Their rising time varies from several microseconds to about 200 microseconds, of which the middle value is 34.3 μs. The distribution of rising time mainly ranges from 10 μs to 50 μs. The half wave time varies from several tens of microseconds to about 350 microseconds, of which the middle value is 105.25 μs. The distribution of half wave time mainly ranges from 40 μs to 160 μs. These features are much different from those of the standard lightning impulse voltage.

(3) Three types of typical switching transient voltages, including switching on the no-load long transmission line, switching on the no-load power transformer, and switching off the no-load power transformer, were measured during the commissioning of 500 kV middle Tibet power grid. A 1.73 p.u. overvoltage was recorded during switching off the no-load power transformer because the breaker cut off the current near its peak value.

(4) Based on the wave transmission theory, the lightning stroke position, the actual transmission velocity of the traveling wave can be estimated with an accurate time of the transient voltage.

The new measuring technology based on optical electric-field sensor supply a more convenient field measuring method for transient voltages.

Author Contributions: Methodology, S.X.; equipment development, S.X., H.Y. (Huaiyuan Yang); field experiment, Y.Z., H.Y. (Huaiyuan Yang), H.Y. (Hao Yu), X.C., and S.X.; project administration, S.C.; writing—original draft preparation, S.X., C.Z.; writing—review and editing, R.H., Z.M.

Funding: This research was funded by the science and technology project of SGCC, grant number 52199918000G.

Acknowledgments: The authors wish to thank the anonymous reviewers, whose comments assisted in improving the clarity of the paper.

Conflicts of Interest: The authors declare no conflict of interest.

References

1. Xie, S.; Su, S.; He, H.; Guo, R.; Liu, F.; Zhang, C. Internal Overvoltage and Insulation Coordination of Electrode Line in HVDC Project. *High Volt. Eng.* **2018**, *22*, 2331–2337.
2. Lu, B.; Shi, Y.; Zhan, H.; Duan, S. VFTO Suppression by Selection of a Combination of Initial Phase Angle and Contact Velocity. *IEEE Trans. Power Deliv.* **2017**, *33*, 1115–1123. [CrossRef]
3. Bamne, S.; Kinhekar, N. Estimation and Analysis of VFTO at Various Locations in 132kV GIS Substation. In Proceedings of the 2018 International Conference on Computation of Power, Energy, Information and Communication (ICCPEIC), Chennai, India, 28–29 March 2018.
4. Zhao, L.; Ye, L.; Wang, S. Research on Very Fast Transient Overvoltage During Switching of Disconnector in 550kV GIS. In Proceedings of the 2018 IEEE 3rd International Conference on Integrated Circuits and Microsystems (ICICM), Shanghai, China, 24–26 November 2018.
5. International Electrotechnical Commission. *IEC 60071-1 Insulation Co-Ordination—Part 1: Definitions, Principle and Rules*; IEC Central Office: Geneva, Switzerland, 2006; p. 8.
6. Feng, Y.; Wang, X.; Chen, X.M.; Wu, S.P.; Mao, A.L. Influence of Circuit Parameters of Capacitor Voltage Transformer on Grid Harmonic Voltage Measurements. *Proc. CSEE* **2014**, *34*, 4968–4974.
7. Xiao, Y.; Fu, J.; Hu, B.; Li, X.; Deng, C. Problems of Voltage Transducer in Harmonic Measurement. *IEEE Trans. Power Deliv.* **2004**, *19*, 1483–1487. [CrossRef]
8. Ghassemi, F.; Gale, P.; Cumming, T.; Coutts, C. Harmonic Voltage Measurements using CVTs. *IEEE Trans. Power Deliv.* **2005**, *20*, 443–449.
9. Sima, W.X.; Lan, H.T.; Du, L.; Sun, C.X.; Yao, C.G.; Yang, Q. Study on Response Characteritics of Voltage Sensor Mounted at the Tap of Transformer Bushing. *Proc. CSEE* **2006**, *26*, 172–176.

10. Malewski, R.; Douville, J.; Lavallee, L. Measurement of switching transients in 735 kV substations and assessment of their severity for transformer insulation. *IEEE Trans. Power Deliv.* **1988**, *3*, 1380–1390. [CrossRef]
11. Ma, G.M.; Li, C.R.; Quan, J.T.; Jiang, J. Measurement of VFTO Based on the Transformer Bushing Sensor. *IEEE Trans. Power Deliv.* **2011**, *26*, 684–692. [CrossRef]
12. Hanson, L.K.; Fajer, J.; Thompson, M.A.; Zerner, M.C. Electrochromic effects of charge separation in bacterial photo synthesis: Theoretical models. *J. Am. Chem. Soc.* **1987**, *109*, 4728–4730. [CrossRef]
13. Fernandez-Valdivielso, C.; Matias, I.R.; Gorraiz, M.; Arregui, F.J.; Bariain, C.; Lopez-Amo, M. Low Cost Electric Field Optical Fiber Detector. In Proceedings of the Optical Fiber Sensors Conference Technical Digest, Portland, OR, USA, 6–10 May 2002.
14. Shimizu, R.; Matsuoka, M.; Kato, K.; Hayakawa, N.; Hikita, M.; Okubo, H. Development of Kerr electro-optic 3-d electric field measuring technique and its experimental verification. *IEEE Trans. Dielectr. Electr. Insul.* **1996**, *3*, 191–196. [CrossRef]
15. Zahn, M. Optical, electrical and electromechanical measurement methodologies of field, charge and polarization in dielectrics. *IEEE Trans. Dielectr. Electr. Insul.* **1998**, *5*, 627–650. [CrossRef]
16. Zeng, R.; Wang, B.; Niu, B.; Yu, Z. Development and Application of Integrated Optical Sensors for Intense E-Field Measurement. *Sensors* **2012**, *12*, 11406–11434. [CrossRef] [PubMed]
17. Hidaka, K. Progress in Japan of space charge field measurement in gaseous dielectric using a pockels sensor. *IEEE Electr. Insul. Mag.* **1996**, *12*, 17–28. [CrossRef]
18. Kuhara, Y.; Hamasaki, Y.; Kawakami, A.; Murakami, Y.; Tatsumi, M.; Takimoto, H.; Tada, K.; Mitsui, T. BSO/fibre-optic Voltmeter with Excellent Temperature Stability. *Electron. Lett.* **1982**, *18*, 1055–1056. [CrossRef]
19. Mitsui, T.; Hosoe, K.; Usami, H.; Miyamoto, S. Development of Fiber-optic Voltage Sensors and Magnetic-field Sensors. *IEEE Trans. Power Deliv.* **1987**, *2*, 87–93. [CrossRef]
20. Kanda, M.; Masterson, K.D. Optically Sensed EM-field Probes for Pulsed Fields. *Proc. IEEE* **1992**, *80*, 209–215. [CrossRef]
21. Bulmer, C.H.; Burns, W.K.; Moeller, R.P. Linear Interferometric Waveguide Modulator for Electromagnetic-Field Detection. *Opt. Lett.* **1980**, *5*, 176–178. [CrossRef] [PubMed]
22. Cooray, G.V. *The Lightning Flash*, 2nd ed.; The Institution of Engineering and Technology: London, UK, 2008; p. 191.
23. Xie, S.; Wang, H.; Zeng, R. Overvoltage Measurement Technique Based on Integrated Optical Electric Field Sensor. *High Volt. Eng.* **2016**, *42*, 2929–2935.
24. International Electrotechnical Commission. *IEC 60060-3 High Voltage Test Techniques—Part 3 Definitions and Requirements for On-Site Tests*; IEC Central Office: Geneva, Switzerland, 2006; p. 17.

Article

A Traceable High-Accuracy Velocity Measurement by Electro-Optic Dual-Comb Interferometry

Bin Xue [1,*], Haoyun Zhang [1], Tuo Zhao [2] and Haoming Jing [3]

1 School of Marine Science and Technology, TianJin University, TianJin 300072, China; zhanghaoyun@tju.edu.cn
2 Beijing Institute of Aerospace Control Devices, Beijing 100854, China; zhaotuo1993@163.com
3 Department of Electrical and Computer Engineering, UC Santa Barbara, CA 93106, USA; h_jing@ucsb.edu
* Correspondence: xuebin@tju.edu.cn

Received: 20 August 2019; Accepted: 19 September 2019; Published: 2 October 2019

Abstract: An electro-optic dual-comb Doppler velocimeter for high-accuracy velocity measurement is presented in this paper. The velocity information of the object can be accurately extracted from the change of repetition frequency, which is in the microwave frequency domain and can be locked to an atomic clock. We generate two optical combs by electro-optic phase modulators and trace their repetition frequencies to the rubidium clock. One functions as the measurement laser and the other the reference. Experimentally, we verify its high accuracy in the range of 100–300 mm/s with a maximum deviation of 0.44 mm/s. The proposed velocimeter combines the merits of high accuracy and wide range. In addition, since the repetition frequency used for the measurement is traceable to the rubidium clock, its potential superiority in traceability can be utilized in velocity metrology.

Keywords: electro-optic dual-comb interferometry; laser Doppler velocimetry; Traceability

1. Introduction

The laser Doppler velocimeter (LDV) is a well-known equipment used to measure motions, fluid, and airflow [1–6]. By illuminating the flow or object with a laser and measuring the scattering caused by movement, it is possible to calculate its speed [7,8]. LDV is capable of providing high spatial resolution and high response speed and is characterized by non-contact measurement. In recent years, the development of the laser Doppler velocimeter has been more and more mature, and various techniques for the LDV have been explored in practical applications [9–12]. Zeyuan Kuang et al. have designed a dual-polarization fiber grating laser-based laser Doppler velocimeter and achieved a velocity measurement range of 2–37 m/s [13]. Recently, the mode-locked laser is applied in absolute distance measurement and its precision is in the nanometer magnitude. Mohammad U. Piracha et al. utilize a train of oppositely chirped pulses to probe a fast-moving target at >91 m/s [1]. Yan Bai et al. utilize a mode-locked laser to measure a target, whose speed is dozens of m/s, through the method of heterodyne Doppler velocimetry, with the measurement error being only 0.4 m/s [14]. However, these technologies are all used for the field of high-speed measurement. Hongbin Zhu et al. have exploited a birefringent dual-frequency laser Doppler velocimeter in the low-velocity area [15]. In their work, the performance of the developed LDV is evaluated through velocity measurements with a range of 0.159 mm/s to 32.273 mm/s. They all achieved a high resolution and high accuracy but had no thought of metrology. In future, measuring instruments that can be traced directly to the natural standard is a tendency [16].

Recently, optical comb technology has been widely exploited in metrology laboratories and physics research and is starting to become commercially available [17,18]. The optical comb performs as a periodic interval comb in the frequency domain. The stabilized frequency comb is capable of functioning as a high precision wavelength ruler, which offers the unique advantage that the measurement uncertainty can be well traced directly to the atomic clock. Consequently, the measurement accuracy can be improved by several orders of magnitude in comparison to other methods [19–21]. Chih-Hao Li et al. have used a laser frequency comb to calibrate a spectrograph and in their work realize velocity measurement of astronomical objects to a precision of 1 cm/s [22]. Recently, electro-optic (EO) frequency combs have attracted the increasing interest of researchers. In 1994 [23], M. Kourogi generated frequency combs by electro-optical phase modulation. After this, he established the OptoComb Company and gradually applied electro-optic frequency combs on scanners, distance meters, and vibrometers. Distance measurement and vibration measurement technology based on the electro-optical comb has been rapidly developed. However, most applications of the optical comb all make measurements through detecting its phase change. Considering that the repetition frequency of the optical comb is located in the microwave field and can be locked to the atomic clock, by using the Doppler effect of the repetition frequency as a basic principle to measure the target velocity it is possible to trace the velocity directly to the atomic clock. This will also utilize the advantage of high time resolution and frequency resolution of the optical comb.

We are thus motivated to develop a method of an electro-optic dual-comb Doppler velocimeter, making the velocity measurement results able to be traced directly back to a rubidium atomic clock, which directly embodies the metrological thought of shortening the tracing links. In this paper, a narrow line width laser of 1543.5 nm is used as the seed laser. Two optical combs are modulated by the electro-optic modulators and their repetition frequencies are locked to the Rubidium atomic clock. The Doppler frequency shift, which occurs in the repetition frequency of the backward reflected light, is detected after the laser is illuminated on the moving target. By utilizing the Doppler effect formula, the target's velocity can be calculated. Inspired by this idea, we deduce some formulas that prove the correctness of the measurement principle. Then, we propose a measurement method based on it. A verified experiment is set up, and the experimental results show that our idea is feasible and that the proposed measurement method makes full use of the high-accuracy characteristic of the optical comb.

The remainder of this paper is organized as follows. In Section 2 we give original rational deduction towards the Doppler effect of optical comb repetition frequency. In Sections 3 and 4 we introduce the measurement method and our experimental system, followed by a detailed analysis of the experimental data. In Section 5 we conclude with a discussion and future outlook.

2. Measurement Principle

To generate an optical comb, an electro-optic phase modulator is used to modulate the seed laser to produce different frequency sidebands. The modulated laser can be expressed as

$$E = \sum_{m=-16}^{16} A_m \cos[2\pi(f_0 + mf_{rep}) + \varphi_m] \tag{1}$$

where m is an integer (the range of m is [−16, 16], meaning that our experimental device produced 32 sidebands), A_m is the amplitude of the mth sideband, f_0 is the optical frequency of the seed laser, f_{rep} is the repetition frequency of the modulated laser, and φ_m is the initial phase. When this laser beam is incident on the moving target, the corresponding Doppler signal can be written as

$$E_{dop} = \sum_{m=-16}^{16} A_m \cos[2\pi(f_0 + mf_{rep})(1 + \frac{2v}{c}) + \varphi_m] \tag{2}$$

where v is the velocity of the moving target and c is the velocity of light. The repetition frequency of the optical comb changes from f_{rep} to $(1+2v/c)f_{rep}$. Thus, the Doppler shift occurring at the repetition frequency reflects the information of velocity. When the target is close to the laser source, v is positive and the repetition frequency is increased. When the target is moving towards the laser source, v is negative and the repetition frequency is reduced. From Formula (1) and Formula (2), the velocity of the moving target can be expressed as $v{=}\Delta f_{rep}/2f_{rep}{\cdot}c$, where Δf_{rep} is the Doppler shift of the repetition frequency.

As the velocity of the moving target is much smaller than that of light, in order to improve the resolution, it is better to increase the modulation repetition frequency. For the same moving velocity, the higher the repetition frequency, the greater the Doppler shift and the higher the resolution. However, detecting high repetition frequency requires very expensive detection equipment. Hence, we developed a method using a dual optical comb to detect the Doppler shift. One optical comb modulates the repetition frequency to $f_{rep.sig}$ as a signal source and the other optical comb modulates the repetition frequency to $f_{rep.loc}$ as a local oscillator.

The signal source can be expressed as

$$E_{sig\_i} = \sum_{m=-16}^{16} A_m \cos[2\pi(f_0 + f_{AOM} + mf_{rep.sig}) + \varphi_{m1}] \tag{3}$$

The local oscillator can be expressed as

$$E_{loc} = \sum_{m=-16}^{16} A_m \cos[2\pi(f_0 + mf_{rep.loc}) + \varphi_{m2}] \tag{4}$$

Then, the beat signal manifests as a new comb in the frequency domain and can be detected by an oscilloscope. The detected beat signal can be expressed as

$$I \propto \sum_{m=-16}^{16} \cos[2\pi(f_{AOM} + m(f_{rep.sig} - f_{rep.loc}) + (\varphi_{m1} - \varphi_{m2})] \tag{5}$$

After being reflected by the moving target, the signal source changes to

$$E_{sig\_r} = \sum_{m=-16}^{16} A_m \cos[2\pi(f_0 + f_{AOM} + mf_{rep.sig})(1 + \frac{2v}{c}) + \varphi_{m3}] \tag{6}$$

Hence, the detected beat signal can be expressed as

$$I' \propto \sum_{m=-16}^{16} \cos[2\pi((1 + \frac{2v}{c})f_{AOM} + \frac{2v}{c}f_0 + m(1 + \frac{2v}{c})f_{rep.sig} - mf_{rep.loc}) + (\varphi_{m3} - \varphi_{m2})] \tag{7}$$

When comparing Formula (5) and Formula (7), it can be seen that the repetition frequency changes from $f_{rep.sig}$-$f_{rep.loc}$ to $(1+2v/c)f_{rep.sig}$-$f_{rep.loc}$. In this way, the velocity of the moving target can be calculated.

3. Experiment Setup

Figure 1 shows the experimental system. The output of the seed laser (RIO ORION Laser Module, 1543.5 nm, 5.1 kHz line width, 18 mW) is spilt into two parts. One part is modulated by an electro-optic phase modulator (EOM, EOSpace, <5 V half-wave voltage V_π, 2 W RF power) with 10 GHz repetition frequency, which works as the seed of the signal source. The output of the EOM is amplified to 500 mW by an Er-doped fiber amplifier. After the optical circular, the beam size of the signal source is expanded to 20 mm by a large collimator (micro laser system, FC40), which can guarantee that the received reflected light is strong enough. Then, the laser is incident on a moving corner prism, which is mounted on an electrical rail (Zolix, LMA-TR-200, travel 200 mm, resolution 1 μm/s) whose velocity can be precisely controlled. The other part serves as the seed of the local oscillator. An acousto-optic modulator (AOM; AA opto-electronic MT110, 110 MHz frequency shift) is used to avoid the frequency ambiguity. Then, the output of the AOM is modulated by another EOM (EOSpace, <5 V V_π, 2W RF power) with 10.001 GHz repetition frequency. The RF driver1 (Agilent N5173B) and RF driver2 (Agilent E8267D) are locked to a rubidium atomic clock (Microsemi 8040, 2×10^{-11} stability with 1 s averaging time). Finally, the two parts are combined by a beam splitter and detected by a fast photodetector (Menlosystems FD310). The stored waveforms by an oscilloscope (LeCroy Waverunner 610Zi) can be processed to obtain the Doppler shift information.

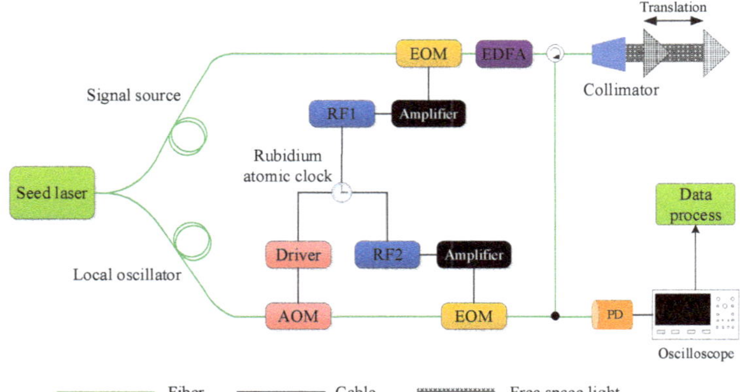

Figure 1. Experimental system for the dual optical comb laser Doppler velocimeter (LDV). Legend: RF1, RF driver1; RF2, RF driver2; EOM, electro-optic modulator; AOM, acousto-optic modulator; EDFA, Er-doped fiber amplifier; PD, photodetector.

4. Results and Discussion

Before moving the target, beat notes between the signal source and the local oscillator were measured by a spectrum analyzer (ROHDE&SCHWARZ FSH8, 3 kHz RBW). Then, the waveform obtained by the oscilloscope was Fourier transformed to obtain the new comb due to beat. As shown in Figure 2, we can see that the Fourier transform can basically reproduce the beat frequency spectrum. Additionally, the center frequency of the beat notes is 110 MHz, which is the driving frequency of the AOM. The frequency interval of the comb teeth is 1 MHz, which is the frequency difference between the driving frequencies of the two RF drivers. Actually, the frequency of RF driver1 is 10 GHz and the frequency of RF driver2 is 10.001 GHz. This means that $f_{rep.sig}$ = 10 GHz and $f_{rep.loc}$ = 10.001 GHz.

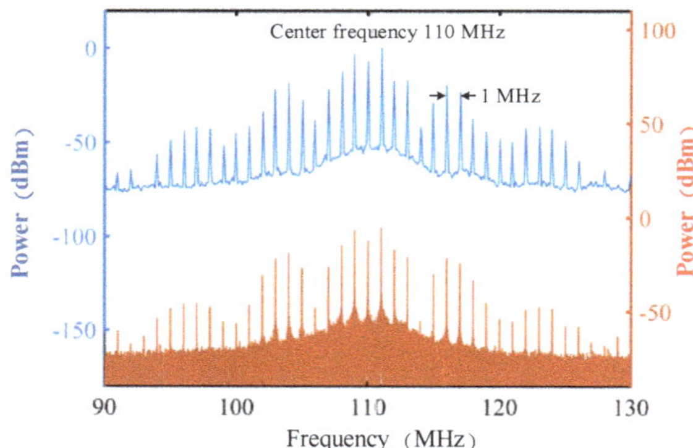

Figure 2. Beat notes between the signal source and the local oscillator. The blue line is measured by a spectrum analyzer with the left vertical axis; the red line is the result of Fourier transforming the waveform obtained by the oscilloscope with the right vertical axis.

To show the feasibility of the electro-optic dual-comb Doppler velocimeter, Figure 3 shows the difference between Fourier transforming the waveforms when the target is stationary and when it is moving. In Figure 3a, the black comb expresses the measured spectrum when the target is stationary and the red comb is that when the target is moving away from the collimator. It can be seen that every comb has a reduced offset due to the Doppler effect. As the target's velocity is much lower than that of light, the Doppler shift generated at 10 GHz repetition frequency is too small to be seen in the spectrum. In order to show that the repetition frequency has changed, we translate the red comb and move its center frequency to 110 MHz in Figure 3b. Hence, the changes of frequency interval can be seen on the high-order comb teeth due to the cumulative effect. It is thus proved that the Doppler effect causes a change in the repetition frequency, which can be used to calculate the velocity of a moving target. To show the directional discriminability, Figure 3c shows the frequency spectrum when the target is stationary, as well as leaving from and approaching the collimator with $v = -100$ mm/s and $v = +100$ mm/s. As can be seen, for the same teeth of the combs, the green line is on the left and red line is on the right. This proves that approaching leads to an increase in the repetition frequency and leaving leads to a reduction in the repetition frequency.

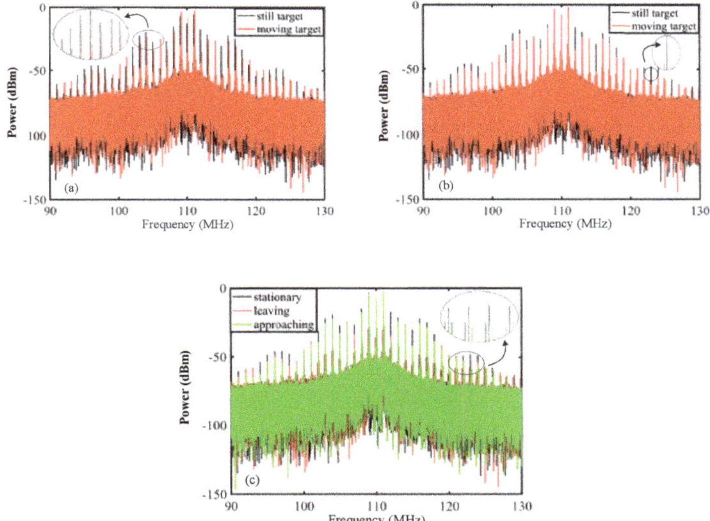

Figure 3. Beat notes when the target is moving and stationary. The black line is the spectrum when the target is stationary; the red line is the spectrum when the target is moving away from the collimator. (**a**) Fourier transforming the waveform obtained by the oscilloscope; (**b**) translating the red comb and moving its center frequency to 110 MHz; (**c**) frequency spectrum of the target in different states.

When processing the data, we selected the two teeth at 110 MHz and 126 MHz to calculate the change in repetition frequency. This means taking the average of 16 repetition frequency data and increasing the resolution by 16 times. The greater the distance between the two teeth involved in the calculation, the higher the resolution. However, we saw only 16 sidebands in the spectrum analyzer; the other sidebands were suppressed due to their weak intensity. Hence, we selected those two teeth to make the calculation. We have conducted experiments to set the moving target at v from 100 to 300 mm/s with an interval of 50 mm/s and in the directions toward and away from the collimator. Figure 4 shows the results of velocity measurement. As shown in Figure 4, the velocity of the target is accurately measured by the dual optical comb LDV with a maximum deviation of 0.44 mm/s. Due to the minimum speed of the electric rail being 100 mm/s, we were not able to measure at a lower speed.

In addition, due to the limit of the length of the electric rail, the sampling time was too short at a higher speed to collect enough data.

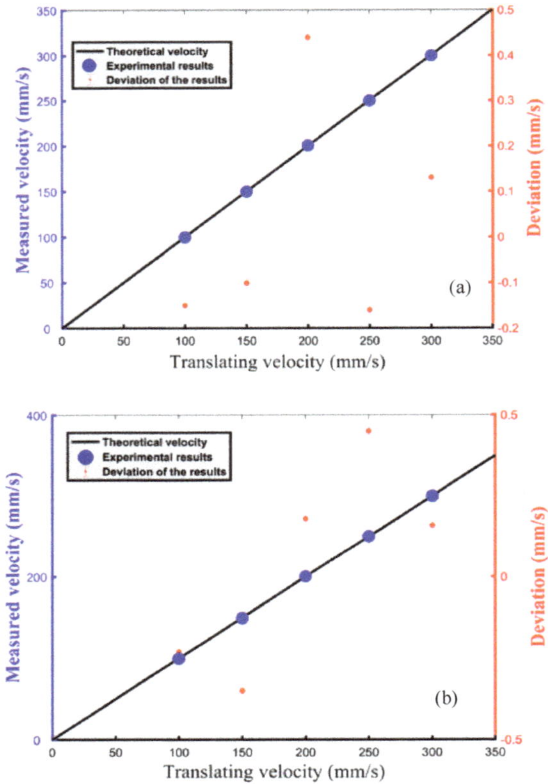

Figure 4. Results of velocity measurement. (**a**) The target is leaving the collimator; (**b**) the target is approaching the collimator.

Theoretically, the upper limit of the speed measurement is limited by the difference between the two optical comb repetition frequencies. In this experiment, a difference of 1 MHz limited the maximum speed of the measurement at 15 km/s. Additionally, the resolution was limited by the frequency stability and FFT resolution. Figure 5 shows the Allan deviation of a 1 MHz frequency, which is mixed with the output of the two RF drivers at 110 MHz and 111 MHz. The Allan deviation is well below 14 mHz, with the averaging time larger than 1 s, which corresponds to a resolution of 0.21 mm/s. In our data processing, the FFT resolution was 5 Hz, which corresponds to a resolution of 75 mm/s. Since we selected 16 teeth to calculate the change of repetition frequency, the resolution could be improved to 4.69 mm/s.

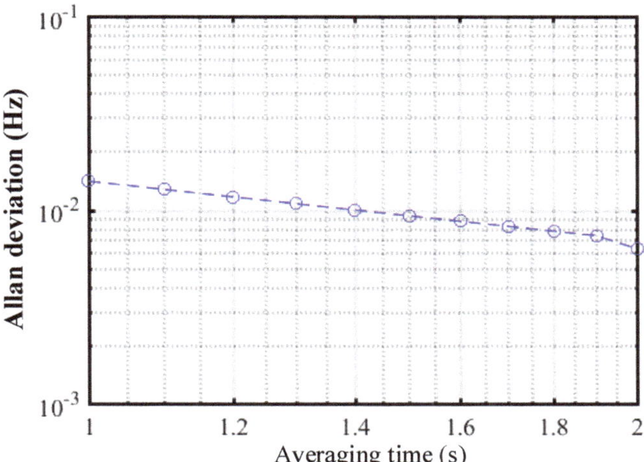

Figure 5. Allan deviation of 1 MHz frequency.

5. Conclusions

In order to achieve a wide range and high accuracy in the field of velocity measurement, an electro-optic dual-comb Doppler velocimeter was proposed and studied in this work. In principle, the proposed velocimeter which was derived is feasible and can further improve velocity measurement accuracy. Experimentally, we locked the repetition frequencies of the two optical combs in the microwave frequency domain and beat them to make the detection. In this way, we verified its high accuracy in the range of 100–300 mm/s. The measured range of 4.69 mm/s to 15 km/s was theoretically derived. The proposed velocimeter combines the merits of high precision, high accuracy, and wide range. In addition, since the repetition frequency used for the measurement is traceable to a rubidium clock, its potentially superior traceability can be utilized in velocity metrology. It is of great meaning and necessity because it helps to provide an available velocimeter with high stability and an extremely compact configuration, making a potential contribution to the velocimeter in practical engineering applications.

Author Contributions: B.X. designed the experiment. H.Z. and T.Z. performed the experiment. H.Z. and T.Z. analyzed the data. H.Z. and T.Z. wrote the original manuscript. B.X. and H.J. did the review and editing.

Funding: This research was funded by the Research Project of Tianjin Education Commission (No.JWK1616).

Acknowledgments: We thank Zhiyang Wang and Kai Zhang for their help in performing the experiment.

Conflicts of Interest: The authors declare no conflict of interest.

References

1. Piracha, M.U.; Nguyen, D.; Ozdur, I.; Delfyett, P.J. Simultaneous ranging and velocimetry of fast moving targets using oppositely chirped pulses from a mode-locked laser. *Opt. Express* **2011**, *19*, 11213–11219. [CrossRef]
2. Kliese, R.; Rakić, A.D. Spectral broadening caused by dynamic speckle in self-mixing velocimetry sensors. *Opt. Express* **2012**, *20*, 18757–18771. [CrossRef]
3. Chen, J.B.; Zhu, H.B.; Xia, W.; Guo, D.M.; Hao, H.; Wang, M. Self-mixing birefringent dual-frequency laser Doppler velocimeter. *Opt. Express* **2017**, *25*, 560–572. [CrossRef]
4. Yoshikawa, E.; Ushio, T. Wind ranging and velocimetry with low peak power and long-duration modulated laser. *Opt. Express* **2017**, *25*, 8845–8859. [CrossRef]

5. Beuth, T.; Fox, M.; Stork, W. Influence of laser coherence on reference-matched laser Doppler velocimetry. *Appl. Opt.* **2016**, *55*, 2104–2108. [CrossRef]
6. Eiji, O.; Yohei, W. Influence of perfusion depth on laser Doppler flow measurements with large source-detector spacing. *Appl. Opt.* **2003**, *42*, 3198–3204.
7. Ojo, A.O.; Fond, B.; Wachem, B.G.V.; Heyes, A.L.; Beyrau, F. Thermographic laser Doppler velocimetry. *Opt. Lett.* **2015**, *40*, 4759–4762. [CrossRef]
8. Otsuka, K. Self-mixing thin-slice solid-state laser Doppler velocimetry with much less than one feedback photon per Doppler cycle. *Opt. Lett.* **2015**, *40*, 4603–4606. [CrossRef]
9. Maru, K.; Watanabe, K. Cross-sectional laser Doppler velocimetry with nonmechanical scanning of points spatially encoded by multichannel serrodyne frequency shifting. *Opt. Lett.* **2014**, *39*, 135–138. [CrossRef]
10. Maru, K.; Fujiwara, T.; Ikeuchi, R. Nonmechanical transverse scanning laser Doppler velocimeter using wavelength change. *Appl. Opt.* **2011**, *50*, 6121–6127. [CrossRef]
11. Moro, E.A.; Briggs, M.E.; Hull, L.M. Defining parametric dependencies for the correct interpretation of speckle dynamics in photon Doppler velocimetry. *Appl. Opt.* **2013**, *52*, 8661–8669. [CrossRef]
12. Maru, K.; Hata, T. Non-mechanical scanning laser Doppler velocimeter with directional discrimination using single transmission path. In Proceedings of the 2013 Conference on Lasers and Electro-Optics Pacific Rim (CLEOPR), Kyoto, Japan, 30 June–4 July 2013; Volume 125, pp. 1–2.
13. Kuang, Z.Y.; Cheng, L.H.; Liang, Y.Z.; Liang, H.; Guan, B.O. Dual-polarization fiber grating laser-based laser Doppler velocimeter. *Chin. Opt. Lett.* **2016**, *14*, 10–13.
14. Bai, Y.; Ren, D.M.; Zhao, W.J.; Qu, Y.C.; Qian, L.M.; Chen, Z.L. Heterodyne Doppler velocity measurement of moving targets by mode-locked pulse laser. *Opt. Express* **2012**, *20*, 764–768. [CrossRef]
15. Zhu, H.B.; Chen, J.B.; Guo, D.M.; Xia, W.; Hao, H.; Wang, M. Birefringent dual-frequency laser Doppler velocimeter using a low-frequency lock-in amplifier technique for high-resolution measurements. *Appl. Opt.* **2016**, *55*, 4423–4429. [CrossRef]
16. Mari, L.; Goodwin, J.; Jacobson, E.; Pendrill, L. The definitions of the base units in the revised SI: A terminological analysis. *Measurement* **2019**, 106976, in press. [CrossRef]
17. Wu, H.Z.; Zhang, F.M.; Liu, T.Y.; Petr, B.; Qu, X.H. Absolute distance measurement by multi-heterodyne interferometry using a frequency comb and a cavity-stabilized tunable laser. *Appl. Opt.* **2016**, *55*, 4210–4218. [CrossRef]
18. Wu, H.Z.; Zhang, F.M.; Liu, T.Y.; Meng, F.; Li, J.S.; Qu, X.H. Absolute distance measurement by chirped pulse interferometry using a femtosecond pulse laser. *Opt. Express* **2015**, *23*, 31582–31593. [CrossRef]
19. Delfyett, P.J.; Gee, S.; Choi, M.T.; Izadpanah, H.; Lee, W.K.; Ozharar, S.; Quinlan, F.; Yilmaz, T. Optical Frequency Combs from Semiconductor Lasers and Applications in Ultrawideband Signal Processing and Communications. *J. Lightwave Technol.* **2006**, *24*, 2701–2719. [CrossRef]
20. Diddams, S.A. The evolving optical frequency comb [Invited]. *J. Opt. Soc. Am. B* **2010**, *27*, B51–B62. [CrossRef]
21. Jones, D.J.; Diddams, S.A.; Taubman, M.S.; Cundiff, S.T.; Ma, L.S.; Hall, J.L. Frequency comb generation using femtosecond pulses and cross-phase modulation in optical fiber at arbitrary center frequencies. *Opt. Lett.* **2000**, *25*, 308–310. [CrossRef]
22. Li, C.J.; Benedick, A.J.; Fendel, P.; Glenday, A.G.; Kärtner, F.X.; Phillips, D.F.; Sasselov, D.; Szentgyorgyi, A.; Walsworth, R.L. A laser frequency comb that enables radial velocity measurements with a precision of 1 cm s-1. *Nature* **2008**, *452*, 610–612. [CrossRef]
23. Kourogi, M.; Enami, T.; Ohtsu, M. A monolithic optical frequency comb generator. *IEEE Photonics Technol. Lett.* **1994**, *6*, 214–217. [CrossRef]

Article

Monolitic Hybrid Transmitter-Receiver Lens for Rotary On-Axis Communications

René Kirrbach [1,2,*], Michael Faulwaßer [1], Tobias Schneider [1], Philipp Meißner [1], Alexander Noack [1] and Frank Deicke [1]

1 Department of Wireless Microsystems, Fraunhofer IPMS, 01109 Dresden, Germany; michael.faulwasser@ipms.fraunhofer.de (M.F.); tobias.schneider@ipms.fraunhofer.de (T.S.); philipp.meissner@ipms.fraunhofer.de (P.M.); alexander.noack@ipms.fraunhofer.de (A.N.); frank.deicke@ipms.fraunhofer.de (F.D.)

2 Faculty of Electrical and Computer Engineering, Technical University of Dresden, 01069 Dresden, Germany

* Correspondence: rene.kirrbach@ipms.fraunhofer.de or rene.kirrbach@mailbox.tu-dresden.de

Received: 29 January 2020; Accepted: 19 February 2020; Published: 24 February 2020

Abstract: High-speed rotary communication links exhibit high complexity and require challenging assembly tolerances. This article investigates the use of optical wireless communications (OWC) for on-axis rotary communication scenarios. First, OWC is compared with other state-of-the-art technologies. Different realization approaches for bidirectional, full-duplex links are discussed. For the most promising approach, a monolithic hybrid transmitter-receiver lens is designed by ray mapping methodology. Ray tracing simulations are used to study the alignment-depended receiver power level and to determine the effect of optical crosstalk. Over a distance of 12.5 mm, the lens achieves an optical power level at the receiver of −16.2 dBm to −8.7 dBm even for misalignments up to 3 mm.

Keywords: hybrid lens; optical wireless communications; Li-Fi; freeform lens; optic design; rotary interfaces; rotary joint; wireless rotary electrical interface; rotating electrical connectors; full-duplex data transfer; Gigabit-Ethernet; industrial communications; real-time

1. Introduction

Reliable, real-time connectivity is the backbone of industrial automation. Data transmission over rotating parts is required in a broad range of applications such as wind turbines [1], industrial communications [2,3], surveillance radars [3–7], military [2,8], aerospace [9] and many more. Table 1 gives an overview of the most important transmission principles used in rotary communication links. Slip rings were widespread in the past. However, due to mechanical contact they suffer wear and tear which limits their durability [1,2,5,6,9–11] and thus increases maintenance costs. Precious brush materials and lubricants are used to extend lifetime [12] at the expense of increased system complexity and higher costs. Therefore, contactless data transfer is favored nowadays [6,10]. Life times of several hundreds of revolutions are common and rotation speeds in the order of magnitude of 10^3 rpm or even 10^4 rpm are reached with contactless methods.

Capacitive-based near-field transmission links are known as reliable and cost-effective [11]. However, the system proposed by Doleschel et al. [13] shows that system complexity of modern solutions is clearly not negligible. Data rates of up to several Gbit/s are possible [11,13]. Practical systems support conventional Gigabit-Ethernet and industrial protocols like ProfiNET and EtherCAT for instance. Thus, devices with data rates ranging from 500 kbit/s to 1 Gbit/s were developed [13]. The maximum transmission distance is in the range of 1 cm. Inductive coupling is mainly employed for power transfer and only rarely used for high-speed data transmission [14,15].

Table 1. Common principles for rotary data transmission and their typical state-of-the-art performance. Values should be understood as orders of magnitude rather than exact values. * Higher values are possible but have not been published yet; ‡ this work is limited to on-axis scenarios.

Group	Contact	Contactless				
Type	**Slip Ring**	**Capacitive**	**RF**	**Fiber (FORJ)**		**OWC**
				Single	**Multi**	
Ref.	[5,6,16–18]	[6,9,11,13]	[1,6,10,11,13]	[4,7,8]	[6,13]	[19]
data rate/Gbit/s	0.1... 3	$5 \cdot 10^{-4}$...>1	0.054...5	10...40	>40	> 10 *
max speed / rpm	10^1...10^4	10^3...10^4	10^3...10^4	10^3... $\cdot 10^4$	10^2...10^3	>1400 *
max revolutions	10^7...$2 \cdot 10^8$	>10^8	>10^8	>10^8	>10^8	>10^8
cost (initial)	medium	low/medium	low/medium	high	very high	low
cost (maintenance)	high	low	low	low	high	low
RF immunity	weak	medium	weak	strong	strong	strong
on-/off axis	off-axis	off-axis	off-axis	on-axis	on-axis	on-axis ‡

Several systems using conventional low-power radio-frequency (RF) technologies were proposed [1,10] with data rates in the range of tens of Mbit/s or below. Standards like 802.11ac and 802.11ad might be able to provide data rates in the range of Gbit/s. Their main problem is reliability and robustness in terms of ensuring a bandwidth and low-latency data transfer [2,3,13]. Future millimeter-wave based communication [20] standards like IEEE 802.11ay might even reach data rates in the range of several Gbit/s to several tens of Gbit/s [11]. However, their practicality and cost-effectiveness has to be proven.

Highest data rates are reached with fiber optical rotary joints (FORJ). Besides their superior data rate in the range of Gbit/s up to multiple tens of Gbit/s per channel [6,7,13], these links provide immunity against RF. Single-fiber systems only consist of an optical transceiver at both sides and optical fibers in-between. However, due to sophisticated mechanical alignment [8], these systems are expensive. Multi-fiber links offer even higher data rates but exhibit a very high complexity [6].

Optical wireless communications (OWC) aim to combine the advantages of rotary FORJ with relaxed mechanical tolerances, reduced system complexity and thus lower costs. Initially only light emitting diode (LED) based systems with data rates in the lower Mbit/s range [2,3] or laser diode based uni-directional links were demonstrated [21]. Faulwaßer et al. [19] introduced a full-duplex link for data rates of up to 10 Gbit/s for on-axis rotary data transfer. Similar to fiber-based communications, data rates are likely to increase in the future. The demonstrated rotation speed from 0 rpm to 1400 rpm [19] was limited by the test equipment. There is no OWC-exclusive factor limiting the speed. The lifetime of OWC links is expected to be similar to other contactless methods, since there are no significant aging effects. The optoelectronic components, i.e., light emitting diodes (LEDs), laser diodes (LDs) and photodiodes (PDs) are known for high reliability and long lifetime [22–24]. The key element of the transceiver in [19] is a monolithic hybrid lens that acts as transmitter (TX) and receiver (RX) optics in parallel and thereby relaxes the mechanical alignment to several millimeters. The form factor of the transceiver is only 5 mm × 5 mm × 5 mm [19].

This article investigates the potential of OWC for bidirectional, full-duplex, on-axis rotary scenarios and describes how to design a monolithic, hybrid TX-RX lens. In Section 2 a channel model is used to derive some adequate figures of merit. The use of a hybrid lens is motivated by discussing several realization approaches of rotary OWC. Next, the design procedure of a hybrid lens is described and the choice of design parameters is discussed. In Section 3, the performance of the lens and a second system is evaluated and compared using optical ray tracing simulations. Thereby, the alignment-depended optical signal power at the receiver and optical crosstalk is studied. The results are discussed in Section 4. Finally, Section 5 provides a short summary.

2. Materials and Methods

2.1. Fundamental Concepts

2.1.1. Channel Model

OWC use optical emitters like LEDs or LDs at the TX to convert an electrical signal into the optical domain. A PD is used for back-conversion at the RX. Similar to FORJ, both transceivers are placed in front of each other [25]. OWC use lenses instead of optical fibers to enable larger mechanical tolerances. The key goal for the designer is to increase the optical power P_{PD} that falls onto the PD in order to improve the RX signal-to-noise ratio (SNR) [26] and to minimize the bit-error-rate (BER). If the SNR is already sufficient, the excess can be converted into a higher data rate by increasing bandwidth or applying a multilevel modulation scheme.

There will always be a misalignment between both transceivers due to positioning, assembly tolerances or vibrations. The rotation axis might even exhibit a nutation, i.e., nonideal motion around the ideal rotation axis. The combination of these nonidealities and the communication distance leads to a minimum field of view (FOV) that is required for robust operation. The FOV sizes are denoted by the half-opening angles $\theta_{\mathrm{TX\ FOV}}$ and $\theta_{\mathrm{RX\ FOV}}$. In order to ensure eye-safe operation, i.e., to classify the system as laser class 1 according to IEC 60825-1:2014 (DIN EN 60825-1:2015-07) [27], the optical transmitter power P_{TX} is limited. Therefore, efficient transceiver design is required to meet link-budget requirements, ensure eye-safety and to maximize data rate. The designer tries to maximize the dynamic range by keeping the TX and RX performance constant within a plane perpendicular to the optical axis [28]. In other words, TX has to provide constant irradiance E_{TX} and RX has to detect the same signal level within this plane.

For a moment we assume the distance z between both transceivers is large compared to their apertures. Although this does not fully apply for short distances, the assumption is useful to show the fundamental dependencies. As a result, we can assume that parallel rays are incident onto RX. For a homogeneous FOV, the optical power P_{PD} can be expressed as product of the irradiance $E_{\mathrm{TX}}(z)$, the effective receiver input aperture $A_{\mathrm{RX,eff}}$ and the geometrical coupling coefficient ϵ_{c} as shown in Equation (1). For the ideal optical link, TX and RX have the same FOV size and both FOVs are placed on the optical axis. In practice, both FOVs may differ or they might be misaligned. The geometrical coupling coefficient ϵ_{c} quantifies this misalignment. It becomes crucial for short-distance communication as we know from the challenging assembly tolerances of FORJ. ϵ_{c} is defined as the fraction of the illuminated area of TX, which is overlapping with the FOV of the RX divided by the total illuminated area $A_{\mathrm{FOV\ TX}}$. For the ideal optical channel, i.e., equally sized TX and RX FOVs with no misalignment, $\epsilon_{\mathrm{c}} = 1$ applies.

$$P_{\mathrm{PD}} = E_{\mathrm{TX}}(z, \theta_{\mathrm{TX\ FOV}}) \cdot A_{\mathrm{RX,eff}}(\theta_{\mathrm{RX\ FOV}}) \cdot \epsilon_{\mathrm{c}} \tag{1}$$

The effective receiver input aperture $A_{\mathrm{RX,eff}}(\theta_{\mathrm{RX\ FOV}})$ is expressed as product of active PD area A_{PD}, optical gain $g(\theta_{\mathrm{RX\ FOV}})$ and efficiency $\eta_{\mathrm{RX}}(\theta_{\mathrm{i}})$ at the angle of incidence θ_{i} as it is shown in Equation (2). A large PD area A_{PD} is favorable for the link budget but goes along with a large PD capacitance, which limits the modulation bandwidth (BW) [26,29]. Consequently, a PD with large A_{PD} but sufficient BW is chosen.

$$P_{\mathrm{PD}} = E_{\mathrm{TX}}(z, \theta_{\mathrm{TX\ FOV}}) \cdot A_{\mathrm{PD}} \cdot g(\theta_{\mathrm{RX\ FOV}}) \cdot \eta_{\mathrm{RX}}(\theta_{\mathrm{i}}) \cdot \epsilon_{\mathrm{c}} \tag{2}$$

The RX FOV should always be chosen as small as possible, since the optical gain $g(\theta_{\mathrm{RX\ FOV}})$ decreases with increasing FOV due to conservation of Ètendue [30]. Moreover, a restricted RX FOV improves the robustness against noise and interchannel interference. Next, $g(\theta_{\mathrm{RX\ FOV}})$ is substituted by the maximum theoretical optical gain [30] as shown in Equation (3). Now $\eta_{\mathrm{RX}}(\theta_{\mathrm{i}})$ is used as a figure of merit for the optical RX efficiency. The ideal lossless receiver achieves $\eta_{\mathrm{RX}}(\theta_{\mathrm{i}}) = 1$ for all $\theta_{\mathrm{i}} \in [0, \theta_{\mathrm{RX\ FOV}}]$. The angle $\theta_{\mathrm{RX,PD,max}}$ denotes the maximum coupling angle from the RX optics to the

PD surface normal. In the next step, the irradiance $E_{TX}(z, \theta_{TX\ FOV})$ is substituted by the product of the optical TX power P_{TX} and the TX efficiency η_{TX} divided by the illuminated spot area $A_{FOV\ TX}$.

$$P_{PD} = \frac{P_{TX} \cdot \eta_{TX}}{A_{FOV\ TX}} \cdot \left(\frac{n_1 \cdot \sin\theta_{RX,PD}}{n_{air} \cdot \sin\theta_{RX\ FOV}} \right)^2 \cdot A_{PD} \cdot \eta_{RX}(\theta_i) \cdot \epsilon_c \tag{3}$$

$A_{FOV\ TX}$ is replaced by the corresponding triangular relationship, which includes the communication distance z and the tangent of the TX FOV $\theta_{TX\ FOV}$. Finally, Equation (4) is a useful expression for the most important geometric dependencies of P_{PD}.

$$P_{PD} = \frac{P_{TX} \cdot \eta_{TX}}{\pi \cdot (z \cdot \tan\theta_{TX\ FOV})^2} \cdot \left(\frac{n_1}{\sin\theta_{RX\ FOV}} \right)^2 \cdot A_{PD} \cdot \eta_{RX} \cdot \epsilon_c \tag{4}$$

The performance merits from Equation (4) are η_{TX}, η_{RX} and ϵ_c. The transmitter efficiency η_{TX} specifies how much of the emitted power P_{TX} is concentrated into the FOV. Its loss mechanisms are $\zeta_{TX\ M}$ and $\zeta_{TX\ F}$. $\zeta_{TX\ M}$ describes rays that strike the target plane outside the FOV. $\zeta_{TX\ F}$ describes back-reflected rays due to Fresnel-reflections.

Since η_{TX} contains no information concerning irradiance homogeneity, we additionally introduce the effective transmitter efficiency $\eta_{TX\ eff}$. It is defined according to Equation (5) by using the minimum irradiance within the FOV E_{min} [28,31]. The difference between η_{TX} and $\eta_{TX\ eff}$ is called the inhomogeneity factor $\zeta_{TX\ H}$.

$$\eta_{TX\ eff} = \frac{E_{min}}{E_{min,ideal}} = \frac{E_{min} A_{FOV\ TX}}{P_{TX}} \tag{5}$$

The sensitivity of RX is typically limited by the interaction of several internal noise mechanisms that exhibit a Gaussian probability function [26]. The signal can additionally be corrupted by optical crosstalk, i.e., adjacent communication channels. This nonGaussian noise has a limited range of variation. This effect introduces a power penalty PP [26] that reduces the usable peak-to-peak amplitude of the signal. The optical power that falls onto the PD consists of a signal part P_{PD} and another part arising from crosstalk $P_{PD\ cross}$. The corrected power value $P_{PD\ PP}$ takes the eye-closure effect into account by subtracting the $P_{PD\ cross}$ from P_{PD} [26]. It holds $P_{PD\ PP} < P_{PD}$ as soon as crosstalk is present. In this work, all optical power values are understood as average values to ensure comparability with literature. When dealing with the power penalty the signal strength is considered in a peak-to-peak manner. It is still valid to consider average values as long as the extinction ratio of the signal and the crosstalk signal is equal. This assumption applies to our case, since both link directions are designed equally and the crosstalk arises from the signal itself.

2.1.2. Ideal Arrangement

The ideal arrangement consists of TX and RX placed at the same position on the optical axis as shown in Figure 1. Both FOVs are equal in size and they fully overlap. The arrangement can be realized by using an LED. On the one hand, applying a forward-bias to the LED causes a forward-current and leads to photon generation. Biasing the PN-junction reversely enables fast photo-detection on the other hand. As a result, the transceiver only needs a single optoelectronic component for TX and RX. Data rates of up to 150 Mbit/s were demonstrated for close distances [32]. However, the LED is a rather bad PD. It has a low responsivity, small area A_{PD} and low bandwidth [32]. Although a bidirectional link is realizable, data transfer is restricted to half-duplex mode, because the link direction is switched in time domain.

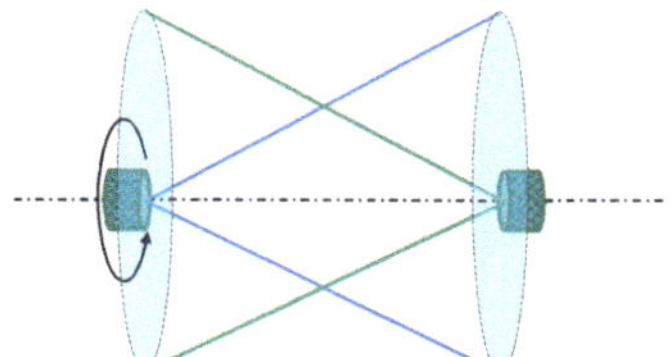

Figure 1. Ideal rotary arrangement: transmitter (TX) (blue) and receiver (RX) (green) of a transceiver are placed on the optical axis at the same position. The field of view (FOV) of TX and RX overlap ideally.

For high-speed bidirectional data transfer in full-duplex mode, TX and RX are separated. The optical arrangement becomes more challenging and design trade-offs have to be met. The following section introduces several geometrical arrangements for optical wireless rotary communication scenarios and compares their performance.

2.1.3. Geometrical Arrangements

Figure 2 shows three geometrical approaches for rotary OWC. In Figure 2a TX and RX are placed next to each other separated by a spacing d. In Figure 2b, TX and RX are radially separated regarding the optical axis. Third, both elements arranged along the optical axis and the front element are transparent as illustrated in Figure 2c.

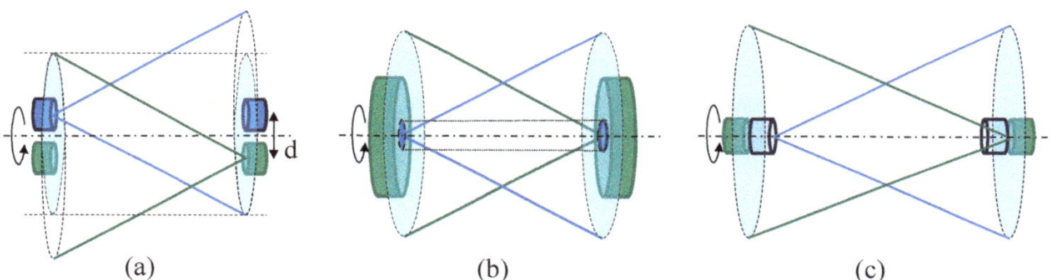

Figure 2. Schematic illustration of arrangements of separated TX (blue) and RX (green). (**a**) TX and RX are placed next to the rotation axis. The FOV of TX and RX overlap only partially. (**b**) TX and RX are placed at the optical axis and they are separated in radial direction. (**c**) TX and RX are placed in a line along the optical axis and the front element is transparent.

Approach (a) is used by commercial low data rate IrDA transceivers like Vishays *TFBS4711* [33] but also by a high-speed transceiver demonstrated by Faulwaßer et al. [34]. The axis of rotation is placed trough one of the elements or between them. This leads to misaligned FOVs. The designer tries to minimize the spacing d. Next, both FOVs are chosen large enough to ensure a sufficient ϵ_c. As we learned from Equation (4), P_{PD} scales with $1/\tan\theta^2_{\text{TX FOV}}$ and $1/\sin\theta^2_{\text{RX FOV}}$. This penalty is typically significant. A numerical example is given in Chapter 3.

This penalty is not present for the radial separation from Figure 2b. The approach directly provides aligned FOVs and a high ϵ_c. TX can be placed in the center surrounded by RX as shown in Figure 2b or vice-versa.

The approach shown in Figure 2c avoids shadowing of the back component by designing the front element transparent. If the front component is TX, it has to have separated emission and absorption bands, i.e., it must exhibit a Stokes shift similar to the fluorescence materials [35]. The issue here is the back plane of the emitter: on the one hand, it has to be reflective to direct the transmitted signal towards the other transceiver. On the other hand, it has to be transparent for the incoming signal. This contradiction does not seem to be easily resolved.

Clearly the radial separation from Figure 2b has superior performance over Figure 2a and has no fundamental concept issue like architecture Figure 2c. However, what is the best way of realizing radial separation? Figure 3 illustrates three principles.

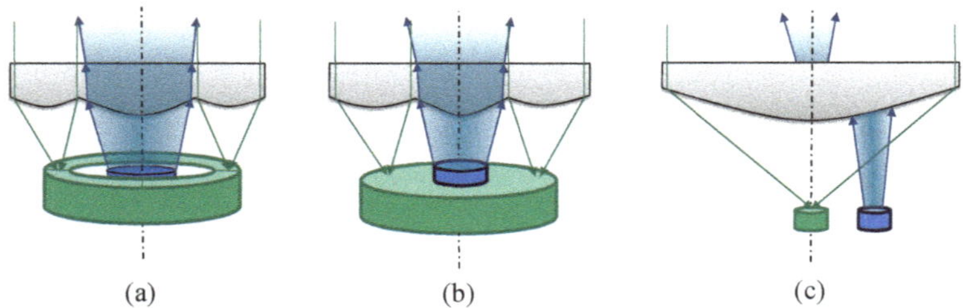

Figure 3. Schematic illustration of different radial separations of TX (blue) and RX (green) with an optical element (grey). (**a**) Direct integration of TX and RX into a plane. (**b**) Stacking an emitter chip onto a PD. (**c**) An optical element is used to homogenize the TX ray bundle and move it to the optical axis.

Designing a transmitter element surrounded by one or more high-speed PDs or vice versa on a single chip, like it is shown in Figure 3a, comes with many design challenges including process compatibility. Thereby, it seems easier to produce the chips separately and stack them afterwards. A small emitter die is bonded onto a large area PD chip as it is depicted in Figure 3b. The emitter is contacted with bonds or directly trough the PD chip. The PD could be separated into multiple parts to ease the contacting and to reduce the transit time of the electrons and holes as they might limit the bandwidth [26]. The main issues of this approach include crosstalk between TX and RX, shadowing of the PD by the emitter and the fact that both chips are custom designs.

By using an optical system like it is shown in Figure 3c, conventional emitter and receiver chips can be used. Those dies are placed next to each other and a hybrid TX-RX lens is used to redirect the rays to achieve radial separation. The spatial separation of both chips is favorable to reduce electrical crosstalk. Injection molding allows the fabrication of the lens in high volume [36] and low cost. Ultra-precise drilling and milling [36] is used to produce the mold tool.

Since a part of the lens is used to direct the TX rays, the maximum theoretical gain $g_{\max}$ cannot be reached. From the PDs point of view, the solid angle of the TX lens part is not used for optical concentration. In Equation (4) this is expressed by a reduced η_{RX}.

In summary, using a hybrid lens for radial separation is most promising: besides the potential for $\epsilon_c \approx 1$, commercial emitter and PD chips can be used. Since the hybrid lens can be fabricated at low cost by injection molding, the system costs are expected to be low.

2.2. Hybrid Lens

2.2.1. Concept

In order to achieve separation in radial direction, the lens is divided into a TX and RX part, i.e., a center part and a surrounding one. Generally, both parts consist of nonrotationally freeform surfaces at the top and bottom of the lens to form a constant irradiance pattern E_{TX} and provide a homogeneous gain g. In order to limit the lens thickness t, Fresnel-structures could be applied to the top and bottom surface. However, it is favorable to keep the top flat to improve reliability, since cavities tend to fill up with particles.

There are two possible arrangements of TX and RX:

1. The emitter is placed centrally and the PD is positioned off-axis as depicted in Figure 4a. In this case only the emission profile of the emitter has to be adjusted. This includes the homogenization of the profile and an adjustment of the angle θ_{TX}. This can ideally be achieved with a single

freeform interface. Therefore, the lower surface can be used for beam shaping and the top surface can be flat. The downside of this approach is a challenging RX lens design: the focus point of the RX lens part is off-axis.

2. The PD is placed centrally and the emitter is positioned off-axis as shown in Figure 4b. The TX lens part fulfills two tasks: first, it compensates the displacement of the ray bundle regarding the optical axis. Second, it reshapes the ray bundle to the anticipated FOV. Because two surfaces are required, the top aperture of the lens cannot be flat. On the other hand, the design of the RX lens part is simplified, because the focal point is on the optical axis. If the TX part is neglected, the RX lens can be designed to be rotationally symmetrical. However, the shadowing effect of the TX part introduces a nonrotationally symmetric factor. Theoretically, this can be partly compensated by a nonrotationally symmetric RX lens part. The shadowing effect is also mitigated by minimizing the size of TX.

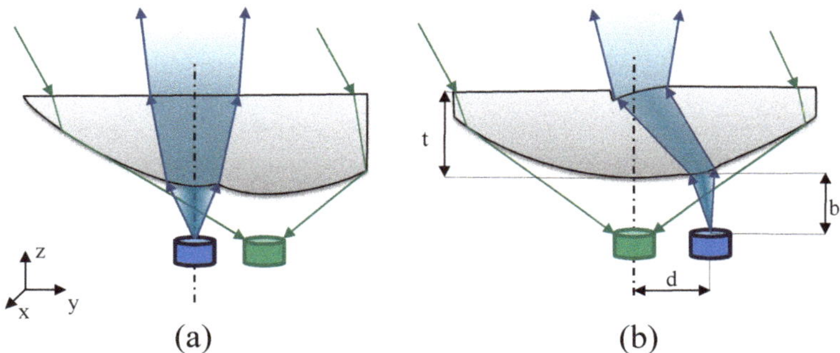

Figure 4. Ray path in the hybrid lens system. (**a**) Center TX (blue) and off-axis RX (green). (**b**) Center RX and off-axis TX.

We choose Approach 2, due to the simplified RX lens design. In this configuration, the lens thickness t typically results from the vertical distance between both TX surfaces. A low t is desirable for size, weight and cost reduction. The costs are reduced, because less material is needed and due to faster processing [36,37]. Nevertheless, a certain thickness t is required to keep the refraction angles at the TX surfaces low in order to limit undesirable Fresnel-reflections. If the RX part determines t, the surface can be split into several Fresnel-grooves to reduce t.

2.2.2. Optic Design Methods

The hybrid lens is a nonimaging optical system. It can be designed by two fundamental approaches [38]: numerical optimization and direct calculation.

Numerical optimization is a straight-forward approach for designing complex optic modules. Modern optic simulation tools like Optic Studio Zemax enable forming and optically simulating arbitrarily shaped optics by overlapping parametric objects. Optimization algorithms like the Levenberg–Marquardt algorithm are used for adjusting parameters of those objects until a sufficiently good result is achieved. Due to the large amount of variables, the optimization is typically inefficient, because of many local minima in the merit function [38]. As a result, it is easy to find a solution, but its performance is very limited, especially if the systems become more complex.

In contrast, direct calculation algorithms follow well-defined design procedures and yield deterministic outcomes. Thereby, they provide better results than numerical optimization methods [38], especially if the systems are complex. A great variety of design methods are known, for example ray mapping [39–46], forming surfaces using Cartesian ovals [47], the simultaneous multiple surface method in 2D [48] and 3D [38,49] or the tailored freeform design method proposed by Ries and Muschaweck [50]. Nowadays ray mapping approach, i.e., the combination of energy mapping in conjunction with geometrical surface construction, is in the focus of illumination research [39–46]. Here, a mass-transfer problem is solved by transforming the source power irradiance E_s into the target power irradiance E_t. This transformation is represented by a mapping $\phi : \Omega_s \rightarrow \Omega_t$ from the source domain Ω_s to the target domain Ω_t. Then, the laws of refraction and reflection are applied for $k \times l$ points to calculate a corresponding vector field $\boldsymbol{N}$ containing the surface normals $\boldsymbol{n}_{i,j}$ with $i \in [1, l], j \in [1, k]$. The challenge is to find a mapping ϕ which yields a vector field $\boldsymbol{N}$ that satisfies the integrability condition for a continuous surface. This condition is shown in Equation (6) [41,42,50]. It states that $\boldsymbol{N}$ has to be curl-free or exhibit at least minimum curl.

$$\boldsymbol{N} \cdot (\nabla \times \boldsymbol{N}) = 0 \tag{6}$$

Nowadays, parametrization and consecutive optimization are widely employed for generating a mapping ϕ [28,31,39,40]. Circular shaped FOV are formed with the mapping shown in Figure 5a. An equi-flux grid around the source in spherical coordinates θ_s, ϕ is mapped onto a target grid in polar coordinates β, r. In 3D, mappings like these typically lead to a normal vector field $\boldsymbol{N}$ with substantial curl. Therefore, an optimization procedure purposely distorts the target grid, for instance by varying r, to improve the performance. It was shown that this approach works well for on-axis scenarios [28,39]. In case of an off-axis placed emitter and two optical surfaces, the scenario is more challenging due to the nonparaxial nature. Hence, the result will deviate from the anticipated irradiance pattern. An additional variable β could be used in lateral direction as illustrated in Figure 5a. The downside is a slower optimization process.

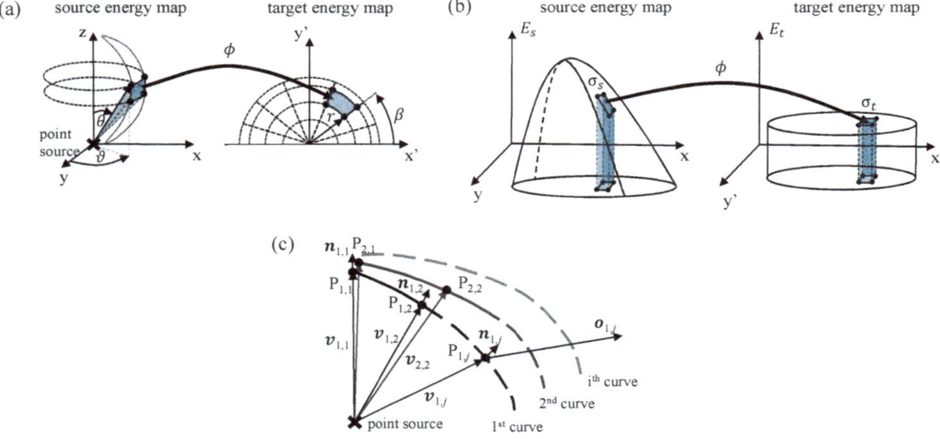

Figure 5. Energy and ray mapping. (**a**) Equi-flux grids at the source and target. Adapted from Wang et al. [39]. (**b**) Mapping from non equi-flux source to anticipated target irradiance. (**c**) Input vectors $\boldsymbol{v}_{i,j}$ and output $\boldsymbol{o}_{i,j}$ vectors are used to construct the surface geometry by subsequently calculating points $P_{i,j}$ and their normals $\boldsymbol{n}_{i,j}$.

For the TX lens part of the hybrid lens, inefficient optimization can be avoided by taking the curl of ϕ directly into account. An initial curl-free mapping ϕ_0 is generated, assuming that the resulting vector field $\boldsymbol{N}$ exhibits minimum curl [42]. Figure 5b illustrates the irradiance in front of the source E_s and the anticipated irradiance at the target plane E_t. The mapping ϕ has to ensure that the infinitesimal area elements at σ_s and σ_t are passed by the same flux. This is achieved by expansion or contraction of the area elements. Mathematically spoken, the mapping has to satisfy Equation (7) for every σ_s of the source grid [42,46].

$$\det(\nabla\phi(\sigma_s))\rho_t(\phi(\sigma_s)) = \rho_s(\sigma_s) \tag{7}$$

The term $\det(\nabla\phi(\sigma))$ represents the expansion and contraction of the area element [46]. Although many mappings ϕ might satisfy Equation (7), there is only a single one that minimizes transport cost [46]. Solving Equation (7) turns out to be nontrivial [42,46]. In this article, the algorithm proposed by Wu et al. [46] is used since it provides good convergence.

For the RX lens part, a rotationally symmetric concentrator lens is designed. It consists of two sections: the center is based on refraction, whereas the outer section works with total internal reflection (TIR). TIR is generally superior over refraction for large angles $\theta_{\text{RX PD i}}$, since it limits the lens diameter and reduces Fresnel-losses [51]. Similar to the TX lens part, the surface is calculated from input and output vectors $\boldsymbol{v}_{i,j}$ and $\boldsymbol{o}_{i,j}$. In the first attempt, the edge ray principle [30] was applied for generating the input vectors $\boldsymbol{v}_{i,j}$. Due to the TX lens part and the discontinuity between both RX lens sections, the edge ray principle is not valid. Therefore, the gain will vary over the FOV and may drop at certain alignments. In the second attempt, this issue is addressed by defining the input vectors for a range of angles of incidence rather than only for the maximum incidence angle. The output rays $\boldsymbol{o}_{i,j}$ are derived from coupling angles $\theta_{\text{RX PD i}}$ to the PD active area. Ideally, they cover the whole half-space in front of the PD.

The normal vector fields $\mathbf{N}$ of both lens parts and an initial point $\mathbf{P}_{1,1}$ for each surface is used to calculate a finite number of surface points $\mathbf{P}_{i,j}$. The result is only a point cloud representation for the optical surfaces. Nonuniform rational B-splines (NURBS) [52] are used for interpolation. NURBS are very flexible and they are supported in the most popular computer aided design formats, which makes them suitable for optical simulation in third-party software and for subsequent fabrication.

2.3. System with Separated TX and RX

In order to show the potential of the methodology and the concept of the hybrid TX-RX lens, a second optical wireless link based on the principle shown in Figure 2a is developed. A concave lens is chosen for TX to widen up the beam. A convex RX lens is used for optical concentration. The distance d between the TX and RX determines the maximum size of both lenses. The RX lens must not be too small to enable a sufficient optical gain g. Therefore, d will be larger compared to the hybrid lens approach. The exact optical parameters are determined by numerical optimization. The hybrid TX-RX lens and the fully separated TX-RX lens system are denoted by H TX-RX and S TX-RX respectively.

2.4. Simulation Parameters

For optical simulations, Monte Carlo ray tracing in *Optic Studio Zemax 17* is used. For each simulation, $5 \cdot 10^5$ rays are traced. Thereby, polycarbonate lenses are used with a refractive index of $n_1 = 1.57$ at $\lambda = 940\,\text{nm}$. We especially investigate Fresnel-reflections which lead to direct optical crosstalk $P_{\text{PD cross}}$ if the reflected rays reach the PD. The detector is a circular shaped PIN-PD with a radius of 100 µm. The LD has an output power of $P_{\text{TX}} = 4\,\text{mW} = 6\,\text{dBm}$ at $\lambda = 940\,\text{nm}$. First, the transmitter part is considered. Then, the full channel is characterized. In order to give a comparable result to other rotary communication technologies, the link distance is set to 12.5 mm.

Conventionally, optical concentrators are characterized by estimating the gain over the angle of incidence by assuming parallel light [53]. However, Figure 6a shows a large divergence of the incident rays. Thus, the assumption of parallel rays does not apply for the present arrangement. The RX performance depends directly on the TX characteristics. Therefore, it is evaluated by the full-channel simulation.

3. Results

Figure 6a shows a render of a cross-section through the lens. It is 6.0 mm in diameter and has a thickness of $t = 2.0\,\text{mm}$ in the center and $t = 2.7\,\text{mm}$ at the groove. The PD is placed on-axis at the origin. The LD position is $(x, y) = (-0.5\,\text{mm}, 0\,\text{mm})$.

For the second system, the numerical optimization leads to a separation between LD and PD of $d = 1.9\,\text{mm}$. The LD is placed at $x = -0.95\,\text{mm}$ and the PD at $x = 0.95\,\text{mm}$. The TX and RX lens have a diameter of 1.3 mm and 2.2 mm and a focal length of −9 mm and 2.7 mm, respectively.

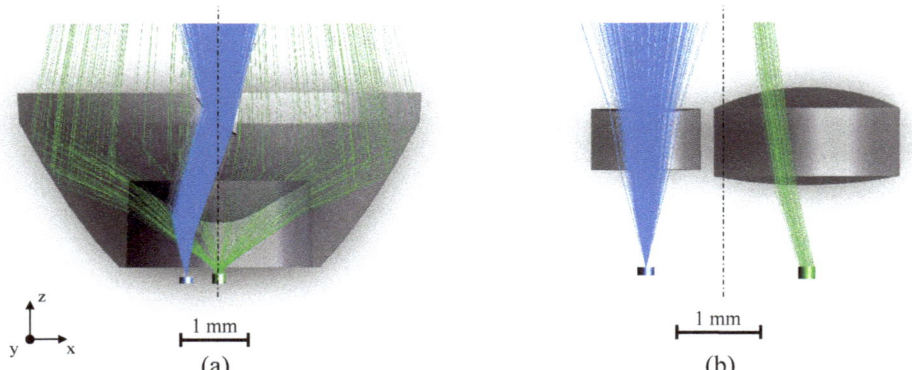

Figure 6. Render of the cross-section through the optical system with TX rays (blue) and RX rays (green). Both devices are placed on the rotation axis (black dotted line) with no misalignment. Only RX rays that hit the detector are shown. (**a**) Hybrid TX-RX lens (H TX-RX). (**b**) Fully separated TX and RX (S TX-RX) in worst-case orientation, i.e., TX lens facing TX lens and RX lens facing RX lens.

3.1. TX Performance

Figure 7 shows the irradiance at a distance of 12.5 mm, measured from the TX aperture. Table 2 lists the detailed merits of both profiles. Figure 7a shows the donut-like shaped profile of the LD without optics. The off-axis placement of the LD is directly observable as profile displacement along the x-axis. Figure 7b shows that the hybrid TX-RX lens is able to remove the displacement and homogenize the profile. The effective transmitter efficiency $\eta_{\text{TX eff}}$ is 57.6 % with a minimum irradiance of $117\,\mu\text{W}/\text{mm}^2$ at $(x, y) = (1.5\,\text{mm}, 1.5\,\text{mm})$. η_{TX} and $\eta_{\text{TX eff}}$ are both reduced by $\zeta_{\text{TX M}} = 24.7\,\%$ and $\zeta_{\text{TX F}} = 12.1\,\%$. Moreover, $\eta_{\text{TX eff}}$ is lowered by another $\zeta_{\text{TX H}} = 5.6\,\%$ due to inhomogeneity within the FOV.

Figure 7c shows the irradiance for the system with fully separated TX and RX as described in Section 2.3. Due to the profile displacement, $\zeta_{\text{TX M}} = 36.0\,\%$ of the power misses the FOV. Moreover, the spherical lenses are not able to correct the donut-shaped profile resulting in a high $\zeta_{\text{TX H}} = 38.6\,\%$. As a result, the irradiance is very inhomogeneous within the FOV and it drops down to $25.9\,\mu\text{W}/\text{mm}^2$ at the right side. This corresponds to $\eta_{\text{TX eff}} = 12.8\,\%$.

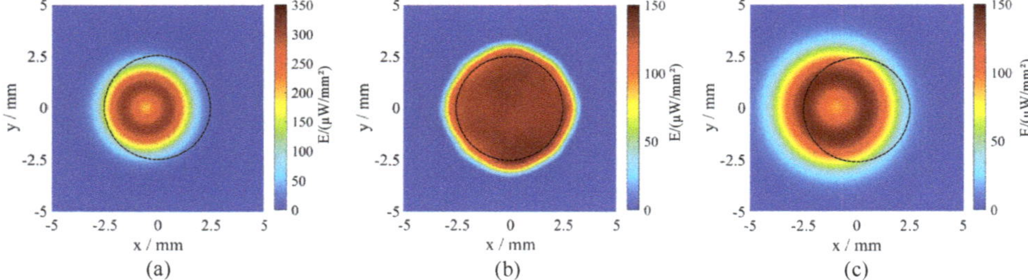

Figure 7. Irradiance over $z = 12.5\,\text{mm}$. (**a**) For the LD without any optics (LD). (**b**) For the LD with the hybrid TX-RX lens. (**c**) For fully separated TX and RX. The black circle highlights the anticipated TX FOV. Note the different color bar scales for the subfigures.

Table 2. TX merits for the laser diode (LD), the hybrid TX-RX system (H TX-RX) and the separated TX-RX lens system (S TX-RX).

Merit	Unit	LD	H TX-RX	S TX-RX
E_{min}	$\frac{\mu W}{mm^2}$	14.6	117	25.9
η_{TX}	%	89.5	63.2	51.4
$\eta_{\text{TX eff}}$	%	7.2	57.6	12.8
$\zeta_{\text{TX M}}$	%	10.5	24.7	36.0
$\zeta_{\text{TX F}}$	%	0	12.1	12.6
$\zeta_{\text{TX H}}$	%	82.3	5.6	38.6

3.2. Full-Channel Performance

Figure 8a–c displays P_{PD}, $P_{\text{PD cross}}$ and $P_{\text{PD PP}}$ at a lens-to-lens distance of 12.5 mm for the hybrid lens. The values are determined by misaligning the receiving transceiver, whereas the transmitting one is placed at the origin. Table 3 shows numerical values for different misalignments in negative x-direction. Figure 8a shows a rotationally symmetric performance in its fundamental structure. However, P_{PD} is not fully homogeneous within the FOV for a revolution. The largest variation ΔP_{PD} over one revolution is reached if the misalignment is between 2 mm and 3 mm. There, ΔP_{PD} is in the range of 3 dB to 3.29 dB. The crosstalk power $P_{\text{PD cross}}$ in Figure 8b is similarly distributed as P_{PD} with a power level which is about 10 dB lower than P_{PD}. As a result, the optical power with applied power penalty $P_{\text{PD PP}}$ in Figure 8c is similar to P_{PD}. This can also be observed in Table 3: the difference between P_{PD} and $P_{\text{PD PP}}$ is in the range of 0.1 dB... 0.9 dB (1.023...1.230). The crosstalk $P_{\text{PD cross}}$ consists of two parts: a constant part $P_{\text{PD cross 1}}$ of -32.4 dBm and an alignment-depended part $P_{\text{PD cross 2}}$.

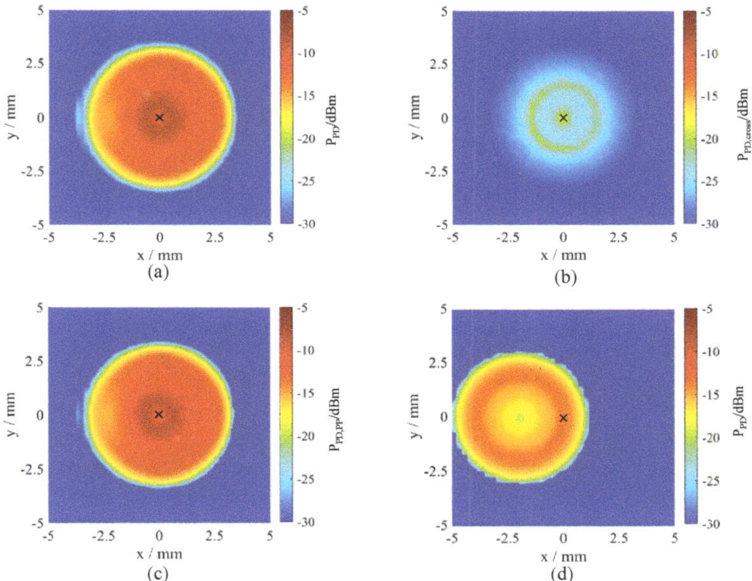

Figure 8. Incident power P_{PD} and $P_{\text{PD cross}}$ over misalignment at $z = 12.5$ mm. The "x" marks the axis of rotation. (**a**) P_{PD}/dBm for the hybrid lens. (**b**) $P_{\text{PD cross}}$/dBm for the hybrid lens. (**c**) $P_{\text{PD PP}}$/dBm for the hybrid lens. (**d**) P_{PD}/dBm for the system with fully separated TX and RX (Section 2.3) in worst-case orientation, i.e., TX lens facing TX lens and RX lens facing RX lens. Note: The graphs are clipped below -30 dBm to highlight features within the FOV. Therefore, $P_{\text{PD cross 1}}$ cannot be seen in Figure 8b.

The system with fully separated TX and RX is shown in Figure 8d. It suffers from a displacement of the communication-area in x-direction. The effect can be seen in Table 3; a communication is only possible for a misalignment of 0.5 mm... 1 mm, depending on the data rate.

Table 3. Minimum P_{PD} depending on the misalignment in negative x-direction ($y = 0$). Values are determined by choosing the minimum value P_{PD} on a circle around the center with a radius of the misalignment. Values lower than 40 dBm are very noisy due to the finite number of simulated rays.

Misalignment/mm	−0	−0.5	−1	−1.5	−2	−2.5	−3	−3.5
H TX-RX: P_{PD}/dBm	−10.1	−8.5	−9.5	−12.0	−14.1	−15.3	−16.0	−26.4
H TX-RX: $P_{\mathrm{PD\ cross}}$/dBm	−19.2	−22.7	−23.9	−22.9	−25.2	−27.4	−28.8	−30.4
H TX-RX: $P_{\mathrm{PD\ PP}}$/dBm	−10.7	−8.7	−9.7	−12.4	−14.5	−15.6	−16.2	−28.6
S TX-RX: P_{PD}/dBm	−13.7	−15.2	−20.6	<−40	<−40	<−40	<−40	<−40

4. Discussion

4.1. TX Performance

Figure 7 proves the suitability of the ray mapping method based on curl-free mapping for the TX part. In order to assess the performance of the design, it is meaningful to have a closer look at the loss mechanism: the main loss is represented by $\zeta_{\mathrm{TX\ M}} = 24.7\,\%$. This magnitude is quite common. It results from the extended source effect from the LD with regards to the TX lens part and the remaining curl in **N**. Furthermore, the overall ray mapping performance depends on the extent of the TX output aperture and the distance to the target plane [41]. The reason for this is that ray mapping is an optical far-field design method that neglects the rays' position vector on the output aperture.

The inhomogeneity within the FOV is with $\zeta_{\mathrm{TX\ H}} = 5.6\,\%$ very low. Generally, a $\zeta_{\mathrm{TX\ H}}$ below 10 % is a good result. The major part of the remaining inhomogeneity results from the FOV edge, where the irradiance starts to decrease. The Fresnel-loss $\zeta_{\mathrm{TX\ F}}$ is with 12.1 % in a common region for two material interfaces. Although an anti-reflection (AR) coating might reduce this effect by a factor of 3 to 4, it is challenging to homogeneously coat the nonplanar TX surfaces.

4.2. Full-Channel Performance

$P_{\mathrm{PD\ PP}}$ is in a sufficient range for data transfer, but how does it correspond to the data rate? For a bit error rate of 10^{-12} Säckinger calculates a sensitivity of −26.5 dBm for a 2.5 Gbit/s link and −20.5 dBm for a 10 Gbit/s link [26]. Tzeng et al. [54] demonstrated a sensitivity of −21.5 dBm for a 10 Gbit/s link. It can be concluded from Table 3 that $P_{\mathrm{PD\ PP}}$ is sufficient for data transmission in the Gbit/s data rate range and even provides a margin for ageing effects and other nonidealities. The optical crosstalk introduces a power penalty of up to 0.9 dB. The magnitude of $P_{\mathrm{PD\ cross}}$ depends directly on P_{TX}. A lower P_{TX} leads to reduced $P_{\mathrm{PD\ cross}}$. The downside is that the SNR cannot be improved by increasing P_{TX} if the crosstalk is the dominant noise factor. The constant crosstalk part $P_{\mathrm{PD\ cross\ 1}}$ is alignment-independent. Therefore, it has to be the internal optical crosstalk. In contrast, the back-reflected signal $P_{\mathrm{PD\ cross\ 2}}$ from the opposite transceiver depends on the transceiver alignment. The crosstalk and thus the power penalty is reduced if z is increased. In the far field, P_{PD} is decreasing with approximately $P_{\mathrm{PD}} \propto z^{-2}$ and the crosstalk scales with $P_{\mathrm{PD\ cross\ 2}} \propto z^{-4}$, since the back-reflected rays have to travel twice the distance. Although this relation is not fully correct for the near field, the trend is still valid.

The crosstalk $P_{\mathrm{PD\ cross\ 2}}$ can be lowered by reducing the Fresnel-reflections $\zeta_{\mathrm{RX,F}}$. The planar top surface is well suited for an AR coating. The problem is that the nonplanar top surface of the TX lens part is also affected by the coating. If one is able to solve the TX coating problem, a single coating is twice as effective, because the ray passes the coating two times: first at TX and second at RX. The link-budget is improved by approx. 6 % per coating instead of only approx. 3 %. Moreover, the crosstalk $P_{PD\ cross\ 2}$ is reduced. Alternatively, the top surface of the RX lens part could be designed nonplanar in a way that back-reflected rays miss the transmitting device.

Although TX exhibits a quite homogeneous performance, $P_{PD\ PP}$ varies about 6.9 dB along radial direction for misalignment $\leq$ 3 mm according to Figure 8a and Table 3. Thus, there is clearly some room for improvement for the RX lens part. The main issue results from the mismatch between near and far field in terms of ray optics. This can be seen in Figure 6a: the incident ray bundle (green) exhibits a large divergence. The situation is different at $z = 50$ mm, for instance. The solid angle of incidence is much smaller and the incident ray bundle exhibits a smaller divergence. The larger the distance, the better the design approach with the edge-ray principle works. Another issue is the nonrotationally symmetric shadowing effect of the TX lens part that manifests itself as a local minimum around $(x, y) = (-2.5\,\text{mm}, 0\,\text{mm})$ in Figure 8a,c. The variation $\Delta P_{PD} = 3.29$ dB is not crucial for data transmission. Assuming a misalignment of -2.5 mm and a rotation speed of 10 000 rpm, the link moves 9.42 µm over the surface in Figure 8a during one bit-duration of 1 ns (1 Gbit/s). The change of P_{PD} over a sequence of bits is small enough and does not influence the transmission drastically. Ideally, the receiver circuit includes a decision-point control mechanism for continuous adaptation of the decision level to improve the BER [26].

4.3. Suitability for Rotary Scenarios

Faulwaßer et al. [19] already reported data rates of up to 10 Gbit/s. However, what data rates are generally possible compared to FORJ? From an electrical point of view, OWC is able to achieve similar data rates as single-FORJ. In contrast to FORJ, the PD has typically a larger area and thus a larger capacitance, which limits the bandwidth. Choosing a smaller PD, will reduce the maximum tolerable misalignment. The hybrid lens from this work exhibits 14.5 dB link loss at $(x, y) = (0\,\text{mm}, 0\,\text{mm})$. This link loss is a part of the active link concept. The signal is directly recovered at RX by amplification and optimally by subsequent analog-to-digital conversion. The magnitude of the OWC link loss depends on the magnitude of tolerable misalignment. Thus, a higher data rate requires a smaller FOV. In summary, the data rate of OWC links might be slightly below single-fiber FORJ due to a higher link loss.

With regards to the communication distance, the hybrid lens is flexible and not restricted to $z = 12.5$ mm. If the lens is designed for larger ranges, the TX beam exhibits a lower divergence and RX should be designed for smaller angles of incidence. Thereby, the hybrid lens can be tailored to the ideal distance of the rotary system.

As mentioned in Section 2.1.3, the hybrid lens approach has the potential to be very cost effective: in contrast to fully separated TX and RX, only a single lens has to be fabricated, potentially coated and assembled.

The proposed hybrid lens works in on-axis configuration like FORJs. Future work will deal with off-axis optical links to improve flexibility. Only data rates in the range of kbit/s to a few Mbit/s have been demonstrated [2,3], which cannot compete with modern capacitive links [6]. Another interesting field of research is the realization of multi-channel optical wireless links similar to multi-FORJs. In nonrotary scenarios, data rates of several hundreds of Gbit/s have already been demonstrated [55].

5. Conclusions

This work has shown the suitability of OWC for full-duplex, high-speed data transfer in on-axis rotary scenarios. The proposed hybrid lens is able to provide a sufficient RX signal level $P_{PD\ PP}$ of more than -16.2 dBm even for misalignments of up to 3 mm at a communication distance of $z = 12.5$ mm. OWC is therefore able to provide a robust data transfer without the strict mechanical tolerances compared to FORJs. The results show a maximum power penalty resulting from optical crosstalk of 0.9 dB within the FOV. The approach is promising since it allows low-cost fabrication. Besides the electronical components, only a single optical component is required that can be fabricated by injection molding in high volume.

6. Patents

The hybrid TX-RX lens and its derivatives are covered by several patents including DE102018205559 B3 [56] (WO19197343A1 [57]). Further patents are submitted.

Author Contributions: Conceptualization, R.K., A.N.; methodology, R.K., T.S.; software, R.K.; validation, R.K. and T.S.; formal analysis, R.K.; investigation, R.K.; data curation, R.K.; writing–original draft preparation, R.K.; writing–review and editing, M.F., T.S., P.M., F.D. and A.N.; visualization, R.K.; supervision, R.K.; project administration, R.K. All authors have read and agreed to the published version of the manuscript.

Funding: This research received no external funding.

Conflicts of Interest: The authors declare no conflict of interest. The funders had no role in the design of the study; in the collection, analyses, or interpretation of data; in the writing of the manuscript, or in the decision to publish the results.

Abbreviations

The following abbreviations are used in this manuscript:

BER	bit error rate
BW	bandwidth
EM	emitter
FORJ	fiber optic rotary joint
FOV	field of view
H TX-RX	hybrid transmitter-receiver lens system
LED	light emitting diode
LD	laser diode
NURBS	nonuniform rational B-splines
OWC	optical wireless communications
PD	photodetector
RF	radio frequency
S TX-RX	separate transmitter-receiver lens system
SNR	signal-to-noise ratio
RX	receiver
TIR	total internal reflection
TX	transmitter

References

1. Astier, R.; Capitaine, T.; Bourny, V.; Dubois, J.; Fortin, J. Technological leap of signal transfer system: CONTACTLESS. In Proceedings of the 38th Annual Conference on IEEE Industrial Electronics Society (IECON), Montreal, QC, Canada, 25–28 October 2012; pp. 3238–3243.
2. Hong, E.; Gregory, D.A.; Krishnamurthy, A. Opto-electronics based wireless slip ring system. *J. Eng. Comput. Arch.* **2007**, *1*, 1–6.
3. Wang, Z.; Gu, H.; Sun, P. Contactless signal transmission for industrial bus at rotary joint. In Proceedings of the 2015 IEEE International Conference on Signal Processing, Communications and Computing (ICSPCC), Ningbo, China, 19–22 September 2015; pp. 1–6.
4. Ghelfi, P.; Laghezza, F.; Scotti, F.; Serafino, G.; Pinna, S.; Onori, D.; Lazzeri, E.; Bogoni, A. Photonics in Radar Systems: RF Integration for State-of-the-Art Functionality. *IEEE Microw. Mag.* **2015**, *16*, 74–83. [CrossRef]
5. Nickel, H.U.; Doleschel, A.; Schmid, M. Multichannel rotary joints for surveillance radars - State-of-the-art and future trends. In Proceedings of the 14th International Radar Symposium (IRS), Dresden, Germany, 19–21 June 2013; pp. 282–287.
6. Nickel, H.U.; Zovo, J.; Schmid, M. Refining Radar Architectures: Multichannel Rotary Joints for Surveillance Radars. *IEEE Microw. Mag.* **2016**, *17*, 60–74. [CrossRef]
7. Vollbracht, D.; Gekat, F.; Hilger, D.; Hille, M. Waveguide fibre optic rotary joint for antenna mounted radar receivers. In Proceedings of the 34th Conference on Radar Meteorology, Poster, Williamsburg, VA, USA, 5–9 October 2009.

8. Mi, L.; Yao, S.-L.; Sun, C.-D.; Sun, B.; Zhang, H.-J. A single-channel fiber optic rotary joint on the basis Of TEC fiber collimators. In Proceedings of the International Conference on Optical Communications and Networks (ICOCN 2010), Nanjing, China, 24–27 October 2010; pp. 437–440.
9. Roberts, G.; Hadfield, P.; Humphries, M.E.; Bauder, F.; Izquierdo, J.M.G. Design and evaluation of the power and data contactless transfer device. In Proceedings of the 1997 IEEE Aerospace Conference, Snowmass, Aspen, CO, USA, 10 Febuary 1996; Volume 3, pp. 523–533.
10. Song, Y.; Wang, S.; Feng, Y. Stable and Secure Contactless Connectivity of Power and Data in Rotational Applications. In Proceedings of the 2018 IEEE Asia-Pacific Conference on Antennas and Propagation (APCAP), Auckland, New Zealand, 5–8 August 2018; pp. 401–402.
11. Patyuchenko, A. *60 GHz Wireless Data Interconnect for Slip Ring Applications*; Technical Report; Analog Devices, Inc: Norwood, MA, USA, 2019.
12. Timsit, R.S.; Antler, M. Tribology of Electronic Connectors: Contact Sliding Wear, Fretting, and Lubrication: 7.4. Lubrication. In *Electrical Contacts: Principles and Applications*; Slade, P.-G., Ed.; CRC Press: Boca Raton, FL, USA, 2014; pp. 481–506.
13. Doleschel, A.; Lege, M. Contactless solutions for radar rotary joint systems. In Proceedings of the 16th International Radar Symposium (IRS), Dresden, Germany, 24–26 June 2015; pp. 451–456.
14. Panhans, C.; Stolle, R. High Bandwidth Contactless Rotary Transmitter Design Optimized for Baseband Transmission. In Proceedings of the 11th International Conference on Information Technology and Electrical Engineering (ICITEE), Pattaya, Thailand, 10–11 October 2019; pp. 1–6.
15. Wageningen, D.V.; Staring, T. The Qi wireless power standard. In Proceedings of the 14th International Power Electronics and Motion Control Conference (EPE-PEMC), Ohrid, Macedonia, 6–8 September 2010; pp. S15-25–S15-32.
16. Dorsey, G.F.; Coleman, D.S.; Witherspoon, B.K. High Speed Data across Sliding Electrical Contacts. In Proceedings of the 2012 IEEE 58th Holm Conference on Electrical Contacts (Holm), Portland, OR, USA, 23–26 September 2012; pp. 1–12.
17. Chen, H; Hsu, S.; Chang, T.; Chi, W.; Ma, S.; Wu, T. Signal integrity design for GHz high-speed differential wires with slip ring capsule. In Proceedings of the 2012 IEEE International Symposium on Electromagnetic Compatibility, Pittsburgh, PA, USA, 6-10 August 2012; pp. 18–21.
18. Slade, P.-G. *Electrical Contacts: Principles and Applications*; CRC Press: Boca Raton, FL, USA, 2014.
19. Faulwaßer, M.; Kirrbach, R.; Schneider,T.; Noack, A. 10 Gbit/s bidirectional transceiver with monolithic optic for rotary connector replacements. In Proceedings of the 2018 Global LIFI Congress (GLC), Paris, France, 8–9 February 2018; pp. 95–102.
20. Rappaport, T,S.; Sun, S.; Mayzus, R.; Zhao, H.; Azar, Y.; Wang, K.; Wong, G. N.; Schulz, J.K.; Samimi, M.; Gutierrez, F. Millimeter Wave Mobile Communications for 5G Cellular: It Will Work!. *IEEE Access* **2013**, *1*, 335–349. [CrossRef]
21. Aizawa, T.; Sakai, M.; Toguchi, S.; Hirohashi, K. Optical Slip Rings for Home Security High-Definition Cameras. In Proceedings of the 2007 Digest of Technical Papers International Conference on Consumer Electronics, Las Vegas, NV, USA, 10–14 January 2007; pp. 1–2.
22. Guenter, J.K.; Hawthorne, R.A.; Granville, D.N.; Hibbs-Brenner, M.K.; Morgan, R.A. Reliability of proton-implanted VCSELs for data communications.In Proceedings of the SPIE 2683, Fabrication, Testing, and Reliability of Semiconductor Lasers, San Jose, CA, USA, 4 April 1996; pp. 102–113.
23. Christou, A.; Unger, B.A. *Semiconductor Device Reliability*; Christou, A., Unger, B.A., Eds.; Springer: Dordrecht, The Netherlands, 1990; p. 60.
24. Vinh, Q.T.; Khanh, T.Q.; Ganev, H.; Wagner, M. Measurement and Modeling of the LED Light Source. In *LED Lighting: Technology and Perception*, 1st ed.; Khanh, T.Q., Bodrogi, P., Vinh, Q.T., Winkler, H., Eds.; John Wiley & Sons, Ltd: Hoboken, NJ, USA, 2014; pp. 214–222.
25. Dorsey, G.F. Fiber Optic Rotary Joints - A Review. *IEEE Trans. Compon. Hybrids Manuf. Technol.* **1982**, *5*, 37–41. [CrossRef]
26. Säckinger, E. *Analysis and Design of Transimpedance Amplifiers for Optical Receivers*, 1st ed.; John Wiley and Sons: Hoboken, NJ, USA, 2018; pp. 46, 48, 56, 120–127, 197–199, 469–474.
27. *Safety of Laser Products - Part 1: Equipment Classification and Requirements*; IEC 60825-1:2014; IEC: Geneva, Switzerland, 2014.

28. Kirrbach, R.; Jakob, B.; Noack, A. Introducing Advanced Freeform Optic Design to Li-Fi Technology. In Proceedings of the 7th International Conference on Photonics, Optics and Laser Technology (PHOTOPTICS 2019), Prague, Czech Republic, 25–27 February 2019; pp. 249–250.
29. Kahn, J.; Barry, J.R. Wireless Infrared Communications. *Proc. IEEE* **1997**, *85*, 265–298. [CrossRef]
30. Welford, W.T.; Winston, R. *High Collection Nonimaging Optics*, 1st ed.; Academic Press Inc.: San Diego, CA, USA, 1989; pp. 1–28, 54.
31. Kirrbach, R.; Faulwaßer, M.; Jakob, B. Non-rotationally Symmetric Freeform Fresnel-Lenses for Arbitrary Shaped Li-Fi Communication Channels. In Proceedings of the 2019 Global LIFI Congress (GLC), Paris, France, 12–13 June 2019; pp. 103–108.
32. Stepniak, G.; Kowalczyk, M.; Maksymiuk, L.; Siuzdak, J. Transmission Beyond 100 Mbit/s Using LED Both as a Transmitter and Receiver. *IEEE Photonics Technol. Lett.* **2015**, *27*, 2067–2070. [CrossRef]
33. Serial Infrared Transceiver (SIR), 115.2 kbit/s, 2.4 V to 5.5 V Operation TFBS4711, Vishay Semiconductors, rev. 3.3, 08.04.2019. Available online: http://www.vishay.com (accessed on 30 November 2019).
34. Faulwaßer, M.; Deicke, F.; Schneider, T. 10 Gbit/s bidirectional optical wireless communication module for docking devices. In Proceedings of the 2014 IEEE Globecom Workshops (GC Wkshps), Austin, TX, USA, 8–12 December 2014; pp. 512–517.
35. Manousiadis, P.; Rajbhandari, S.; Mulyawan, R.; Vithanage, D.; Chun, H.; Faulkner, G.; O'brien, D.; Turnbull, G.; Collins, S.; Samuel,I. Wide field-of-view fluorescent antenna for visible light communications beyond the étendue limit. *Optica* **2016**, *3*, 702–706. [CrossRef]
36. Naumann, H.; Schröder, G.; Löffler-Mang, M. *Handbuch Bauelemente der Optik: Grundlagen, Werkstoffe, Geräte, Messtechnik*, 7th ed.; Carl Hanser Verlag: Munich Vienna, Germany, 2014; pp. 69–70, 81–83.
37. Mäkinen, J.-K. Cost Modeling of Injection-Molded Plastic Optics. In *OECC Handbook of Plastic Optics*, 2nd ed.; Bäumer, S., Ed.; WILEY-VCH Verlag GmbH and Co. KGaA: Weinheim, Germany, 2010; p. 229.
38. Miñano, J.C. SMS 3D Design Method. In *Illumination Engineering: Design of Nonimaging Optics*; Anderson, J., Ed.; Wiley-IEEE Press: Hoboken, NJ, USA, 2013; p. 102.
39. Wang, K.; Liu, S.; Xiaobing, L.; Wu, D. Basic Algorithms of Freeform Optics for LED Lighting. In *Freeform Optics for LED Packages and Applications*, 1st ed.; John Wiley and Sons: Singapore, 2017; pp. 25–69.
40. Ma, D. Exploration of Ray Mapping Methodology in Freeform Optics. Ph.D. Thesis, The University of Arizona, Tucson, AZ, USA, 2015.
41. Bruneton, A.; Bäuerle, A; Wester, R.; Stollenwerk, J.; Loosen, P. Limitations of the ray mapping approach in freeform optics design. *Opt. Lett.* **2013**, *38*, 1945–1947. [CrossRef]
42. Desnijder, K.; Hanslaer, P.; Meuret, Y. Flexible design method for freeform lenses with an arbitrary lens contour. *Opt. Lett.* **2017**, *42*, 5238–5241. [CrossRef]
43. Desnijder, K.; Hanslaer, P.; Meuret, Y. A ray mapping method for off-axis and non-paraxial freeform illumination lens design. *Opt. Lett.* **2019**, *44*, 771–774. [CrossRef] [PubMed]
44. Desnijder, K.; Ryckaert, W; Hanslaer, P.; Meuret, Y. Luminance spreading freeform lens arrays with accurate intensity control. *Opt. Express* **2019**, *27*, 32994–33004. [CrossRef] [PubMed]
45. Wu, R.; Liu, P.; Zhang, Y.; Zheng, Z.; Li, H.; Liu, X. A mathematical model of the single freeform surface design for collimated beam shaping. *Opt. Express* **2013**, *21*, 20976–20989. [CrossRef] [PubMed]
46. Wu, R.; Zhang, Y; Sulman, M.M.; Zheng, Z.; Benítez, P.; Miñano, J.C. Initial design with L2 Monge-Kantorovich theory for the Monge–Ampère equation method in freeform surface illumination design. *Opt. Express* **2014**, *22*, 16161–16177. [CrossRef]
47. Michaelis, D.; Schreiber, P.; Bräuer, A. Cartesian oval representation of freeform optics in illumination systems. *Opt. Lett.* **2011**, *36*, 918–920. [CrossRef]
48. Winston, R.; Minano, J.C.; Benitez, P.G.; Shatz, N.; Bortz, J.C. *Nonimaging Optics*, 1st ed.; Academic Press: Burlington, MA, USA, 2005; pp. 181–218.
49. Benitez, P.G.; Miñano, J.C.; Blen, J.; Arroyo, R.M.; Chaves, J.; Dross, O.; Hernandez, M.; Falicoff, W. Simultaneous multiple surface optical design method in three dimensions. *Opt. Eng.* **2004**, *43*, 1489–1502. [CrossRef]
50. Ries, H.; Muschaweck, J. Tailored freeform optical surfaces. *J. Opt. Soc. Am.* **2002**, *19*, 590–595. [CrossRef]
51. Wallhead, I.; Jiménez, T.M.; Ortiz, J.V.G.; Toledo, I.G.; Toledo, C.G. Design of an efficient Fresnel-type lens utilizing double total internal reflection for solar energy collection. *Opt. Express* **2010**, *20*, A1005–A1010. [CrossRef]

52. Piegl, L.; Tiller, W.J. *The NURBS Book*, 2nd ed.; Springer: Berlin/Heidelberg, Germany, 1995.
53. Ning, X.; Winston, R.; O'Gallagher, J. Dielectric totally internally reflecting concentrators. *Appl. Opt.* **1987**, *26*, 300–305. [CrossRef]
54. Tzeng, L.D.; Mizuhzra, O.; Nguyen, T.V.; Ogawa, K.; Watanabe, I.; Makita, K.; Tsuji, M.; Taguchi, K. A high-sensitivity APD receiver for 10-Gb/s system applications. *IEEE Photonics Technol. Lett.* **1996**, *8*, 1229–1231. [CrossRef]
55. Gomez, A.; Shi, K.; Quintana, C.; Maher, R.; Faulkner, G.; Bayvel, P.; Thomsen, B.C.; O'Brien, D. Design and Demonstration of a 400 Gb/s Indoor Optical Wireless Communications Link. *J. Lightw. Technol.* **2016**, *34*, 5332–5339. [CrossRef]
56. Schneider, T. Optische Sende-/Empfangs-Einheit und Vorrichtung zur Signalübertragung. DE102018205559 B3, 12 April 2018.
57. Schneider, T. Optical Transceiver Unit and Device for Signal Transmission. Patent No. WO19197343A1, 8 April 2019.

Review

Key Roles of Plasmonics in Wireless THz Nanocommunications—A Survey

Efthymios Lallas

General Department of Lamia, University of Thessaly, 35100 Lamia, Greece; elallas@uth.gr

Received: 20 October 2019; Accepted: 10 December 2019; Published: 13 December 2019

Abstract: Wireless data traffic has experienced an unprecedented boost in past years, and according to data traffic forecasts, within a decade, it is expected to compete sufficiently with wired broadband infrastructure. Therefore, the use of even higher carrier frequency bands in the THz range, via adoption of new technologies to equip future THz band wireless communication systems at the nanoscale is required, in order to accommodate a variety of applications, that would satisfy the ever increasing user demands of higher data rates. Certain wireless applications such as 5G and beyond communications, network on chip system architectures, and nanosensor networks, will no longer satisfy speed and latency demands with existing technologies and system architectures. Apart from conventional CMOS technology, and the already tested, still promising though, photonic technology, other technologies and materials such as plasmonics with graphene respectively, may offer a viable infrastructure solution on existing THz technology challenges. This survey paper is a thorough investigation on the current and beyond state of the art plasmonic system implementation for THz communications, by providing in-depth reference material, highlighting the fundamental aspects of plasmonic technology roles in future THz band wireless communication and THz wireless applications, that will define future demands coping with users' needs.

Keywords: wireless NoC (WiNoC); graphene based WiNoCs (GWiNoCs); wireless nanosensor networks (WNSNs); surface plasmon polariton (SPP); GFET; multiple-input-multiple-output (MIMO); graphennas; THz transceiver

1. Introduction

Wireless data traffic has experienced an unprecedented boost in the past years, and it is expected to increase sevenfold up to 2021 [1]. Data traffic forecasts in wireless communication networks will account for more than 60% of the overall internet traffic by then [2]. Current wireless communications handle data rates of tens of Gbps per link or even more, and the prospect for the future demands will be 100 Gbit/s within 10 years [3], with multiplexed rates well beyond 100 Gbit/s, and eventually Tbit/s. Wireless communications seem to be in advance against conventional wired communications. By 2030, wireless data rates will be sufficient enough to compete with wired broadband rates [4]. Therefore, the use of even higher carrier frequency bands in the THz range is required, via adoption of new technologies equipping future THz band wireless communication systems at the nanoscale, in order to accommodate a variety of applications, that would satisfy the ever increasing user demands for higher data rates. Certain wireless applications such as 5G and beyond communications, NoC system architectures and nanosensor networks, will no longer satisfy their speed and latency demands with existing technologies and system architectures. Apart from conventional CMOS technology, and the already tested, still promising, photonic technology, other technologies and materials such as plasmonics with graphene material, may offer a viable solution on existing THz technology challenges.

At the moment, wireless traffic in the access 5G networks exploits millimeter wave (mmW) bands. In order to accommodate the continuously increasing traffic demands of 5G and beyond communications, researchers have been focused on taking advantage of higher regions in the radio spectrum, pointing to the THz band communication and infrastructure, as a promising solution to equip 5G plus networks, thus enabling efficient operation of bandwidth hungry applications, that are not feasible at the moment with current infrastructure.

Wireless NoC (WiNoC) [5] with its inherent broadcast capability, appears as a promising approach to overcome all abovementioned bottlenecks of ancestor technologies. Ultra small miniature sizes of plasmon based antennas and other nanolink components as well, with considerably much less wiring, are the desired features of this technology, in order to enable the integration of one or multiple antennas per core, paving the way for dense, scalable NoC schemes, as required by future applications. Graphene based WiNoCs (GWiNoCs) is probably the most updated promising approach for THz nanoscale wireless communications, and it is therefore considered to be the basis for implementing future on chip network architectures. Alternatively, hybrid optical wireless schemes, may be also proved to be a promising NoC solution [6], by combining the best assets of these two worlds: low loss dielectric waveguide media, and miniature sized plasmonic material oscillating at THz rates.

Last, wireless nanosensor networks (WNSNs) is another established THz nanoscale application, with basic similarities as in WiNoCs, such as core to core or to memory communication, but also with other unique characteristic types of communication, mainly between nanosensors and nanomachines in the THz band [7]. Such EM communication in the THz band, is usually enabled by plasmonic materials, as graphene for implementing plasmonic nanotransceivers and nanoantennas, as in the WiNoC case. These three important wireless THz nanocommunication applications can be seen altogether in Figure 1.

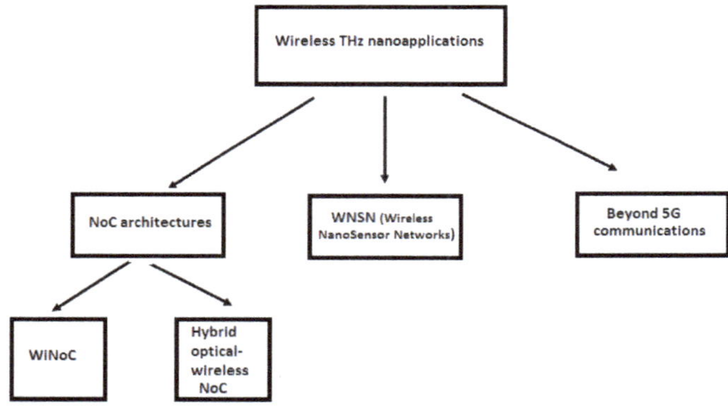

Figure 1. Applications for wireless THz nanocommunications.

Nonetheless, the THz band as the last undiscovered frontier of the total EM spectra range, and THz wireless nanoscale communications, are still urging for an efficient, compact and standardized interconnect solution for generating, transmitting, propagating, and detecting the THz wave information. Despite the fact that new efficient methods have been introduced based on modern system architectures via the utilization of new technologies for manipulating THz Band signals by the research academia, still there are many challenges to face, such as the very high propagation signal loss, the impedance mismatch between THz link components, the limited size restrictions along with integration potentials, associated with high bandwidth availability and ultrafast operating data rates with minimum latency requirements. Conventional CMOS based electronic interconnects are definitely far from the target to meet THz speed, low propagation signal loss, and the impedance match between THz link components. Current CMOS scaling technology growth has restricted the cut-off frequency and the maximum oscillation frequency of the device to several hundreds of a GHz. Traditionally, the operating frequency should be much below the cut-off frequency when designing mixed-signal circuits at high frequencies, and the CMOS device technology applied, should be capable of continuously scaling down the size and values of on-chip components (e.g., transistors, capacitors, and inductors), in order to achieve higher throughput and reduced circuit footprints [8]. Evidently, at THz frequencies this approach has weak potential for improvements, which are further limited by the loss encountered in on-chip metal structures [9]. The alternatives for such a case, would be either to search for a new implementation technology that offers better scaling prospects and a higher intrinsic transit frequency, or to exploit the non-linearities of the device for efficient power generation at higher-order harmonics. As concerns the latter, there are approaches that use novel circuit to generate, radiate, and control THz frequencies widely adopted in academia and industry [10], however it is high time for other technologies to take this on, at this critical stage.

Artificial intelligence (AI) computing with its hot AI chip topic may be an alternative for CMOS wired interconnection and NoC architectures, against CMOS process and device bottleneck and Von Neumann and memory wall bottlenecks. In a broader sense, the AI chip is the adoption of AI principles in computing processing systems in the form of accelerators, in order to boost their computing performance. AI approaches such as processing-in-memory (PIM), machine learning (ML), and especially neural network (NN)-based accelerators, such as FPGA, GPU and ASIC, are considered as mature solutions for speeding up computing performance [11]. PIM or near data computing (NDC), is a promising solution to tackle with the memory wall bottleneck. PIM architectures put additional computation logic in or near memory by leveraging 3D memory technologies to integrate computation logic with memory, thus speeding up NN computations on larger memory capacity and bandwidth, via in-memory data communication, at the same time. The metal-oxide resistive random access memory (ReRAM) has showed great potential to be used for main memory, with its crossbar array structure, capable for performing matrix vector multiplication efficiently, and accelerating NN computations [12].

However, as processing system scalability increases, AI processing loads are getting more and more data-intensive and demand higher bandwidth and heavy data movement between computing logic and memory. Hence, when the scale of the NN computation and accelerator increases, the NoC-based data communication within NN accelerators would evidently have to deal with a performance bottleneck. In addition to performance, the energy consumption of the NoC in an NN accelerator may be also a big challenge to deal with. Particularly, when all this AI processing is done at the edge level, embedded in sensors, smartphones or general IOT equipment, there are more strict power requirements and the need for much more specific hardware implementations [13]. Compared with cloud applications, the application requirements and constraints of edge devices are much more complex, and special architecture design may be needed for handling different situations. Among them, the most important feature and at the same time, request for current edge IOT devices, is their ability to locally perform "inference", relieving thus processing burden from cloud servers and reducing delay [13]. However, in such a case, the demand for training in edge embedded devices is not very clear, given that in the future, all these wearable IOT devices should be capable to perform efficient inference computing, which in turn, requires them to have sufficient inference computing ability, so as to achieve a certain intelligence threshold under the strict power and cost constraints of the edge area, in order to meet the challenges of various different AI application scenarios. Efforts from the research community have been made towards the direction of locally reducing accuracy, and computational complexity, by combining some data structure transformations, such as FFTs to reduce the multiplication in matrix operations, or table lookup to simplify the implementation of multiply-and-accumulate (MAC) operations. Moreover, various low power methods have been applied to AI chips of edge devices to further reduce the overall power consumption, such as the clock-gating applied to MAC. Nowadays, industry has been focused in developing specialized AI chips and all kinds of IOT devices with enhanced inference capabilities at low power and costs. The collaborative training and inference among cloud and edge devices would be an interesting direction to be explored by research academia [13].

As current NN training accelerators relied on conventional wired NoCs, seem to have to deal with certain limitations, especially as time goes by, WiNOC on the other hand, with its inherent features, as broadcast support and multiple access to the shared medium by beamforming and antenna beam narrowing, spatial multiplexing within package, reduced latency, as wireless channel is a distance independent communication means, flexibility in a sense of virtually mapping different topologies within a cycle, and scalability potential as systems scale linearly with the number of cores, may be also considered as an alternative, for being exploited as an interconnection means for these accelerators [14]. Since NN training accelerator parallelized computation nature in a many-core-like environment, is similar to one-to-all, or all-to-all WiNoC communication nature, focusing on exploring the matching points between these two pillars, may be an interesting direction to be investigated by research academia. In this way, the improvement of the efficiency of a WiNoC architecture should be sought not only on the implementation of the miniature antennas and transceiver wireless equipment, but also on the proper design of the NN architecture, as concerns the intensive data movement between processing core and memory units.

In general, the AI chip concept is a complex multi-variable issue, lying in the middle of a whole layer stack, with demanding tasks ranging from providing efficient support for higher layer cloud applications and algorithms, up to orchestrating entities based on AI principles in low level architectures, consisting of devices and circuits, processes and materials, and hence there are a lot of unsolved issues and unanswered uncertainties that may be set under consideration [13], which is out of the scope of this work. Despite the fact that AI chips have made significant progress in the area of ML and NN computations, it is still in its infancy stage, and there seems to be a long way to go, before achieving a generic standardized AI framework; the so called artificial general intelligence (AGI), capable for solving out heterogeneous nature AI applications, especially at the edge network [13].

The photonic based interconnect solution is undoubtedly a viable approach for providing high data rates at low propagation losses, still, their component size is one with two orders of magnitude larger than what is required for the THz band case. Plasmon based THz link components on the other hand, due to their extremely small size and their ability to operate at ultra-high rates, may be a promising approach for equipping wireless THz nanoscale communication systems. Moreover, they could be perfectly combined with photonic technology, and particularly with dielectric waveguiding, as plasmonic waveguiding is quite lossy for long interconnect distances.

To this end, there is still an urge to have a comprehensive view on the current progress and recent advances in the wireless THz communications field, that would help researchers to have a reference point, and based on that, to expand their own ideas and directions, and find motivations to further develop research in this field. This work is a thorough investigation on current and beyond state of the art plasmonic system implementation for THz communications, by identifying the target nanoscale applications and major open research challenges, as well as the recent research achievements. It is the aim of this comprehensive survey then, to highlight the key roles of plasmon based technologies on equipping future competitive THz nanoscale communication systems hosting wireless THz nanoapplications, namely NoCs, WNSNs and beyond 5G communications. This survey paper may be well considered as a complementary work of [15], which had emphasized on key roles of plasmonics and silicon photonics, on equipping wired ultra-high bit rate interconnects, ranging from nanoscale intra and inter-chip interconnections, up to board to board and rack to rack interconnections between data centers (DCs).Particularly, the current work aims to complete the fundamental roles of plasmonic elements and mechanisms referred in the previous work, by associating the currently under investigation, THz system infrastructure in wireless communications. The potential of the THz communications is highlighted by illustrating the basic design issues in equipping these three important THz applications, that will define future wireless application demands coping with users' needs. Moreover, key roles of plasmonics for equipping each single, individual part of a future wireless THz nanocommunication link, namely the antennas and the transceiver parts, are also highlighted.

The rest of this paper is structured according to these two pillars, wireless THz nanoapplications and the implementation of future THz transceiver components. Sections 2–4 are each dedicated accordingly on these three major wireless THz nanoapplications, namely NoCs, WNSNs and beyond 5G communications. In each of these sections, in-depth reference material is provided, which includes the latest literature findings regarding the fundamental aspects of plasmonic technology roles and accompanied photonic technologies whenever required, for each one of these wireless THz nanoapplications, respectively. In Section 5, we focus individually on each critical plasmon based, or hybrid component part of a wireless THz nanocommunication link, namely the antennas and the transceiver parts. Finally, we conclude the paper in Section 6, by also providing two summary tables, with all the information aggregated, including all, state of the art and beyond state of the art characteristic plasmon based and hybrid achievements, for equipping future competitive THz nanoscale communication systems and wireless THz nanoapplications, accordingly.

2. WiNoCs

2.1. WiNoC Architectures Potential

As the number of processing cores within a chip area increases in pace with the increased requirements of running applications, communication needs increase as well, and given the fact that the chip area is limited, the node architecture complexity increases accordingly. More complex and sophisticated system designs are required, not only for cores, but also for memory parts to communicate with each other efficiently. Evidently, multi-level memory hierarchy communication needs with multi-core architectures, may be causing a communication bottleneck, as they grow in size.

Nowadays, state of the art multi-core architectures are based on wired NoC paradigm designs [16]. The first multi-processing core interconnections were shared bus architectures [17] which later on were replaced by on-chip CMOS electrical wired interconnections, according to the NoC framework. These were originally implemented via metal traces over a substrate forming a PCB [18]. In the last decade, many alternative fabrics and technologies have been progressively proposed in order to deal effectively with the NoC communication bottleneck, such as 3D NoC [19], RF signals over on-chip transmission lines [20], FSO communication systems at IR frequencies and above [21], photonic NoC [22], nanophotonic NoC [23], and recently WiNoC [24,25], or hybrid WiNoC [26,27]. Three dimensional NoCs are definitely an advantageous network architecture with desired features such as low distance, multiple variety horizontal or vertical interconnections, allowing integration of different technologies at different layers. This technology, however, requires thermal management to deal with the increased heat density due to the superposition of active layers and complex alignment methodologies for the precise positioning of the vertical interconnects. On chip RF schemes allow the interconnection of multiple cores over the same channel with dynamic bandwidth allocation, but they don't have much scalability potential, as they require an increased area and power overhead for the implementation of complex multi signal transceivers, and they also have to deal with the energy reflections at the line terminals. FSO systems on the other hand, may be a promising solution for providing high data rates at large bandwidth and operating at high frequencies accordingly, and they still have to tackle with a few issues such as the low transmission power budget due to eye-safety limits, the impact of several atmospheric effects (e.g., fog, rain, etc.) on signal propagation, and the strict alignment between transmitter and receiver that limits the achievable data rates [28]. Photonic and nanophotonic NoCs are definitely suggested for providing ultra-high data rates and bandwidths, they are CMOS compatible, but there are some parts within a NoC chip area, that are difficult to be implemented all optically, such as buffers, memories, and header controllers.

In general, it seems that as the number of cores on a chip increases and hence the communication performance requirements increase accordingly, all conventional wired interconnection and NoC schemes are inadequate to provide at the same time guaranteed desired latency, throughput, bandwidth, and energy efficiency, while wireless or hybrid wireless optical NoC solutions may be proved to be more promising alternatives. Specifically, WiNoCs, due to its inherent broadcast and multicast features, should be capable of providing improved performance in terms of scalability, flexibility and area overhead for multi-core systems. Only a single wireless transceiver along with integrated antennas and considerably less wiring equipment are required for interconnecting and sharing resources, among all the chip components, instead of many individual wired connections that would otherwise be required in conventional wired NoCs. It is a critical target for WiNoCs, to be able to manage efficiently wireless communication requirements at the core level, by exploiting miniature sizes of plasmon based antennas and other transceiver parts, in order to enable the integration of one or multiple antennas per core, as seen in Figure 2.

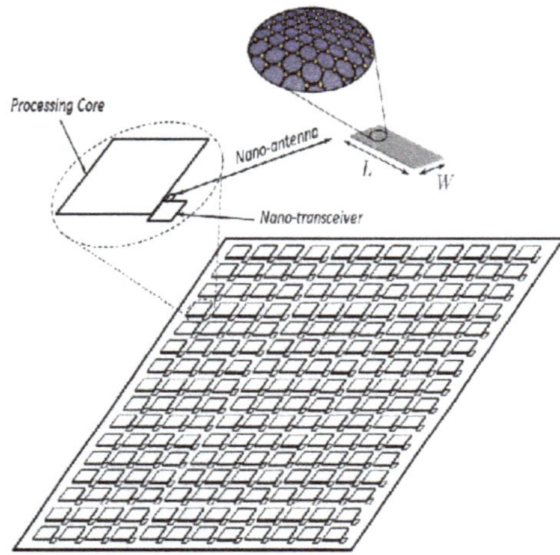

Figure 2. Wireless NoC(WiNoC) critical target—one nanoantenna per core.

Such antennas are mostly graphene based planar antennas, which radiate signals at the THz band, and utilize the minimum chip area than other conventional metallic counterparts [29]. Evidently, wireless interconnects are feasible to reduce wire equipment and parasitic and area occupation, as well as the power dissipation of long global, or multi hop short wires that would normally be required at wired competitors, providing the same high bandwidth and low latency communication [5]. Moreover, as wireless schemes natively enable all-to-all communication, they deal effectively with many other interconnection related issues such as multi-core interconnection with single memory, data coherency, consistency, and synchronization. Indeed, memory ordered execution operations, and cache coherence operations which involve a single memory image accessible to all processors, are critical in terms of latency, especially as the number of cores on a chip increases, in which case, traditional wired NoCs would be insufficient for guaranteeing such latency conditions, while wireless NoCs may do offer a promising solution [5].

2.2. GWiNoCs

WiNoC's main enabler is considered to be its on chip antenna, which it is integrated with a proper transceiver. Originally, WiNoC implementations were based on millimeter on-chip antennas radiating in the GHz band, integrated with adequate high frequency transceivers [30]. Nowadays, the research community has been focused on advanced wireless communication at the THz era. By increasing the communication frequency from GHz to THz domain, first, we anticipate for smaller footprints of the transceiver and the antenna, thus improving the integration potential of the system, and second, we anticipate for larger available transmission bandwidth and higher achievable data rates. At this critical crossroad there seem to be two reasonably strong trends to act jointly as a promising solution:

the considerable reduction of the size of the current metallic antennas and other transceiver components, so as to operate at very high resonant THz frequencies, which are mostly implemented via graphene material [31], and the adoption of an hybrid, wireless optical approach for WiNoCs, based on seamless integration of optical and wireless links on chip, enabling wireless multicasting and broadcasting of data, at optical frequencies [32].

GWiNoC, is a relatively recent approach that relies on graphene material for implementing not only nanoantennas [33], but also any other THz wireless transceiver part, for fully equipping the interconnection between the cores of a multiprocessor. As mentioned above, it is not feasible to reach extremely high resonant THz band frequencies by simply scaling down current metallic antennas [34], at the expected size of a nanosystem (a few μm) [35], as it would result in a huge channel attenuation. Graphene based nanoantennas on the other hand, are inherently just a few micrometers in size, i.e., two orders of magnitude below the dimensions of future metallic on-chip antennas, and hence they could provide inter-core communication in the THz band (usually between 0.1–10 THz). These graphene inherent features would offer both size compatibility with each continuously shrunk processor core, as well as adequate bandwidth for massively parallel processing.It seems that the ultimate WiNoC design target has been already set, consisting of a single graphene based nanoantenna and a nanotransceiver interconnected for each individual processing core, for managing the data of outgoing and incoming transmissions to the antenna respectively.

Graphene based antennas or graphennas, have shown excellent behavior as far as concerns the propagation of surface plasmon polariton (SPP) waves in the THz band. SPPs are coupled electron-light oscillations at the interface between a dielectric and a metal, that can propagate at the speed of light. SPPs in graphene are confined much more strongly than those in conventional noble metals, and they are electrically and chemically tunable by electrical gating and doping [8]. Hence, graphene can be considered as an appropriate THz tunable material for building THz resonator devices [36]. Graphene based nanoantennas and transceivers are a hundred times smaller in size than conventional microstrip antennas, with equal or higher bandwidth and gain [37]. Its long plasmon lifetime and the very high propagation velocity characterize it as an ideal material for implementing plasmonic waveguides for on-chip communication. Moreover, graphene has been found to be an appropriate material to enable the elaboration of GFET, providing higher speed and lower energy than conventional CMOS devices [38], and what's more, it is CMOS compatible. All graphene THz transceiver components can be combined with graphene-based THz antenna arrays, to achieve dynamic beam forming and steering enabling all wireless NoC scenarios. Therefore, apart from antennas, graphene has been equivalently proposed to build all types of THz transceiver components, as described in [39], and will be discussed in the next subsection.

2.3. Hybrid Optical Wireless NoCs

On the other hand, optical technology when combined with chip scale wireless interconnections may be as well considered as a promising hybrid NoC solution to overcome the performance bottlenecks of the current state of the art NoC architectures. Unfortunately, all plasmonic based solutions proposed in the literature for wireless applications do not overcome the problem of integration with SOI based NoC platforms. Moreover, plasmonic waveguides display high propagation losses and, therefore, they are not suitable for implementing long range on chip interconnections. An appropriate solution would be based on the adoption of the hybrid combination of plasmonic resonators as nanoantennas, while keeping dielectric waveguides as the feeding elements [6]. Hence, the employment of plasmonic nanoantennas adjusted to dielectric waveguides for building nano-optical wireless links instead of conventional plasmonic waveguide links, with short range propagation limitations would be a promising solution.

Another key design issue for building successfully such hybrid NoC architectures, is the implementation of the perfect coupling of plasmonic antennas with conventional silicon waveguides, guaranteeing full compatibility with Si photonic and nanophotonics circuitry standards. Waveguide coupled plasmonic antennas may become a drastic solution for a successful coupling without losses [40,41], enabling a hybrid optical wireless approach in the NoC design. The efficient coupling between plasmonic antennas and SOI waveguides is a non-trivial issue, as an on-chip, point-to-point connection normally requires matched directive nanoantennas. The nanoantenna shape and size should be properly designed so as to ensure impedance matching to the waveguide, and directional emission in the desired direction [42].

Moreover, hybrid wireless optical on chip communication takes advantage of the entire WDM spectrum when propagating in the optical wired links, guaranteeing even higher multiple capacities, as required by intra-chip communications [43]. Various configurations of plasmonic nanoantennas for supporting wireless-optical on chip communication have been proposed in the literature, such as plasmonic horn nanoantennas [44], a directional plasmonic Yagi-Uda nanoantenna placed on a dielectric waveguide [45], or a plasmonic nanoantenna array on a dielectric waveguide [46], or various configurations of plasmonic Vivaldi antennas (double, or an array of them) to name but a few [47,48]. Plasmonic antennas will be described more analytically in the upcoming section.

3. Wireless Nano Sensor Networks

WNSNs is another established THz nanoscale application under the internet of nanothings (IoNT) framework [49], which encourages, not only the core to core or to memory communication as in the WiNoC case, but also the interconnection between other nanoscale components, mainly nanosensors and nanomachines. These nanoscale networks rely on the THz band communication between its different components, which as mentioned, could be either nanosensors or nanomachines [7]. Nanosensors are capable of detecting events with unprecedented accuracy, while nanomachines are dedicated to tasks ranging from computing and data storing to sensing and actuation [50]. Hence, WNSNs are composed by integrated nanomachines and nanosensors, which interact with each other through EM communication [51]. EM communication in the THz band are mostly enabled by graphene based plasmonic nanotransceivers and nanoantennas, as in the WiNoC case.

Main features of the WNSNs are: (i) the size of nano-devices, which range from one to a few hundred nanometers, (ii) the exploitation of graphene based nanoantennas for THz band communication, (iii) extremely high bit rates (Tbit/s), and (iv) very short transmission ranges (tens of millimeters) [51]. Evidently, the THz band is considered as the natural domain for the operation of nanosensor components, as this frequency range supports very high transmission bandwidths within a short range. Alternatively, in the event of transmitting at lower frequencies (e.g., the MHz range), nanosensor devices would have to communicate over longer distances, but the energy efficiency of such a process to mechanically generate EM waves for remote control of these devices would be very low, and hence, communication by using the MHz frequencies wouldn't be an appropriate solution. Consequently, nanosensor devices would properly communicate with each other in the THz band [35].

As mentioned, apart from graphene-based THz antennas, graphene is also preferred for the development of other transceiver components in a scale ranging from one to a few hundreds of nanometers, such as: nanoscale FET transistors, nanosensors, nanoactuators, and nanobatteries. With the exploitation of graphene material, the integration of these nano-components in a single device of just a few micrometers in size is feasible, and will result in implementing autonomous nano-devices, able to perform specific tasks at the nanoscale, such as computing, data storing, sensing or actuation [35].

Depending on the measured parameters, nanosensors could be categorized in three types; namely physical, chemical and biological nanosensors [35]. Physical nanosensors such as pressure nanosensors [52], force nanosensors [53] or displacement nanosensors [54], are used to measure magnitudes such as mass, pressure, force, or displacement accordingly. Chemical nanosensors are used to measure magnitudes such as the concentration of a given gas, the presence of a specific type of molecules, or the molecular composition of a substance. Their working principle of both types is more or less the same and it is usually based on the change of the electronic properties of nanotubes and nanoribbons when they are used in a FET configuration, whose on/off threshold voltage changes as well by alteration of the value of each measured magnitude.

Last, biological nanosensors are used to monitor biomolecular processes such as antibody/antigen interactions, DNA interactions, enzymatic interactions or cellular communication processes. A biological nanosensor is usually composed of a biological recognition system or bioreceptor, such as an antibody, an enzyme, a protein or a DNA strain, and a transduction mechanism, e.g., an electrochemical detector, an optical transducer, or an amperometric, voltaic or magnetic detector [55]. There are mainly two subtypes of biological nanosensors based on their working principle: electrochemical biological nanosensors which work in a similar way to chemical nanosensors and photometric biological nanosensors. The latter subtype working principle is based on the use of noble metal nanoparticles and the excitation using optical waves of surface plasmons.

More specifically, a typical generic architecture of a WNSN node as seen in Figure 3 [35], would be consisted of: (i) Sensing unit: graphene material and its derivatives, namely, graphene nanoribbons (GNRs) and carbon nanotubes (CNTs) [56], provide outstanding sensing capabilities and they are the basis for implementing many types of sensors [57]. (ii) Actuation unit: an actuation unit will allow nanosensors to interact with their close environment. Several nanoactuators have also been designed and implemented so far [58]. (iii) Processing unit: nanoscale processors are being enabled by the development of different forms of miniature FET transistors in the nanometer scale. They were mostly implemented via CNTs and GNRs nanomaterials. (iv) Storage unit: graphene has shown excellent performance in a number of applications from supercapacitors [59] to photomechanical actuators [60], however, so far, its potential in nanomemory construction has not been adequately explored. (v) Power Unit: there are two types of nanobatteries [61] for feeding nanomachines: (a) harvesting the energy from the environment via nanoscale energy harvesting systems [62] and (b) wireless energy induced from an external power source [63]. (vi) Communication unit: this consists of nanoantennas and transceivers for guaranteeing EM communication between nanosensors. The working principle of energy harvesting is based on the conversion of mechanical or vibrational or hydraulic energy into electrical energy. The mechanical energy is produced by the human body movements, or muscle stretching, the vibrational energy is generated by acoustic waves or structural vibrations of buildings, and finally the hydraulic energy is produced by body fluids, or the blood flow. This energy conversion is achieved by the piezoelectric effect seen in zinc oxide (ZnO) nanowires, as they are bent, when a voltage appears in the nanowires (Figure 4) [35].

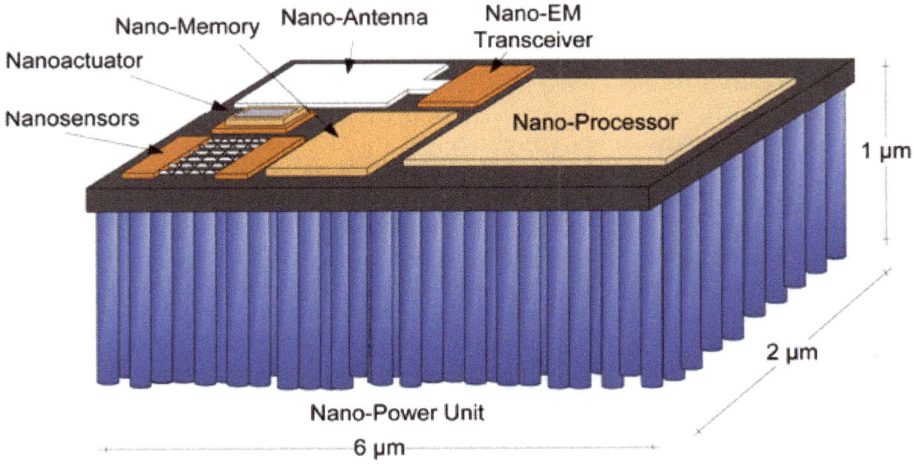

Figure 3. Wireless nanosensor networks (WNSN) node architecture.

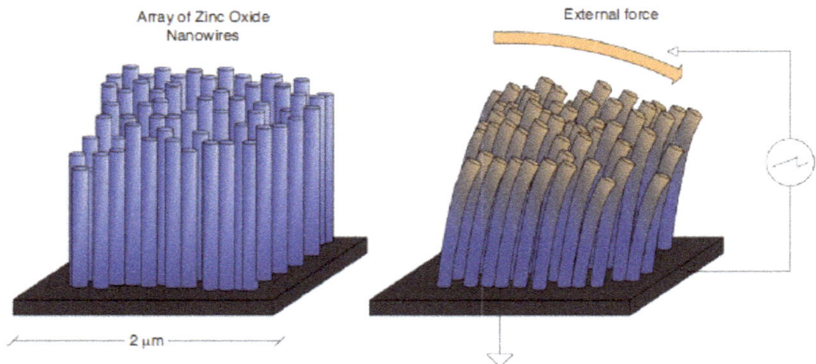

Figure 4. Energy harvesting is based on the piezoelectric effect seen in zinc oxide (ZnO) nanowires.

As mentioned, nanoantennas are mostly implemented via CNTs and GNRs nanomaterials. Concerning the latter case, the propagation of EM waves on a graphene sheet was first analyzed in [64] while in [34], nano-patch antennas based on GNRs and nano-dipole antennas based on CNTs were quantitatively compared. EM transceivers of nanosensors are embedded systems that include all the necessary circuit equipment which processes the transmitted or received signals from the free space through the nanoantenna with proper functioning such as frequency conversion, filtering and power amplification. Several GFET transistors capable of such functioning and operating in the sub-THz or THz band have been demonstrated so far [65]. Other materials, such as Au and Ag have been successfully used for plasmonic sensors in the visual [66,67] and near infrared [68], as well. Ge plasmonic material in the mid-IR, has been also proposed as a promising material for replacing silicon, as a substrate for MOS devices [68].

According to [35], the applications of WNSNs can be classified in four main groups: biomedical, environmental, industrial, and military applications. As far as concerns in the environmental and industrial application domains, various THz nanosensors have been used for the detection of pollutants and empowering the technology of food preservation and food processing [69]. Nanowire based nanosensors are suggested for sensing ambient intelligence, such as vertically bridged nanowires, laterally bridged nanowires, ultra-sharp Ga2O3 nanowire, the nanowire FET chem-biosensor, and the CNT biosensor [70]. As far as concerns military applications domain, various THz nanosensors, detectors and cameras have been suggested for security applications, and specifically for the detection of weapons, explosives, as well as chemical and biological agents [71].

As far as concerns in the biomedical applications domain, graphene based nanoantennas have been used for wireless communication between nanosensors, deployed inside and over the human body, resulting in many bio-nanosensing applications [72]. In-vivo wireless nanosensor networks (iWNSNs) at the THz band, is a characteristic application for providing fast and accurate disease diagnosis and treatment [73]. These networks, fully equipped with nanoscale components (e.g., nodes, routers, gateways, links and interfaces) as in regular networks, as seen in Figure 5, are operating inside the human body in real time, achieving precise and real time medical monitoring and medical implant communication. Apart from iWNSNs, graphene plasmons have also been used in biological sensing applications, monitoring the rotational and vibrational modes of DNA molecules and many large proteins in the THz and far-IR [74,75]. Moreover, as far as concerns imaging applications, and especially the THz band spectroscopy, nanosensors have been used for backscattering techniques for monitoring the dynamics of large biomolecules [76]. A THz biosensor communicating with the biological agents through a graphene plasmonic waveguide is described in [77]. In this sensor, SPPs in the waveguide must be launched by either an external near-field source or an antenna. In [78], another biosensor system has been proposed, where all the required elements (the sensor, frequency modulator, and antenna for energy harvesting) are packaged in a single module [78]. This structure has been designed as a dual resonant antenna part of a GFET.

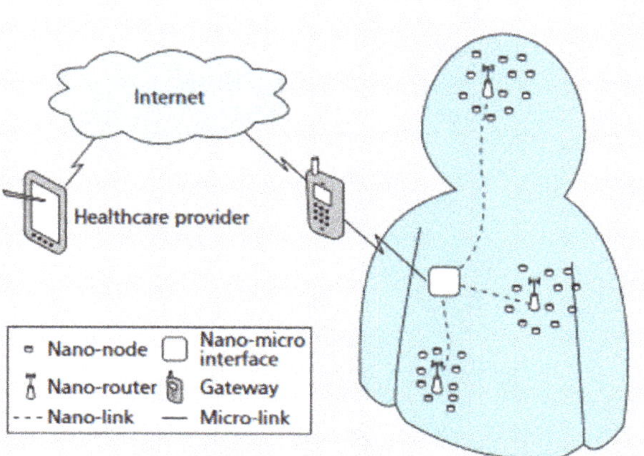

Figure 5. In-vivo wireless nanosensor networks (iWNSNs) for healthcare applications.

4. Beyond 5G Networks: Towards to THz Band Communications

At the moment, wireless traffic in the access 5G networks exploits wide radio bands such as the mmW frequencies and systems, respectively. The short term roadmap for 5G and beyond communication would anticipate the establishment of many high-rate small cells forming the access link, operating in the mm wave spectrum, which are in the order of several Gb/sec, thus, in total, the aggregation capacity of the fronthaul/backhaul link should be several times higher, so as to guarantee reliable and fast data delivery from multiple users, which are connected to the small cell. Hence, in such circumstances, the THz band available bandwidth resource would be highly appreciated, while the high propagation loss of THz band fronthaul/backhaul links would be compensated by the extremely high antennas directivity [79]. In order to accommodate the continuously increasing wireless traffic demands of 5G communications and even beyond, researchers have been focused on taking advantage of higher regions in the radio spectrum (above 300 GHz), pointing to the THz band communication and infrastructure, thus enabling efficient operation of bandwidth hungry applications, that are not feasible for these systems at the moment. Hence, in the case of THz band communication, the supplied bandwidth required, would be one order of magnitude above current mmW systems, offering faster data transfer and download speeds, lower latency, and higher link directionalities [80] with non-line-of-sight (NLoS) propagation [81], as THz waves could penetrate thin objects, able to carry the data, such as the data transmission/reception by a smartphone in a pocket.

Despite the fact that current research works have been focused on various 5G scenarios based on photonic assisted wireless communication systems with very high data rates, the capacity demand of THz wireless systems has not been achieved yet. Uni-travelling photodiodes (UTC-PDs) and comb sources, [82,83], in the W-band (75–110 GHz) incorporating optical polarization division multiplexing (PDM) [84], and spatial multiple-input-multiple-output (MIMO) techniques, are considered as state of the art photonic sources, achieving data rates of 100 Gbit/s and beyond [85,86].

However, as mentioned, higher regions in the radio spectrum at higher operating rates are required, in order to reach THz communication system goals, seeking for brand new drastic solutions. Reference [21] is a survey dedicated to hybrid radio frequency, FSO systems with THz/O links, considering it as a viable approach for equipping future THz wireless communication. Integrated microwave photonics (IMWP) in the THz range [87], have been also proposed as an enabling approach for equipping 5G wireless systems, by providing optical signal generation and distribution of mm waves towards antenna terminals [88], dynamic filtering [89], optical control of antenna arrays [90], and many more other functions [91]. 5G applications, however, pose very stringent requirements on the speed of IMWP circuits, for processing mm waves or even sub-THz frequencies, in order to access the multi-GHz bandwidths required for high data rates [91].

In practice, there are a few more challenges a designer has to consider, when applying these suggested photonic based technologies in THz wireless communication systems. Specifically, given that free space path loss increases as the square of the frequency, directive antennas and directional transmitters focused on individual users, such as high gain steerable phased arrays, will be required in order to compensate for the large free space path losses of EM waves. Beam steering techniques [92] and massive MIMO antenna array elements arranged on terminal devices [93] may be promising solutions for increasing antenna directivity gain. There are also other seamless integration issues of wireless links combined with photonic infrastructures that might appear, such as the connection of optical fibers to THz transmitter and receiver front ends, which consequently requires optical to THz (O/THz) and THz/O converters with high bandwidths (well above 300 GHz). As far as concerns O/THz conversion, UTC-PDs integrated with sub-THz waveguides are considered an established solution [82,94] but the opposite THz/O conversion is still an issue that needs reconsideration, as it requires modulators with electro-optic bandwidth well above 300 GHz, high power management, and very high linearity. Plasmonic modulators may at this point prove to be a viable solution, as they are characterized for their ultra-compact footprints [95] (10's μm^2), ultra-low power consumption (2.8 fJ/bit at 100 GBd) [96], and flat frequency responses up to 170 GHz [97] and 325 GHz [98]. In [99], the experimental demonstration of a plasmonic MZ modulator with sub-THz frequency responses (up to 500 GHz), high power handling, and high linearity is described. The first demonstration of a THz link seamlessly integrated into a fiber optic network using direct THz/O conversion at the wireless receiver, is described in [100]. An ultra-broadband silicon POH modulator, is used for THz/O conversion of WDM signals in [100]. Alternatively, graphene based THz components have shown very promising results in terms of generating, modulating as well as detecting THz waves [101], and hence they may be considered as appropriate THz band transceivers [102].

5. Plasmonic THz Wireless Nanoscale Link Components

The main components of a typical WiNoC layout are basically the THz band antenna along with transceiver components, used as a feeding element. Specifically, the transceiver is held responsible for preparing the information for outgoing transmissions to antenna, and for demodulating incoming transmissions from antenna respectively.

5.1. Plasmonic THz Antennas

5.1.1. Design Issues

There are many challenges when designing a THz band plasmonic antenna, such as the material parameters and properties, the size of the antenna, the impedance mismatch with the coupling waveguides, and its integration potential with processing cores and transceiver parts. As far as concerns the basic parameters that characterize the antenna material, such as the dynamic complex conductivity and permittivity, as well as the propagation properties of SPP waves on the nanoantenna, such as the confinement factor and propagation length, and finally the antenna geometry parameters, such as the length and radius, all these design issues should be taken into consideration before ending up to the proper antenna implementation choice, such as state of the art metallic or hybrid antenna structures [103].

The conductivity of plasmonic materials such as graphene, gold or silver is a complex-valued parameter which affects the global oscillations of electrical charge in close proximity to the surface of the antenna, resulting into the excitation of electromagnetic SPP waves. The frequency at which SPP waves are excited depends on the material conductivity of the antenna components. For example, graphene supports SPP waves at frequencies as low as in the THz band (0.1–10 THz), whereas in noble metals such as gold or silver, SPP waves are only observed at tens of THz and above [104]. There are many analytical models describing the conductivity of metals, among them, the Kubo formula is considered the most appropriate model, associating the complex conductivity and permittivity as functions of the frequency. Particularly for terahertz frequencies, the Drude model contribution is applicable, which takes into account the intra-band electron transitions within the metal energy band structure. According to this model, plasmon material conductivity acquire a different response at THz band, because of its intrinsic kinetic inductance, associated to the imaginary part of the conductivity, that plays the role of negative real permittivity in a bulk material. At these high frequencies, moderate changes in the chemical potential can significantly alter graphene's conductivity and change the sign of its imaginary part [64]. Apart from frequency dependence and band structure, graphene and other metals conductivity also depends on a set of parameters such as chemical doping, Fermi energy (chemical potential), electron mobility, and relaxation time. The relaxation time is the interval required for a material to restore a uniform charge density after a charge distortion is introduced, while the chemical potential refers to the level in the distribution of electron energies at which a quantum state is equally likely to be occupied or empty. These two variables have a strong impact on the resonant frequency and radiation efficiency of the antennas. The value of the chemical potential can be altered by applying an electrostatic bias or chemical doping, providing also significant reconfigurability potential. The bias injects electrons or holes on the active area of the structure, modifying graphene's chemical potential [29].

An example of graphene's dispersive conductivity is shown in Figure 6. In general, the chemical potential variation is directly related to the applied voltage variation. As implied in Figure 6, higher chemical potential leads to better performance through an increase of the conductivity. Figure 7 shows the resonance characteristics of the antenna as a function of the chemical potential. The increase of chemical potential leads to a significant shift of the resonant frequency and to an enhancement of the antenna response without changes in the radiation pattern [105].

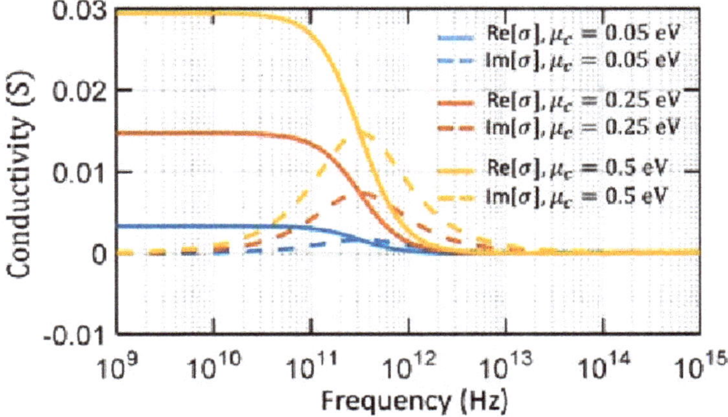

Figure 6. Graphene's dispersive conductivity at THz frequencies.

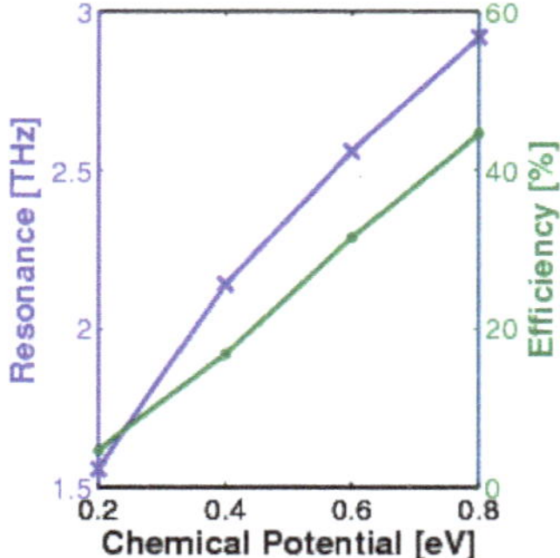

Figure 7. Nanoantenna resonance characteristics as a function of the chemical potential.

Absorption enhancement based on metal nanoparticle dispersive properties of metal nanoantenna structures is also a critical design issue that needs also consideration. These metal nanoparticle dispersive properties response of different metals at THz range are also described by the Drude–Lorentz model by relative complex permittivity of metal nanoantennas as a function of frequency as seen in Figure 8 [106], with ε1 the real part of the relative permittivity, and ε2 is the imaginary part of it. Different metal nanoantennas like gold, silver, copper and aluminum are considered.

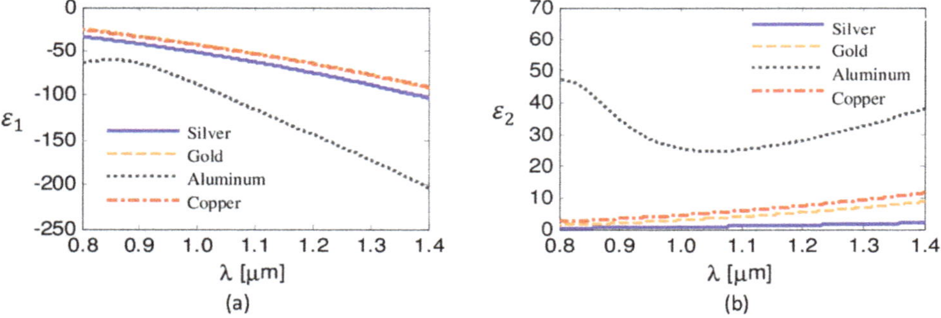

Figure 8. Complex permittivity of different metal nanoantennas at the THz range. (**a**) real part and (**b**) imaginary part of relative permittivity.

An approximate 20% increase in absorption performance is obtained. The frequency that corresponds to the maximum absorption is dependent on the nanoparticle material. The maximum absorption is observed in gold nanoparticle material, as seen in Figure 9 [106]. The frequency of the maximum absorption moves toward lower frequencies as physical dimensions of the nanoparticles increase. Hence, the choice of the best plasmonic material for a given application is a subject of discussion and research. It has been also proved that the radiation frequency of the graphene nanoantenna can be tuned in a wide spectral range by adjusting the dimensions and particularly its length. According to [107], the absorption cross section increases up to a given limit with the substrate size, and hence, a larger substrate improves the performance of the graphene nano-antenna. Moreover, the absorption cross section increases as the graphene patch is located closer to the side of the substrate, while the resonant frequency becomes higher when the patch is farther from the center [107].

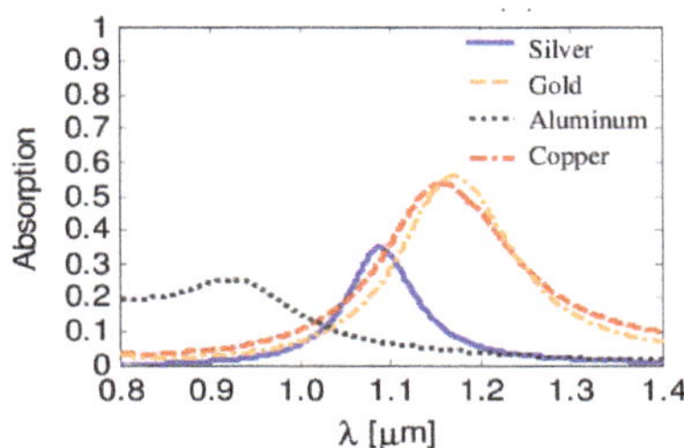

Figure 9. Absorption response at THz range for different metal nanoantennas.

Therefore, the optimal location for on chip graphene nano-antennas may be near the edge of the substrate, in order to maximize their efficiency. As it concerns chemical potential, there is a tradeoff between the amount of power the graphene nano-antenna can absorb and its resonant frequency. Specifically, graphene nano-antennas with zero chemical potential, resonate at a low frequency but with a small absorption cross section, thus limiting their radiation efficiency. On the other hand, nano-antennas with a higher chemical potential experience greater absorption performances with an increased resonant frequency. Thus, there exists a compromise between these two parameters to be taken into consideration.

In [108] there is an exhausting investigation and comparison of the performance in transmission and reception of metallic nano-dipole antennas, implemented via various metals such as Cu, Al, Ag and Au. Taking into account each metal property, such as its dynamic complex conductivity and permittivity, the propagation properties of SPP waves and the antenna geometrical features, as length and radius, a mathematical framework is developed in order to analytically derive critical transmission and reception performance parameters such as the generated plasmonic current in reception and the total radiated power and efficiency in transmission.

As far as concerns the THz antenna size future requirements, and given that the available bandwidth is inversely proportional to the antenna size, it requires only a few micrometers in size, in order to build an appropriate nanoantenna, which is almost two orders of magnitude below the dimensions of current on-chip antennas. Hence, the approach of integrating one antenna per core seems rather unfeasible for CMOS technology, especially as the core sizes continue to shrink to a few hundreds of micrometers [109].

Silicon integrated antennas could be considered as a mature option for on chip antennas, but their size ranges from a few to ten mm's, such as the zig-zag monopole antenna of axial length 1–2 mm proposed in [110], or the demonstration of a miniature on-chip antenna operating at the range of 100–500 GHz in [111]. CNT based antennas have been also considered as an alternative approach for implementing THz antennas. Normally the CNT antennas are capable for equipping WiNoCs, as they have small size, low power losses, high transmission power and they operate at very high rates. Unfortunately, they are characterized by significant manufacturing difficulties for implementation [112]. Ultra-wide broadband (UWB) and multi-band antennas have been also proposed for on-chip wireless communication, due to their multi-channel capability, that can be shared among multiple nodes. A CMOS UWB based antenna for WiNoC has been proposed in [113], its short transmission range, however, does not give much prospect for long distance communications. Horn and paraboloid antennas have been also proposed for transmission at 300 GHz, with a radiation bandwidth in the order of 10 percent of their center frequency, their geometry however, makes them not suitable for mobile and personal devices [28]. The very small size of a THz Band antenna is an uncompromising necessity required for future wireless THz communications, so as to allow the integration of a very large number of antennas with very small footprints, forming very large antenna arrays. New antenna array patterns, such as massive MIMO schemes based on graphene plasmonic material for doping, may be a promising approach [93].

5.1.2. Graphennas

Graphennas are just a few micrometers in size, thus enabling size compatibility with each processor core, and providing enough bandwidth, and hence, they are appropriate candidates for inter-core communication in the THz band. Graphene is a one-atom thick layer of carbon atoms in a honeycomb crystal lattice, and it is considered as an attractive solution, due to its unique electrical and optical characteristics [114]. Graphennas show excellent behavior, as far as concerns the propagation of SPP waves in the THz band. As known, SPPs are electromagnetic waves guided along a metal-dielectric interface and generated by means of an incident high frequency radiation [115]. A graphene layer supports transverse magnetic SPP waves with an effective mode refractive index. A basic configuration of graphene THz nanoantenna is shown in Figure 10. The nanoantenna is basically composed of a graphene layer, which is the active element, along with a metallic flat surface which is the ground layer, and a dielectric material layer in between the former two layers. An antenna feeding mechanism is also required in order to complete the nanoantenna layout [115]. By adjusting the dimensions of the graphene nanoantenna, the radiation frequency can be tuned in a wide spectral range.

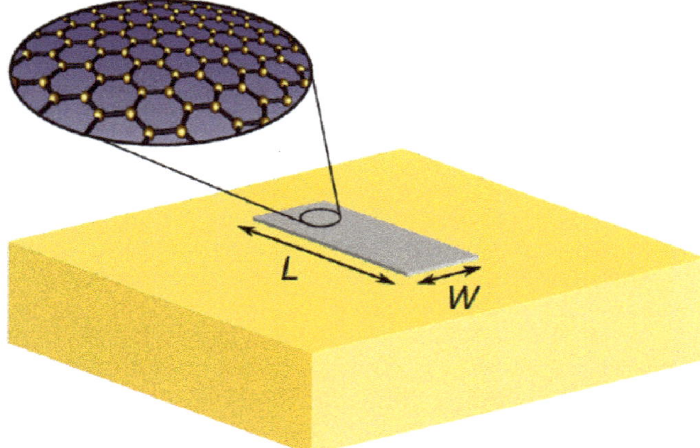

Figure 10. Graphene THz nanoantenna configuration.

As mentioned in the previous section, all plasmon based THz solutions are characterized by relatively high losses of the supported SPP modes, especially during waveguide propagation, hence hybrid combinations of plasmonic resonators with dielectric waveguides appear to be an attractive alternative solution. We may at this point, consider a broader generic hybrid combination of surface plasmon with dielectric wave modes, providing better results between mode confinement and propagation loss in the THz band, as seen in Figure 11 [116]. In other words, graphene may be used either for the development of plasmonic waveguides, or antennas in hybrid combinations with dielectric material. Specifically, as far as concerns graphene based waveguide structures, and given that THz plasmons can be confined laterally in a graphene sheet [117], this property has led to an opportunity of implementing many different graphene based waveguide structures, consisting of a number of graphene layers, mixed with dielectric materials [118], or with dielectric-metal structures [119], obtaining excellent field confinement results. Other approaches involve the use of graphene layers to form wedges [120], or coat grooves carved on the dielectric [121], or a waveguide consisting of a dielectric with high permittivity, on top of a low index dielectric-graphene-dielectric stack [122].

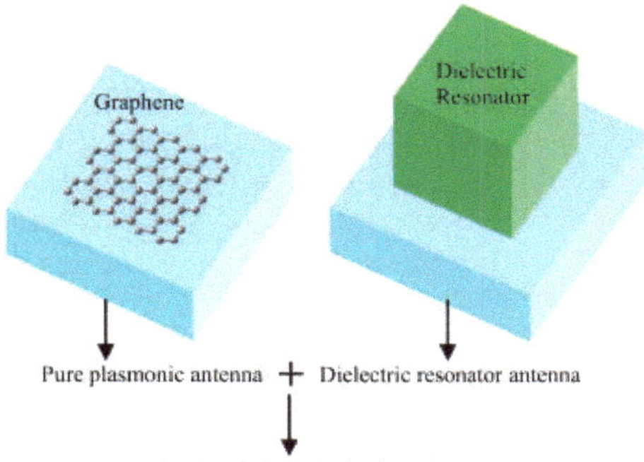

Figure 11. Hybrid combination of surface plasmon with dielectric wave modes for THz nanoantenna implementation.

As far as it concerns graphennas, such structures are based on a number of graphene layers, on a metallic flat surface, a dielectric material in between, and a feed to drive the signals to the antenna, as seen in Figure 10. Other antenna structures, such as patch antenna, and dipole designs, where the feeding mechanism lies in between the two identical graphene patches, have been also proposed in [107,123], respectively. These structures are either based on an ideal photomixer with high impedance as the feeding mechanism, or on more advanced THz sources based on photoconductive materials [124], or on high electron mobility transistors [102]. Furthermore, graphene antennas in MIMO configurations with a considerable number of radiating elements, may be also considered as an updated efficient THz solution [125]. Alternatively, other works are based on graphene potential of tuning ability, rather than acting as a radiating antenna element. Specifically, in [126], graphene sheets are employed between the source and a metallic radiating element to retain the tunability, while in [127], a novel antenna design is proposed, based on hybrid graphene-metal structure for enhancing reconfigurability capabilities of THz antenna. Different hybrid graphene-dielectric antenna structures have been also proposed in [128], such as a two graphene monolayers separated by a thin dielectric structure, or an hybrid structure with two graphene monolayers (H2G), consisting of a layer with a high index material (HIM) for a dielectric mode, close to the graphene layer for a plasmonic mode, and separated by a spacer with a low index material (LIM), as seen in Figure 12d.

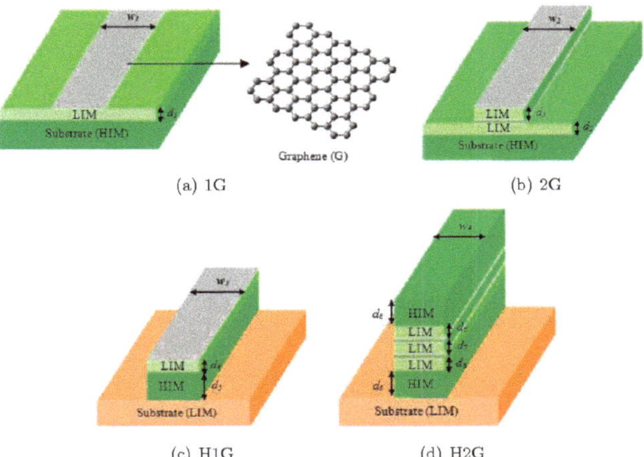

Figure 12. Pure plasmonic (**a**,**b**) and hybrid graphene-dielectric antenna structures (**c**,**d**).

5.1.3. Other Plasmonic Nanoantennas

Other plasmon based nanoantennas have also been proposed, based on hybrid wireless-optical on chip communication. When designing such hybrid structures, large impedance mismatches between the resonant nanoantennas and the waveguides should be completely minimized. Proper impedance matching elements of certain permittivities are needed to be carefully employed as wireless transceiver parts, and positioned within the connecting gap between the nanoantenna and the waveguide. Classic antenna layouts applied in RF communications such as dipole, Yagi-Uda and phased array configurations [129–131], have been also applied in optical wireless nanolinks, using plasmonic waveguides as matching elements, between pure or hybrid plasmonic antennas and their feeding silicon waveguides. Recently, in [132], a dipole loop plasmonic nanoantenna has shown an increased operating bandwidth compared with a single loop antenna. Plasmonic horn nanoantennas proposed in [41], are impedance matched to the feeding waveguides, and show superior performance against those using dipole nanoantennas, for point-to-point optical wireless nanolink communications. A typical plasmonic horn nanoantenna is shown in Figure 13.

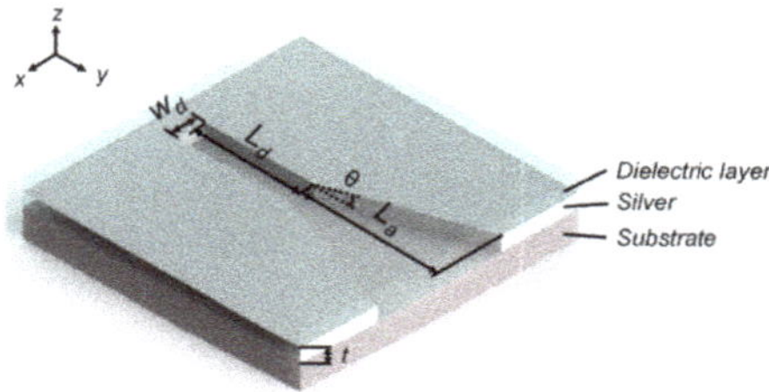

Figure 13. A typical plasmonic horn nanoantenna.

Vivaldi antennas have been also considered as an updated solution for such hybrid communications. They are usually applicable in the microwave and radio frequencies, but also in infrared/optical frequency domain. Vivaldi plasmonic antenna is formed by a slotted microstrip deposited above a silica substrate, along with a hybrid Si-plasmonic coupler, as an impedance matched element to feed silicon waveguides [43]. By increasing the number of Vivaldi antennas, an increase in the directivity and the gain is anticipated too. Hence, double Vivaldi broadside antenna [47], and antenna array configuration based on tilted plasmonic Vivaldi antennas [48], are improving the total antenna radiation performance. A double Vivaldi antenna and its coupling details are shown in Figure 14a,b respectively.

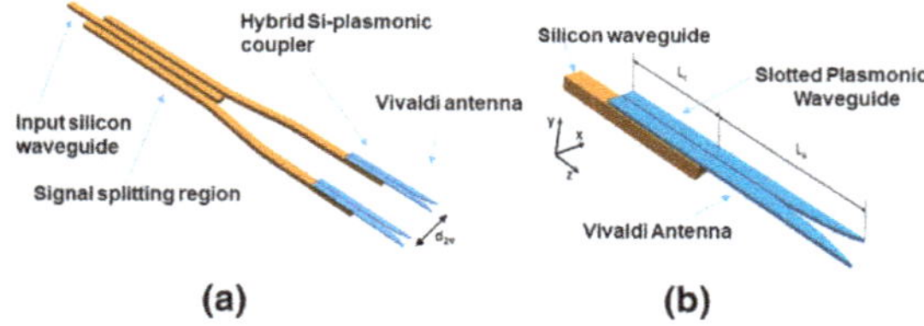

Figure 14. A double Vivaldi antenna and its coupling details ((**a**,**b**) respectively).

All these Vivaldi hybrid structures are characterized as SOI integrated as the optical signals could be propagated through silicon waveguides and plasmonic nanoantennas wireless links, thus avoiding integration of electronic devices and electro-optical conversions, and reducing complexity and energy costs [47]. However, efficient coupling between plasmonic antennas and SOI waveguides is a non-trivial issue, and proper plasmon based impedance matched elements are required to tackle with this issue. Other SOI integrated structures are proposed in [133], such as an antenna array consisting of a series of hybrid plasmonic nanoantennas with subwavelength footprint, that is highly compatible with a low loss silicon waveguide, which feeds light from the bottom of the nanoantenna.

Alternatively, plasmonic nanoantennas may be fed by hybrid plasmonic waveguides, instead of pure silicon waveguides, such as the plasmonic nanopatch antenna, fed by a hybrid metal insulator metal (HMIM) multilayer plasmonic waveguide, for achieving proper impedance matching, as seen in side and top views of Figure 15 [134]. Such proposed hybrid plasmonic waveguide (HPW) based nanoantennas, show high efficiency and directivity, and improve the efficiency by minimizing losses [40].

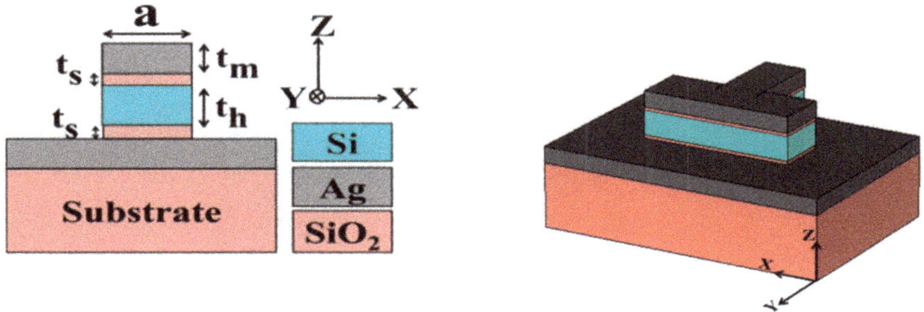

Figure 15. Hybrid metal insulator metal (HMIM) multilayer hybrid plasmonic nano patch antenna.

5.2. THz Band Nanotransceivers

In general, as known, a transceiver is a set of components which is responsible for generating and modulating the outgoing information to the antenna, via appropriate transmitters at the front end of the nanolink, and also responsible for detecting the incoming from the antenna information via appropriate receivers at the back end of the nanolink. A block diagram of a typical plasmonic transceiver architecture is seen in Figure 16 [102]. Apart from the basic transceiver components, namely, the signal generator, the transmitter which is normally a modulator source, and the detector-receiver, there may be other components acting complementary, in order to propagate the information signal across the link with the minimum loss, such as interconnects, switches, filters, high tunability phase shifters, mixers, frequency multipliers, and impedance matched elements, placed between the antenna and the waveguide of the link. As can be seen from Figure 16, at the front end, there is an electric signal generator that generates the electric signal that will later on, feed the transmitter source. Then the outgoing signal, which in our case of plasmonic THz sources is normally a modulated SPP wave, is feeding the plasmonic nanoantenna, which converts the SPP wave into an EM wave. At the receiver end, there is another plasmonic nanoantenna with a similar role, that inversely converts the EM wave again into an SPP wave. Finally, a plasmonic nanoreceiver converts the SPP wave into an electric signal, which will be demodulated to the original information data by the signal detector [102].

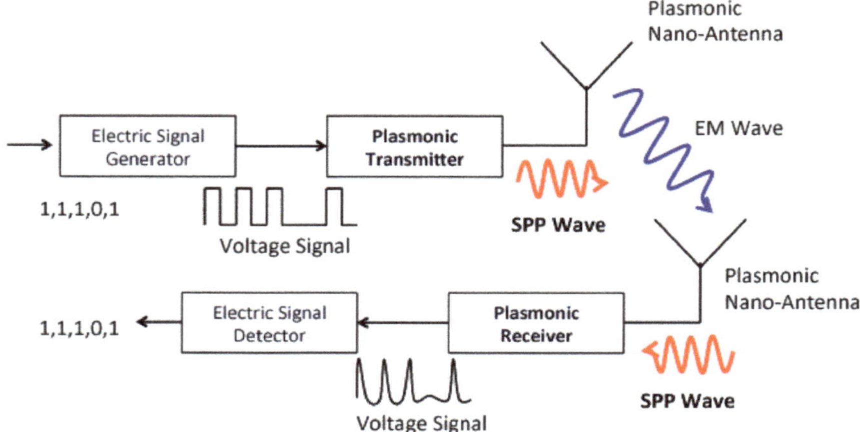

Figure 16. Plasmonic transceiver block diagram.

5.2.1. THz Band Transmitters

For the last few years, silicon photonic devices have been widely adopted for equipping current THz band transceivers, with desired features such as large bandwidth, high transmission power and high detection sensitivity. However, high path losses at the THz band is still considered a major unsolved challenge. As far as concerns THz band transmitters, photonic III/V (InP) devices, such as UTC-PDs with mW power levels at 300 GHz [135], and quantum cascade lasers (QCLs) [136], are considered as state of the art sources. The latter, however, requires an external laser for optical electron pumping, thus limiting the area overhead, while also performing poorly at room temperature [137]. SiGe-based heterojunction bipolar transistors (HBTs) have been also proposed in [138], for equipping THz transmitters operating at 820 GHz.

QCLs at the THz band, such as plasmonic QCLs, or THz QCLs [139] with metallic cavities [140], which belong to the plasmonic laser family, may be a promising candidate for THz transmission, but researchers have to tackle with tunability potential at these high frequencies. Recently, a new tuning mechanism called the antenna feedback mechanism, has been developed for single mode metal-clad plasmonic lasers, based on the principle that the refractive index of the laser's surrounding medium affects the resonant cavity mode as much as the refractive index of gain medium inside the cavity [141]. This mechanism leads to the generation of hybrid SPPs propagating outside the cavity of the laser with a large spatial extent. The emission frequency of the plasmonic laser and its tunability potential are dependent strongly on the effective propagation index of the SPP mode in the surrounding medium, by coupling the resonant SPP mode to a highly directional far field radiation pattern, and integrating it, with the hybrid SPPs of the surrounding medium [142].

Such an antenna feedback principle is shown in Figure 17. Specifically, the general principle of a DFB can be implemented in a THz QCL by introducing periodic slits or holes in its metallic cladding. A parallel-plate metallic cavity is illustrated (Figure 17a). Due to the periodicity of conventional DFBs there is a phase-mismatch between successive apertures for SPP waves on either side of the cladding (Figure 17b). The designed grating period of antenna-feedback effectively couples a single-sided SPP wave that travels in the surrounding medium with the SPP wave traveling inside the active medium (Figure 17c). The antenna-feedback scheme leads to a buildup of in-phase condition at each aperture between counter propagating SPP waves on the either side of metal-cladding. Emission from each aperture adds up coherently and constructively to couple to far-field radiation in the z direction (Figure 17d).

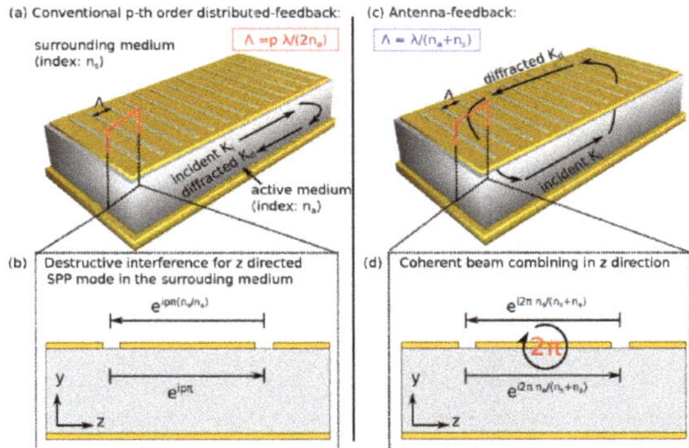

Figure 17. Antenna feedback principle of a DFB laser.

5.2.2. THz Band Receivers

As far as concerns state of the art THz band receivers, waveguide integrated detectors using GaAs Schottky barrier diodes (SBDs) [143], are considered the most common implementation choice. However, they pose limitations on their size and available bandwidth, and behave poorly at room temperature, as QCLs [144]. Silicon CMOS technology has also been proposed in [145], for equipping oscillators at 870 GHz, and sub-harmonic detectors between 790 and 960 GHz. These structures require power amplifiers implemented with InP or GaN technologies, in order to challenge path losses, such as the InP based high electron mobility transistors (HEMT) amplifiers with 10 dB gain at 640 GHz [146], or GaN, GaAs or InGaAs HEMTs [147–149]. Unfortunately, all these components require more chip space and they also pose performance limitations when operating above 1THz.

As far as concerns plasmon based THz wave detectors, hybrid plasmonic schemes consisting of plasmonic waveguides along with silicon waveguides, all integrated with conventional nanoantenna structures, such as dipole or bowtie antennas are considered to be promising, beyond state of the art approaches. A POH slot waveguide integrated with a bowtie antenna is proposed in [150], as a THz wave detector. In this structure, a taper is used to connect silicon strip waveguide with the electro-optical polymer refilled plasmonic slot waveguide, all integrated with a bowtie antenna. Plasmonic slot waveguides are capable of guiding electromagnetic waves at subwavelength scale modes, thus bypassing the diffraction limit bottleneck of conventional waveguides. Consequently, they are functioning as nanocouplers, and hence they can be used as excellent matching elements with the state of the art on chip sources such as nanoLEDs [151] and nanolasers [152], which are all belong to the quantum emitters family. Recently, plasmon based nanocouplers, such as plasmonic grating couplers [153], plasmonic waveguide tapers [154], and nanoantennas [155] have been proposed for integrating silicon photonic chip components with plasmon based components, providing a promising approach for manufacturing plasmonic integrated circuits (PICs) [156].

Alternatively, plasmonic internal photoemission detectors (PIPED) structures, which are actually transmitter and receiver packages, monolithically integrated on a common silicon photonic chip platform, have been also proposed in [157], for THz wave signal generation and coherent detection at frequencies of up to 1 THz. Finally, other more novel approaches are related to the employment of a FET operating at THz frequencies (TeraFET), as a THz detector. Specifically in [158], a TeraFET operation with identical radiation amplitudes at the source and drain antennas but with a phase shift induced asymmetry is proposed, based on the principle that a phase difference between THz signals coupled to the gate and source, and gate and drain terminals of a FET accordingly, enhances device plasmonic resonances. Such a TeraFET structure operates between 200–600 GHz band, and could be used for 5G and beyond, communication systems.

5.2.3. Graphene Based THz Transceiver Components

Apart from graphennas, graphene has been also considered as a promising technology for building THz band transceivers operating at 1–10THz band and even at higher frequencies. In fact, graphene may be used for implementing not only the basic transceiver components, namely sources and detectors, but also for other components such as interconnects, modulators, switches, filters, and phase shifters with high degrees of tunability [101]. Moreover, by applying the same graphene nanomaterial for all these THz transceiver components, the integration potentials of packaging compact THz band nanotransceivers can be highly boosted. Specifically, all graphene THz components can be combined with arrayed THz graphene antennas, so as to enhance total antenna directivity [125]. Evidently, the adoption of the same building material is expected to minimize mismatches, and consequently, the losses between the interconnected components. Many graphene-based components have been proposed for feeding THz link nanoantennas, such as plasmonic switch based on GFET [159], THz LPF [160], BPF [77], and phase-shifters [161]. All these feeding components are based on the same GFET principle, being modeled as a tunable transmission line for the propagation of SPPs, with adjustable operation at different lengths and bias levels each time, depending on the application.

Graphene can be efficiently applied for the implementation of HEMT used as THz source. Specifically, in [102], the proposed THz source is based on a III-V semiconductor based HEMT, enhanced with grapheme, so as to generate the necessary SPP waves that drive a plasmonic nanoantenna with satisfactory impedance matching, thus enabling compatibility between all participating plasmonic THz interconnection components. Moreover, as graphene is a material characterized by ultrafast carrier relaxation recombination dynamics, low pumping threshold level, and wide THz tuning range at room temperatures, it appears to be a promising technology for implementing new types of laser sources, against current semiconductor lasers [162]. It may be used as a THz modulator as well. Specifically, in [163], a single layer graphene THz modulator is proposed, based on a monolayer graphene sheet, lying on a SiO_2/p-Si substrate and biased with metal gates. A voltage applied between the gates, injects carriers on graphene, thus modulating the chemical potential at THz rates. Other modulators, based on the same principles, have been proposed at these rates [164–166], as well as modulators at infrared frequencies [167,168]. Finally, another graphene based plasmonic waveguide phase modulator proposed in [169], is also based on the principle of electronically control the propagation speed of an SPP wave, by modifying the chemical potential of the graphene layer. Last, GFETs operating at room temperatures, have been also proposed and demonstrated as ultrafast THz detectors [170]. A graphene-based THz detector based on a log-periodic circular toothed antenna between the source and the gate of a GFET, is proposed in [171]. The THz oscillating electric field is fed between the gate and the channel of the GFET, inducing a DC signal between source and drain, that is proportional to the received optical power. Finally, a compact graphene slot photodetector on SOI with high responsivity, is proposed in [172].

Apart from graphene other materials, such as black phosphorus (BP) has been characterized by attractive features, such as the high carrier mobility, the in-plane anisotropy, and the tuning capability via electrical gating, 0.3 to 1.7 eV. It can be exploited as a perfect THz detector when being integrated in a microscopic FET for a wide range of THz frequencies from 0.26 THz to 3.4 THz. Such a BP-based FET can be used as a plasma-wave rectifier, a thermoelectric sensor, or a thermal bolometer [173]. Chalcogenides compounds containing at least one of the chalcogen elements, namely, sulfur, selenium or tellurium, and specifically, chalcogenide glasses are characterized as phase-change active plasmonic devices appropriate for active plasmonic switching/modulation functionality for future nanophotonic device applications. Experimental demonstrations employing gallium anthanum sulfide as a photo-active medium, implement that CMOS/SOI-compatible, chalcogenide glasses used for such functionalities [174]. They may be also used as topological insulators for protecting metallic surfaces of plasmonic structures, as they are immune to scattering from disorder and defects, they can be dynamically controlled via external electric, magnetic or optical excitations, and their optoelectronic response is highly sensitive to the polarization state of incident light.

6. Summary and Conclusions

The THz band has been characterized as the last undiscovered frontier of the total EM spectra range that urges for exploration and investigation, since current data traffic and bandwidth hungry applications will no longer satisfy their speed and latency demands with existing technologies and system architectures. On the other hand, wireless communications seem to be in advance against conventional wired communications. Therefore, the migration to higher carrier frequency bands and specifically in the THz band is required, via adoption of new technologies, equipping future THz wireless communication systems at the nanoscale, in order to accommodate a variety of applications that would satisfy the ever increasing user demands for higher data rates. Hence, wireless THz band communications and modern THz wireless nanoscale applications, such as beyond 5G communications, NoC system architectures and WNSNs, are still urging for an efficient, compact and standardized interconnect solution for generating, transmitting, propagating, and detecting the THz wave information.

In this paper, a comprehensive survey has been presented for THz wireless communications and applications, as an attempt to identify unsolved issues and challenges in THz region, such as the very high propagation signal loss, the impedance mismatch between THz link components, limited size restrictions along with integration potentials, associated with high bandwidth availability and ultra-fast operating data rates and minimum latency requirements. Pointing to this direction, the most efficient compact technology, or the hybrid combination of competitive technologies, such as conventional CMOS electronics (photonic and plasmonic), is under investigation, in order to properly equip future THz nanoscale communication systems, hosting modern wireless THz nanoapplications. Among competitive technologies, CMOS-based electronic interconnects are definitely out of the competition, in order to meet THz speed, low propagation signal loss, and the impedance match between THz link components. A photonic solution is indeed; a viable approach for providing high data rates at low propagation losses, still the component size is one with two orders of magnitude larger than what required for THz band case. Plasmon based THz link components, on the other hand, due to their extremely small size and their ability to operate at ultra-high data rates, seems to be a promising approach for equipping wireless THz nanoscale communication systems. Moreover, they could be perfectly combined with photonic technology and particularly with dielectric waveguiding, as plasmonic waveguiding is quite lossy concerning relatively long interconnect distances. Therefore, this survey work, has provided in-depth reference material of the current fundamental aspects of plasmonic technology and hybrid combinations, highlighting plasmon future roles in THz band wireless communication. It is a thorough investigation on current and beyond state-of-the-art plasmonic layouts, implementing THz nanoscale communication systems, and wireless THz nanoapplications accordingly.

Specifically, as far as it concerns NoC architectures, many alternative technologies have been progressively proposed, in order to deal effectively with the NoC communication bottleneck, such as 3D, RF signals over on-chip transmission lines, FSO systems at IR, and photonic and nanophotonic NoC—all these considered to be state of the art technologies, as seen in Table 1. THz band communications however, anticipate for smaller footprints of the transceiver and the antenna for more efficient integration, and for larger available transmission bandwidths, and higher achievable data rates. Given these tight requests, a promising, beyond state-of-the-art solution would be based on reduced size plasmon nanoantennas and other plasmon based THz transceiver components as well, so as to operate at very high resonant THz frequencies. This can be achieved, as it seems, mostly via the use of graphene material supporting graphene based WiNoCs architectures, or alternatively, based on the combination of plasmonic resonators with dielectric waveguiding, supporting wireless-optical on chip communication, and hybrid optical-wireless NoC architectures respectively (Table 1).

Table 1. SoA and beyond SoA technologies for wireless THz applications.

Wireless THz Application	*SoA	Beyond *SoA
NoCs	PCB, 3D, FSO, IR	Graphene based WiNoCs
	Si photonic/nanophotonics	Hybrid optical-wireless NoCs (plasmonic resonators with dielectric waveguide)
WNSNs (wireless nanosensor networks)	Graphene nanoribbon (GNR)/carbon nanotube (CNT) nanosensors, nanoprocessors, nanoantennas, nanotransceivers	Graphene (GFET) based THz antennas and transceiver parts
	Nanoscale energy harvesting systems	
	Au and Ag plasmon sensors	Nanomemories
Beyond 5G communications	Si photonics based uni-travelling photodiodes (UTC-PDs) and comb sources at mm waves	FSO THz/optical links,
		Integrated microwave photonics (IMWP) in THz,
		Plasmon or POH modulator (THz/O)
		Graphene multiple-input-multiple-output (MIMO) antennas structures

*SoA—state-of-the-art.

GNR and CNT based nanosensors, nanoprocessors, nanoantennas, and nanotransceivers, or other material nanosensors such as Au and Ag plasmon sensors, and nanoscale energy harvesting systems, are considered to be the state of the art for implementing todays WNSNs. Basically, in the near future, as far as concerns the communication unit of a WNSN system, it is likely that graphene nanoantennas and other graphene-based THz transceiver components would be required, so as the system to operate at these high resonant THz frequencies (Table 1). At the moment, beyond 5G communication systems are mostly equipped via silicon photonic technology, and particularly via Si photonics-based UTC-PDs, and comb sources operating at mmW. Many research attempts have been proposed in order such systems to operate, well above 300GHz frequencies, at the THz band. These approaches include FSO systems with THz/O links, IMWP in the THz range, graphene MIMO antennas structures, and plasmonic MZ modulators with sub-THz frequency responses, or ultra-broadband silicon POH modulators, required for THz/O signal conversions (Table 1).

Consequently, plasmonics play a critical role for equipping each single, individual component part of a future wireless THz nanocommunication link, namely the antennas and the transceiver parts, as also can be seen in more detail in Table 2. As far as concerns antenna implementation, silicon integrated antennas, CNT based antennas, UWB and multi-band antennas, have been considered as mature options for on chip antennas, but the tight uncompromising limitation of a very small size of a THz band antenna requires other technologies, such as plasmon or hybrid combinations. Hence, graphene based nanoantennas (patch antenna, dipole, MIMO), or hybrid graphene-dielectric antennas(H2G), or plasmonic antennas with dielectric waveguides feeding elements (dipole loop, horn, Vivaldi, hybrid plasmon-dielectric array), or inversely, plasmonic nanopatch antenna, with HMIM plasmonic waveguide as feeding element, are strong and promising candidates for implementing THz nanoantennas, as also seen in Table 2.

As far as concerns other THz transceiver component parts, silicon photonic devices have been widely adopted for equipping current THz band transceivers, with photonic III/V (InP) devices, such as UTC-PDs and QCLs to be considered as state-of-the-art sources. SiGe-based HBTs have been also proposed for equipping THz transmitters. However, high path losses of these technologies, at the THz band is still considered a major unsolved challenge. THz plasmonic lasers such as plasmonic QCLs, THz QCLs with metallic cavities, and single mode metal-clad plasmonic lasers with their antenna feedback tuning mechanism, may be a promising candidate for THz transmission sources. Hybrid transmitters, such as III-V semiconductor based HEMT, enhanced with graphene, may be as well considered to be a promising THz transmitter approach. Moreover, plasmonic slot waveguides may be used as a perfect nanocoupler matching element with current state of the art on chip sources (nanoLEDs, nanolasers) (Table 2).

Table 2. SoA and beyond SoA technologies for THz band transceiver components.

THz Band Transceiver Components	*SoA	Beyond *SoA
Antenna	Silicon integrated antennas, CNT based antennas, ultra-wide broadband (UWB) and multi-band antennas	Graphene based nanoantennas (patch antenna, dipole, MIMO)
		Hybrid graphene-dielectric antennas (H2G- two graphene monolayers separated by a thin dielectric)
		Plasmonic antenna with dielectric waveguides (single/dipole loop plasmon nanoantenna, plasmonic horn nanoantennas, single/double/array Vivaldi plasmon antenna/hybrid Plasmon dielectric array)
		Plasmonic nanopatch antenna, with hybrid metal insulator metal (HMIM) plasmonic waveguide
THz band transmitters	Silicon photonic THz band transmitters (UTC-PDs III/V (InP), QCLs, SiGe-heterojunction bipolar transistors, HBTs)	THz plasmonic lasers (plasmonic quantum cascade lasers (QCLs), THz QCLs with metallic cavities, single-mode metal-clad plasmonic lasers)
		Hybrid THz transmitters III-V semiconductor based HEMT, enhanced with graphene
		Plasmonic slot waveguides-nanocouplers with on chip sources (nanoLEDs, nanolasers)
THz band receivers	GaAs Schottky barrier diodes (SBDs), CMOS with high electron mobility transistors (HEMTs)	Hybrid plasmon THz wave detectors (POH slot waveguide with a bowtie-antenna, graphene slot photodetector on SOI)
		Plasmonic teraFET, graphene-FET
THz band transceiver package		plasmonic internal photoemission detectors (PIPED) package (Tx, Rx) integrated on si-photonic chip platform
THz band transceiver other parts		Graphene (GFET) based switch, LPF, BPF, phase shifter, graphene THz modulator

*SoA—state-of-the-art.

Lastly, as far as concerns state of the art THz band receivers, waveguide integrated detectors using GaAs SBDs are considered as the most common implementation choice. However, they pose limitations on their size and available bandwidth, and behave poorly at room temperature. Si-CMOS technology with HEMTs has also been proposed for equipping detectors between 790 and 960 GHz, as they pose performance limitations above 1THz. Hybrid plasmon structures, such as POH slot waveguide integrated with a bowtie antenna, or graphene slot photodetector on SOI, or GFET based structures such as plasmonic teraFET, are considered to be promising, beyond state of the art, THz wave detectors (Table 2). GFET based structures can be also used for equipping other THz transceiver parts, such as switches, LPFs, BPFs, phase shifters and modulators. Alternatively, PIPED structures, which are actually end to end, transmitter and receiver packages, monolithically integrated on a common silicon photonic chip platform, have been also proposed for THz wave signal generation and coherent detection, at frequencies of up to 1 THz (Table 2).

Tables 1 and 2 include, in summary, all state of the art and beyond state of the art, plasmon based technology, exploited for the implementation of future THz band nanocommunication systems, as have been exhaustively presented in this work. Apparently it seems that, in order to fully equip future THz nanocommunication applications, miniature size transceiver components are required, with uncompromising features such as low propagation signal loss, impedance matching between link components, strong integration potentials and compatibility with ancestor technologies as CMOS, with high bandwidth availability and ultra-fast operating data rates and minimum latency.

This comprehensive survey paper has highlighted such an objective, by qualitatively presenting in such detail, the latest plasmonic and accompanied photonics technologies on equipping future competitive THz nanoscale communication systems, hosting wireless THz nanoapplications, namely NoCs, WNSNs and beyond 5G communications, providing at the same time, motivation for research academia to seek for efficient solutions towards this direction.

Conflicts of Interest: The author declares no conflict of interest.

References

1. Index, C.V.N. *Cisco Visual Networking Index: Forecast and Methodology 2015–2020*; White Paper; CISCO: San Jose, CA, USA, 2015.
2. Index, C.V.N. *Cisco Visual Networking Index: Forecast and Methodology 2016–2021*; White Paper; CISCO: San Jose, CA, USA, 2017.
3. Kürner, T.; Priebe, S. Towards THz communications—Status in research, standardization and regulation. *J. Infrared Millim. Terahertz Waves* **2014**, *35*, 53–62. [CrossRef]
4. Li, R.R. Towards a new internet for the year 2030 and beyond. In *The 3rd ITU IMT-2020/5G Workshop*; International Telecommunication Union (ITU): Geneva, Switzerland, 2018.
5. Abadal, S.; Alarcon, E.; Cabellos-Aparicio, A.; Lemme, M.C.; Nemirovsky, M. Graphene-Enabled Wireless Communication for Massive Multicore Architectures. *IEEE Commun. Mag.* **2013**, *51*, 137–143. [CrossRef]
6. Rodriguez-Fortuno, F.J.; Espinosa-Soria, A.; Martinez, A. Exploiting metamaterials, plasmonics and nanoantennas concepts in silicon photonics. *J. Opt.* **2016**, *18*. [CrossRef]
7. Mohammad, H.; Shubair, R.M. Nanoscale Communication: State-of-Art and Recent Advances. *arXiv* **2019**, arXiv:1905.07722.
8. Petrov, V.; Moltchanov, D.; Komar, M.; Antonov, A.; Kustarev, P.; Rakheja, S.; Koucheryavy, Y. Terahertz Band Intra-Chip Communications: Can Wireless Links Scale Modern x86 CPUs? *IEEE Access* **2017**, *5*, 6095–6109. [CrossRef]
9. Han, R.; Afshari, E. Filling the terahertz gap with sand: High-power terahertz radiators in silicon. In Proceedings of the 2015 IEEE Bipolar/BiCMOS Circuits and Technology Meeting-BCTM, Boston, MA, USA, 26–28 October 2015; pp. 172–177.
10. Seok, E.; Shim, D.; Mao, C.; Han, R.; Sankaran, S.; Cao, C.; Knap, W.; Kenneth, K.O. Progress and challenges towards terahertz CMOS integrated circuits. *IEEE J. Solid-State Circuits* **2010**, *45*, 1554–1564. [CrossRef]
11. Mittal, S. A Survey of ReRAM-Based Architectures for Processing-In-Memory and Neural Networks. *Mach. Learn. Knowl. Extr.* **2019**, *1*, 75–114. [CrossRef]
12. Chin, P.; Li, S.; Xu, C.; Zhang, T.; Zhao, J.; Liu, Y.; Wang, Y.; Xie, Y. PRIME: A novel processing-in-memory architecture for neural network computation in ReRAM-based main memory. In Proceedings of the 2016 ACM/IEEE 43rd Annual International Symposium on Computer Architecture (ISCA), Seoul, Korea, 18–22 June 2016; pp. 27–39. [CrossRef]
13. *White Paper on AI Chip Technologies*; Beijing Innovation Center for Future Chips (ICFC): Beijing, China, 2018.
14. Choi, W.; Duraisamy, K.; Kim, R.G.; Doppa, J.R.; Pande, P.P.; Marculescu, D.; Marculescu, R. On-Chip Communication Network for Efficient Training of Deep Convolutional Networks on Heterogeneous Manycore Systems. *IEEE Trans. Comput.* **2018**, *67*, 672–686. [CrossRef]
15. Thraskias, C.A.; Lallas, E.N.; Neumann, N.; Schares, L.; Offrein, B.J.; Henker, R.; Plettemeier, D.; Ellinger, F.; Leuthold, J.; Tomkos, I. Survey of Photonic and Plasmonic Chip-Scale Interconnect Technologies for Intra-Datacenter and High-Performance Computing Networks. *IEEE Commun. Surv. Tutor.* **2018**, *20*, 2758–2783. [CrossRef]
16. Carloni, L.P.; Pande, P.; Xie, Y. Networks-on-chip in emerging interconnect paradigms: Advantages and challenges. In Proceedings of the 2009 3rd ACM/IEEE International Symposium on Networks-on-Chip, Washington, DC, USA, 10–13 May 2009; pp. 93–102.
17. Kumar, R.; Zyuban, V.; Tullsen, D.M. Interconnections in Multi-core Architectures: Understanding Mechanisms, Overheads and Scaling. In Proceedings of the 32nd International Symposium on Computer Architecture (ISCA'05), Madison, WI, USA, 4–8 June 2005. [CrossRef]

18. Kash, J.A.; Pepeljugoski, P.; Doany, F.E.; Schow, C.L.; Kuchta, D.M.; Schares, L.; Budd, R.; Libsch, F.; Dangel, R.; Horst, F. Communication Technologies for Exascale Systems. In Proceedings of the SPIE OPTO: Integrated Optoelectronic Devices, San Jose, CA, USA, 24 January 2009. [CrossRef]
19. Agyeman, M.O.; Ahmadinia, A.; Shahrabi, A. Heterogeneous 3d network-on-chip architectures: Area and power aware design techniques. *J. Circuits Syst. Comput.* **2013**, *22*. [CrossRef]
20. Socher, E.; Chang, M.-C.F. Can RF Help CMOS Processors? *IEEE Commun. Mag.* **2007**, *45*, 104–111. [CrossRef]
21. Khalighi, M.A.; Uysal, M. Survey on free space optical communication: A communication theory perspective. *IEEE Commun. Surv. Tuts.* **2014**, *16*, 2231–2258. [CrossRef]
22. Shacham, A.; Bergman, K.; Carloni, L.P. Photonic networks-on-chip for future generations of chip multiprocessors. *IEEE Trans. Comput.* **2008**, *57*, 1246–1260. [CrossRef]
23. Xu, Y.; Pasricha, S. Silicon nanophotonics for future multicore architectures: Opportunities and challenges. *IEEE Des. Test* **2014**, *31*, 9–17. [CrossRef]
24. Deb, S.; Ganguly, A.; Pande, P.P.; Belzer, B.; Heo, D. Wireless NoC as interconnection backbone for multicore chips: Promises and challenges. *IEEE J. Emerg. Sel. Top. Circuits Syst.* **2012**, *2*, 228–239. [CrossRef]
25. Matolak, D.W.; Kodi, A.; Kaya, S.; Ditomaso, D.; Laha, S.; Rayess, W. Wireless networks-on-chips: Architecture, wireless channel, and devices. *IEEE Wirel. Commun.* **2012**, *19*, 58–65. [CrossRef]
26. Ganguly, A.; Chang, K.; Deb, S.; Pande, P.P.; Belzer, B.; Teuscher, C. Scalable hybrid wireless network-on-chip architectures for multicore systems. *IEEE Trans. Comput.* **2011**, *60*, 1485–1502. [CrossRef]
27. Agyeman, M.O.; Vien, Q.T.; Ahmadinia, A.; Yakovlev, A.; Tong, K.F.; Mak, T.S. A resilient 2-D waveguide communication fabric for hybrid wired-wireless NoC design. *IEEE Trans. Parallel Distrib. Syst.* **2017**, *28*, 359–373. [CrossRef]
28. Akyildiz, I.F.; Jornet, J.M.; Han, C. TeraNets: Ultra-broadband communication networks in the terahertz band. *IEEE Wirel. Commun.* **2014**, *21*, 130–135. [CrossRef]
29. Correas-Serrano, D.; Gomez-Diaz, J.S. Graphene-based Antennas for Terahertz Systems: A Review. *arXiv* **2017**, arXiv:1704.00371.
30. Samaiyar, A.; Ram, S.S.; Deb, S. Millimeter-Wave Planar Log Periodic Antenna for On-Chip Wireless Interconnects. In Proceedings of the 8th European Conference on Antennas and Propagation (EuCAP 2014), The Hague, The Netherlands, 6–11 April 2014. [CrossRef]
31. Llatser, I.; Cabellos-Aparicio, A.; Alarcón, E.; Jornet, J.M.; Mestres, A.; Lee, H.; Pareta, J.S. Scalability of the Channel Capacity in Graphene-enabled Wireless Communications to the Nanoscale. *IEEE Trans. Commun.* **2015**, *63*, 324–333. [CrossRef]
32. Nafari, M.; Feng, L.; Jornet, J.M. On-chip Wireless Optical Channel Modeling for Massive Multi-Core Computing Architectures. In Proceedings of the 2017 IEEE Wireless Communications and Networking Conference (WCNC), SanFransisco, CA, USA, 19–22 March 2017. [CrossRef]
33. Llatser, I.; Kremers, C.; Cabellos-Aparicio, A.; Jornet, J.M.; Alarcón, E.; Chigrin, D.N. Graphene-based Nano-Patch Antenna for Terahertz Radiation. *Photonics Nanostruct.-Fundam. Appl.* **2012**, *10*, 353–358. [CrossRef]
34. Jornet, J.M.; Akyildiz, I.F. Graphene-Based Nano-Antennas for Electromagnetic Nanocommunications in the Terahertz Band. In Proceedings of the Fourth European Conference on Antennas and Propagation, Barcelona, Spain, 12–16 April 2010.
35. Akyildiz, I.F.; Jornet, J.M. Electromagnetic wireless nanosensor networks. *Nano Commun. Netw.* **2010**, *1*, 3–19. [CrossRef]
36. Luo, X.; Qiu, T.; Lu, W.; Ni, Z. Plasmons in graphene: Recent progress and applications. *Mater. Sci. Eng. R Rep.* **2013**, *74*, 351–376. [CrossRef]
37. Rakheja, S.; Sengupta, P. Gate-voltage tunability of plasmons in single and multi-layer graphene structures: Analytical description and concepts for terahertz devices. *arXiv* **2015**, arXiv:1508.00658. [CrossRef]
38. Wu, Y.; Jenkins, K.A.; Valdes-Garcia, A.; Farmer, D.B.; Zhu, Y.; Bol, A.A.; Dimitrakopoulos, C.; Zhu, W.; Xia, F.; Avouris, P.; et al. State-of-the-art graphene high-frequency electronics. *Nano Lett.* **2012**, *12*, 3062–3067. [CrossRef]
39. Chen, P.-Y.; Alu, A. All-graphene terahertz analog nanodevices and nanocircuits. In Proceedings of the 2013 7th European Conference on Antennas and Propagation (EuCAP), Gothenburg, Sweden, 8–12 April 2013; pp. 697–698.

40. Yousefi, L.; Foster, A.C. Waveguide-fed optical hybrid plasmonic patch nano-antenna. *Opt. Express* **2012**, *20*, 18326–18335. [CrossRef]
41. Yang, Y.; Zhao, D.; Gong, H.; Li, Q.; Qiu, M. Plasmonic sectoral horn nanoantennas. *Opt. Lett.* **2014**, *39*, 3204–3207. [CrossRef]
42. Yang, Y.; Li, Q.; Qiu, M. Broadband nanophotonic wireless links and networks using on-chip integrated plasmonic antennas. *Sci. Rep.* **2016**, *6*, 19490. [CrossRef]
43. Bellanca, G.; Calo, G.; Kaplan, A.E.; Bassi, P.; Petruzzelli, V. Integrated Vivaldi plasmonic antenna for wireless on-chip optical communications. *Opt. Express* **2017**, *25*, 16214–16227. [CrossRef]
44. Afridi, A.; Kocabas, S.E. Beam steering and impedance matching of plasmonic horn nanoantennas. *Opt. Express* **2016**, *24*, 25647–25652. [CrossRef] [PubMed]
45. Arango, F.B.; Kwadrin, A.; Koenderink, A.F. Plasmonic Antennas Hybridized with Dielectric Waveguides. *ACS Nano* **2012**, *6*, 10156–10167. [CrossRef] [PubMed]
46. DeRose, C.T.; Kekatpure, R.D.; Trotter, D.C.; Starbuck, A.; Wendt, J.R.; Yaacobi, A.; Watts, M.R.; Chettiar, U.; Engheta, N.; Davids, P.S. Electronically controlled optical beam-steering by an active phased array of metallic nanoantennas. *Opt. Express* **2013**, *21*, 5198–5208. [CrossRef] [PubMed]
47. Calo, G.; Bellanca, G.; Kaplan, A.E.; Bassi, P.; Petruzzelli, V. Double Vivaldi antenna for wireless optical networks on chip. *Opt. Quant. Electron.* **2018**, *50*, 261. [CrossRef]
48. Calo, G.; Bellanca, G.; Kaplan, A.E.; Kaplan, A.E.; Bassi, P.; Petruzzelli, V. Array of plasmonic Vivaldi antennas coupled to silicon waveguides for wireless networks through on-chip optical technology–WiNOT. *Opt. Express* **2018**, *26*, 30267–30277. [CrossRef] [PubMed]
49. Nayyar, A.; Puri, V.; Le, D.N. Internet of Nano Things (IoNT): Next Evolutionary Step in Nanotechnology. *Nanosci. Nanotechnol.* **2017**, *7*, 4–8. [CrossRef]
50. Akyildiz, I.F.; Jornet, J.M. The Internet of Nano-Things. *IEEE Wirel. Commun. Mag.* **2010**, *17*, 58–63. [CrossRef]
51. Rupani, V.; Kargathara, S.; Sureja, J. A Review on Wireless Nanosensor Networks Based on Electromagnetic Communication. *Int. J. Comput. Sci. Inf. Technol.* **2015**, *6*, 1019–1022.
52. Stampfer, C.; Helbling, T.; Obergfell, D.; Schöberle, B.; Tripp, M.K.; Jungen, A.; Roth, S.; Bright, V.M.; Hierold, C. Fabrication of single walled carbon nanotube based pressure sensors. *Nano Lett.* **2006**, *6*, 233–237. [CrossRef]
53. Stampfer, C.; Jungen, A.; Hierold, C. Fabrication of discrete nanoscaled force sensors based on single-walled carbon nanotubes. *IEEE Sens. J.* **2006**, *6*, 613–617. [CrossRef]
54. Stampfer, C.; Linderman, R.; Obergfell, D.; Jungen, A.; Roth, S.; Hierold, C. Nano electromechanical displacement sensing based on single walled carbon nanotubes. *Nano Lett.* **2006**, *7*, 1449–1453. [CrossRef] [PubMed]
55. Vo-Dinh, T.; Cullum, B.M.; Stokes, D.L. Nanosensors and biochips: Frontiers in biomolecular diagnostics. *Sens. Actuators B Chem.* **2001**, *74*, 2–11. [CrossRef]
56. Avouris, P. Carbon nanotube electronics and photonics. *Phys. Today* **2009**, *62*, 3440. [CrossRef]
57. Yonzon, C.R.; Stuart, D.A.; Zhang, X.; McFarland, A.; Haynes, C.L.; Van Duyne, R.P. Towards advanced chemical and biological nanosensors an overview. *Talanta* **2005**, *67*, 438–448. [CrossRef]
58. Li, C.; Thostenson, E.T.; Chou, T.-W. Sensors and actuators based on carbon nanotubes and their composites: A review. *Compos. Sci. Technol.* **2008**, *68*, 1227–1249. [CrossRef]
59. Bose, S.; Kim, N.H.; Kuila, T.; Lau, K.-T.; Lee, J.H. Electrochemical performance of a graphene-polypyrrole nanocomposite as a supercapacitor electrode. *Nanotechnology* **2011**, *22*, 295202. [CrossRef]
60. Loomis, J.; King, B.; Burkhead, T.; Xu, P.; Bessler, N.; Terentjev, E.; Panchapakesan, B. Graphene-nanoplatelet-based photomechanical actuators. *Nanotechnology* **2012**, *23*, 045501. [CrossRef]
61. Ji, L.; Tan, Z.; Kuykendall, T.; An, E.J.; Fu, Y.; Battaglia, V.; Zhang, Y. Multilayer nanoassembly of sn-nanopillar arrays sandwiched between graphene layers for high-capacity lithium storage. *Energy Environ. Sci.* **2011**, *4*, 3611–3616. [CrossRef]
62. Wang, Z.L. Towards self-powered nanosystems: From nanogenerators to nanopiezotronics. *Adv. Funct. Mater.* **2008**, *18*, 3553–3567. [CrossRef]
63. Wang, Z.L.; Song, J. Piezoelectric nanogenerators based on zinc oxide nanowire arrays. *Science* **2006**, *312*, 242–246. [CrossRef]

64. Hanson, G.W. Dyadic green's functions and guided surface waves for a surface conductivity model of graphene. *J. Appl. Phys.* **2008**, *103*, 064302. [CrossRef]
65. Wu, Y.; Lin, Y.-M.; Bol, A.A.; Jenkins, K.A.; Xia, F.; Farmer, D.B.; Zhu, Y.; Avouris, P. High-frequency, scaled graphene transistors on diamond-like carbon. *Nature* **2011**, *472*, 74–78. [CrossRef] [PubMed]
66. Tong, L.; Wei, H.; Zhang, S.; Xu, H. Recent advances in plasmonic sensors. *Sensors* **2014**, *14*, 7959–7973. [CrossRef] [PubMed]
67. Huang, F.; Drakeley, S.; Millyard, M.J.; Murphy, A.; White, R.; Spigone, E.; Kivioja, J.; Baumberg, J.J. Zero-Reflectance Metafilms for Optimal Plasmonic Sensing. *Adv. Opt. Mater.* **2016**, *4*, 328–335. [CrossRef]
68. Centeno, A.; Aid, S.R.; Xie, F. Infra-Red Plasmonic Sensors. *Chemosensors* **2018**, *6*, 4. [CrossRef]
69. Omanovic-Miklicanin, E.; Maksimovic, M. Nanosensors applications in agriculture and food industry, Technical paper. *Bull. Chem. Technol. Bosnia Herzeg.* **2016**, *47*, 59–70.
70. Islam, M.S.; Vj, L. Nanoscale Materials and Devices for Future Communication Networks. *IEEE Commun. Mag.* **2010**, *48*, 112–120. [CrossRef]
71. Federici, J.F.; Schulkin, B.; Huang, F.; Gary, D.; Barat, R.; Oliveira, F.; Zimdars, D. THz imaging and sensing for security applications explosives, weapons and drugs. *Semicond. Sci. Technol.* **2005**, *20*, S266. [CrossRef]
72. Yang, K.; Hao, Y.; Alomainy, A.; Abbasi, Q.H.; Qaraqe, K. Channel modelling of human tissues at terahertz band. In Proceedings of the IEEE Wireless Communications and Networking Conference Workshops (WCNCW), Doha, Qatar, 3–6 April 2016; pp. 218–221. [CrossRef]
73. Nafari, M.; Jornet, J.M. Metallic plasmonic nano-antenna for wireless optical communication in intra-body nanonetworks. In Proceedings of the 10th EAI International Conference on Body Area Networks, Sydney, Australia, 28–30 September 2015; pp. 287–293. [CrossRef]
74. Alexandrov, B.S.; Gelev, V.; Bishop, A.R.; Usheva, A.; Rasmussen, K. DNA breathing dynamics in the presence of a terahertz field. *Phys. Lett. A* **2010**, *374*, 1214–1217. [CrossRef]
75. Rodrigo, D.; Limaj, O.; Janner, D.; Etezadi, D.; García de Abajo, F.J.; Pruneri, V.; Altug, H. Mid-infrared plasmonic biosensing with graphene. *Science* **2015**, *349*, 165–168. [CrossRef]
76. Plusquellic, D.F.; Heilweil, E.J. Terahertz spectroscopy of biomolecules. In *Terahertz Spectroscopy*; CRC Press: Boca Raton, FL, USA, 2007; pp. 293–322.
77. Chen, P.-Y.; Huang, H.; Akinwande, D.; Alu, A. Graphene-Based Plasmonic Platform for Reconfigurable Terahertz Nanodevices. *ACS Photonics* **2014**, *1*, 647–654. [CrossRef]
78. Huang, H.; Sakhdari, M.; Hajizadegan, M.; Shahini, A.; Akinwande, D.; Chen, P.Y. Toward transparent and self-activated graphene harmonic transponder sensors. *Appl. Phys. Lett.* **2016**, *108*, 173503. [CrossRef]
79. Petrov, V.; Pyattaev, A.; Moltchanov, D.; Koucheryavy, Y. Terahertz band communications: Applications, research challenges, and standardization activities. In Proceedings of the 2016 8th International Congress on Ultra Modern Telecommunications and Control Systems and Workshops (ICUMT), Lisbon, Portugal, 18–20 October 2016. [CrossRef]
80. Federici, J.; Moeller, L. Review of terahertz and subterahertz wireless communications. *J. Appl. Phys.* **2010**, *107*, 6. [CrossRef]
81. Moldovan, A.; Ruder, M.A.; Akyildiz, I.F.; Gerstacker, W.H. Los and nlos channel modeling for terahertz wireless communication with scattered rays. In Proceedings of the 2014 IEEE Globecom Workshops (GC Wkshps), Austin, TX, USA, 8–12 December 2014; pp. 388–392. [CrossRef]
82. Koenig, S.; Lopez-Diaz, D.; Antes, J.; Boes, F.; Henneberger, R.; Leuther, A.; Tessmann, A.; Schmogrow, R.; Hillerkuss, D.; Palmer, R.; et al. Wireless sub-THz communication system with high data rate. *Nat. Photonics* **2013**, *7*, 977–981. [CrossRef]
83. Shams, H.; Shao, T.; Fice, M.J.; Anandarajah, P.M.; Renaud, C.C.; Van Dijk, F.; Barry, L.P.; Seeds, A.J. 100 Gb/s Multicarrier THz Wireless Transmission System with High Frequency Stability Based on A Gain-Switched Laser Comb Source. *IEEE Photonics J.* **2015**, *7*, 1–11. [CrossRef]
84. Pang, X.; Caballero, A.; Dogadaev, A.; Arlunno, V.; Borkowski, R.; Pedersen, J.S.; Deng, L.; Karinou, F.; Roubeau, F.; Zibar, D.; et al. 100 Gbit/s hybrid optical fiber-wireless link in the W-band (75–110 GHz). *Opt. Express* **2011**, *19*, 24944–24949. [CrossRef]
85. Nagatsuma, T.; Ducournau, G.; Renaud, C.C. Advances in terahertz communications accelerated by photonics. *Nat. Photonics* **2016**, *10*, 371–379. [CrossRef]

86. Elayan, H.; Amin, O.; Shubair, R.M.; Alouini, M.S. Terahertz communication: The opportunities of wireless technology beyond 5G. In Proceedings of the International Conference on Advanced Communication Technologies and Networking, Marrakech, Morocco, 2–4 April 2018. [CrossRef]
87. Burla, M.; Bonjour, R.; Salamin, Y.; Abrecht, F.C.; Haffner, C.; Heni, W.; Hoessbacher, C.; Baeuerle, B.; Josten, A.; Fedoryshyn, Y.; et al. Plasmonic Modulators for Microwave Photonics Applications. In Proceedings of the 2017 Asia Communications and Photonics Conference (ACP), Guangzhou, China, 10–13 November 2017.
88. Waterhouse, R.; Novak, D. Realizing 5G: Microwave Photonics for 5G Mobile Wireless Systems. *IEEE Microw. Mag.* **2015**, *16*, 84–92. [CrossRef]
89. Liu, Y.; Marpaung, D.; Choudhary, A.; Eggleton, B.J. Highly selective and reconfigurable Si_3N_4 RF photonic notch filter with negligible RF losses. In Proceedings of the 2017 Conference on Lasers and Electro-Optics (CLEO), San Jose, CA, USA, 14–19 May 2017.
90. Da Costa, I.F.; Rodriguez, S.; Puerta, R.; Vegas Olmos, J.J.; Cerqueira, J.A.; Da Silva, L.G.; Spadoti, D.; Monroy, I.T. Photonic Downconversion and Optically Controlled Reconfigurable Antennas in mm-waves Wireless Networks. In Proceedings of the 2016 Optical Fiber Communications Conference and Exhibition (OFC), Ananheim, CA, USA, 20–24 March 2016.
91. Marpaung, D.; Roeloffzen, C.; Heideman, R.; Leinse, A.; Sales, S.; Capmany, J. Integrated microwave photonics. *Laser Photonics Rev.* **2013**, *7*, 506–538. [CrossRef]
92. Bonjour, R.; Singleton, M.; Gebrewold, S.A.; Salamin, Y.; Abrecht, F.C.; Baeuerle, B.; Josten, A.; Leuchtmann, P.; Hafner, C.; Leuthold, J. Ultra-Fast Millimeter Wave Beam Steering. *IEEE J. Quantum Electron.* **2016**, *52*, 1–8. [CrossRef]
93. Elayan, H.; Amin, O.; Shihada, B.; Shubair, R.M.; Alouini, M.S. Terahertz Band: The Last Piece of RF Spectrum Puzzle for Communication Systems. *arXiv* **2019**, arXiv:1907.05043. [CrossRef]
94. Jia, S.; Pang, X.; Ozolins, O.; Yu, X.; Hu, H.; Yu, J.; Guan, P.; Da Ros, F.; Popov, S.; Jacobsen, G.; et al. 0.4 THz photonic-wireless link with 106 Gb/s single channel bitrate. *IEEE J. Lightwave Technol.* **2018**, *36*, 610–616. [CrossRef]
95. Haffner, C.; Chelladurai, D.; Fedoryshyn, Y.; Josten, A.; Baeuerle, B.; Heni, W.; Watanabe, T.; Cui, T.; Cheng, B.; Saha, S.; et al. Low-loss plasmon-assisted electro-optic modulator. *Nature* **2018**, *556*, 483–486. [CrossRef]
96. Baeuerle, B.; Heni, W.; Fedoryshyn, Y.; Josten, A.; Haffner, C.; Watanabe, T.; Elder, D.L.; Dalton, L.R.; Leuthold, J. Plasmonic-organic hybrid modulators for optical interconnects beyond 100G/λ. In Proceedings of the 2018 Conference on Lasers and Electro-Optics (CLEO), San Jose, CA, USA, 13–18 May 2018; pp. 1–2.
97. Hoessbacher, C.; Josten, A.; Baeuerle, B.; Fedoryshyn, Y.; Hettrich, H.; Salamin, Y.; Heni, W.; Haffner, C.; Kaiser, C.; Schmid, R.; et al. Plasmonic modulator with >170 GHz bandwidth demonstrated at 100 GBd NRZ. *Opt. Express* **2017**, *25*, 1762–1768. [CrossRef]
98. Ummethala, S.; Harter, T.; Köhnle, K.; Muehlbrandt, S.; Kutuvantavida, Y.; Kemal, J.N.; Schaefer, J.; Massler, H.; Tessmann, A.; Garlapati, S.K.; et al. Terahertz-to-optical conversion using a plasmonic modulator. In Proceedings of the CLEO-Conference on Lasers and Electro-Optics, OSA Technical Digest (Optical Society of America, 2018), paper STu3D.4, San Jose, CA, USA, 13–18 May 2018.
99. Burla, M.; Hoessbacher, C.; Heni, W.; Haffner, C.; Fedoryshyn, Y.; Werner, W.; Watanabe, T.; Massler, H.; Elder, D.L.; Dalton, L.R.; et al. 500 GHz plasmonic Mach-Zehnder modulator enabling sub-THz microwave photonics. *APL Photonics* **2019**, *4*, 056106. [CrossRef]
100. Ummethala, S.; Harter, T.; Koehnle, K.; Li, Z.; Muehlbrandt, S.; Kutuvantavida, Y.; Kemal, J.; Schaefer, J.; Tessmann, A.; Kumar Garlapati, S.; et al. THz-to-Optical Conversion in Wireless Communications Using an Ultra-Broadband Plasmonic Modulator. *Nat. Photonics* **2019**, *13*, 519–524. [CrossRef]
101. Hasan, M.; Arezoomandan, S.; Condori, H.; Sensale-Rodriguez, B. Graphene terahertz devices for communications applications. *Nano Comm. Netw.* **2016**, *10*, 68–78. [CrossRef]
102. Jornet, J.M.; Akyildiz, I.F. Graphene-based plasmonic nanotransceiver for terahertz band communication. In Proceedings of the 8th European Conference on Antennas and Propagation (EuCAP 2014), Hague, The Netherlands, 6–11 April 2014; pp. 492–496. [CrossRef]
103. Elayan, H.; Shubair, R.M.; Kiourti, A. On Graphene-based THz Plasmonic Nano-Antennas. In Proceedings of the 16th Mediterranean Microwave Symposium (MMS), Abu Dhabi, United Arab Emirates, 14–16 November 2016. [CrossRef]
104. Mohammadi, A.; Sandoghdar, V.; Agio, M. Gold, copper, silver and aluminum nanoantennas to enhance spontaneous emission. *J. Comput. Theor. Nanosci.* **2009**, *6*, 2024–2030. [CrossRef]

105. Abadal, S.; Hosseininejad, S.E.; Cabellos-Aparicio, A.; Alarcón, E. Graphene-based Terahertz Antennas for Area-Constrained Applications. In Proceedings of the 40th International Conference on Telecommunications and Signal Processing (TSP), Barcelona, Spain, 5–7 July 2017. [CrossRef]
106. Dahab, M.A.; Dawoud, S.; Zainud-Deen, H.; Malhat, H.A. Tapered Metal Nanoantenna Structures for Absorption Enhancement in GaAs Thin-Film Solar Cells. *J. Adv. Res. Appl. Mech.* **2018**, *44*, 1–7.
107. Llatser, I.; Kremers, C.; Chigrin, D.; Jornet, J.M.; Lemme, M.C.; Cabellos-Aparicio, A.; Alarcón, E. Radiation Characteristics of Tunable Graphennas in the Terahertz Band. *Radioeng. J.* **2012**, *21*, 946–953.
108. Nafari, M.; Jornet, J.M. Modeling and Performance Analysis of Metallic Plasmonic Nano-Antennas for Wireless Optical Communication in Nanonetworks. *IEEE Access* **2017**, *5*, 6389–6398. [CrossRef]
109. Abadal, S.; Mestres, A.; Iannasso, M.; Paretta, J.C.; Alarcón, E.; Cabellos-Aparicio, A. Evaluating the Feasibility of Wireless Networks-on-Chip Enabled by Graphene. In Proceedings of the NoCArc'14 International Workshop on Network on Chip Architectures, Cambridge, UK, 13–14 December 2014; pp. 51–56. [CrossRef]
110. Lin, J., Jr.; Wu, H.; Su, Y.; Gao, L.; Sugavanam, A.; Brewer, J.E.; Kenneth, K.O. Communication using antennas fabricated in silicon integrated circuits. *IEEE J. Solid-State Circuits* **2007**, *42*, 1678–1687. [CrossRef]
111. Lee, S.B.; Tam, S.W.; Pefkianakis, I.; Lu, S.; Chang, M.F.; Guo, C.; Reinmann, G.; Peng, C.; Naik, M.; Zhang, L.; et al. A scalable micro wireless interconnect structure for cmps. In Proceedings of the MobiCom'09 15th annual international conference on Mobile computing and networking, Beijing, China, 20–25 September 2009; pp. 217–228. [CrossRef]
112. Huang, Y.; Yin, W.Y.; Liu, Q.H. Performance prediction of carbon nanotube bundle dipole antennas. *IEEE Trans. Nanotechnol.* **2008**, *7*, 331–337. [CrossRef]
113. Zhao, D.; Wang, Y. Sd-mac: Design and synthesis of a hardware efficient collision-free qos-aware mac protocol for wireless network-on-chip. *IEEE Trans. Comput.* **2008**, *57*, 1230–1245. [CrossRef]
114. Geim, A.K. Graphene: Status and prospects. *Science* **2009**, *324*, 1530–1534. [CrossRef] [PubMed]
115. Jornet, J.; Akyildiz, I. Graphene-based plasmonic nano-antenna for terahertz band communication in nanonetworks. *IEEE J. Sel. Areas Commun.* **2013**, *31*, 685–694. [CrossRef]
116. Hosseininejad, S.E.; Alarcón, E.; Komjani, N.; Abadal, S.; Lemme, M.C.; Bolívar, H.P.; Cabellos-Aparicio, A. Study of Hybrid and Pure Plasmonic Terahertz Antennas Based on Graphene Guided-wave Structures. *Nano Commun. Netw.* **2017**, *12*, 34–42. [CrossRef]
117. Nikitin, A.Y.; Guinea, F.; Garcıa-Vidal, F.J.; Martın-Moreno, L. Edge and waveguide terahertz surface plasmon modes in graphene microribbons. *Phys. Rev. B-Condens. Matter Mater. Phys.* **2011**, *84*, 1–4. [CrossRef]
118. Gan, C.H.; Chu, H.S.; Li, E.P. Synthesis of highly confined surface plasmon modes with doped graphene sheets in the mid infrared and terahertz frequencies. *Phys. Rev. B-Condens. Matter Mater. Phys.* **2012**, *85*, 125431. [CrossRef]
119. Gu, X.; Lin, I.T.; Liu, J.M. Extremely confined terahertz surface plasmon-polaritons in graphene-metal structures. *Appl. Phys. Lett.* **2013**, *103*, 071103. [CrossRef]
120. Liu, P.; Zhang, X.; Ma, Z.; Cai, W.; Wang, L.; Xu, J. Surface Plasmon modes in graphene wedge and groove waveguides. *Opt. Express* **2013**, *21*, 32432–32440. [CrossRef] [PubMed]
121. Xing, R.; Jian, S. The Field Enhancement of the Graphene Triple-Groove Waveguide. *IEEE Photonics Technol. Lett.* **2016**, *28*, 2649–2652. [CrossRef]
122. Zhou, X.; Zhang, T.; Chen, L.; Hong, W.; Li, X. A graphene-based hybrid plasmonic waveguide with ultra-deep subwavelength confinement. *IEEE J. Lightwave Technol.* **2014**, *32*, 4199–4203. [CrossRef]
123. Tamagnone, M.; Diaz, J.S.G.; Perruisseau-Carrier, J.; Mosig, J.R. High-impedance frequency-agile THz dipole antennas using graphene. In Proceedings of the 2013 7th European Conference on Antennas and Propagation (EuCAP), Gothenburg, Sweden, 8–12 April 2013; pp. 533–536.
124. Cabellos-Aparicio, A.; Llatser, I.; Alarcon, E.; Hsu, A.; Palacios, T. Use of THz Photoconductive Sources to Characterize Tunable Graphene RF Plasmonic Antennas. *IEEE Trans. Nanotechnol.* **2015**, *14*, 390–396. [CrossRef]
125. Akyildiz, I.F.; Jornet, J.M. Realizing Ultra-Massive MIMO (1024 × 1024) communication in the (0.06–10) Terahertz band. *Nano Commun. Netw.* **2016**, *8*, 46–54. [CrossRef]
126. Tamagnone, M.; Diaz, J.S.G.; Mosig, J.; Perruisseau-Carrier, J. Hybrid graphene-metal reconfigurable terahertz antenna. In Proceedings of the 2013 IEEE MTT-S International Microwave Symposium Digest (MTT), Seattle, WA, USA, 2–7 June 2013; pp. 9–11. [CrossRef]

127. Hosseininejad, S.E.; Komjani, N. Waveguide-fed Tunable Terahertz Antenna Based on Hybrid Graphene-metal structure. *IEEE Trans. Antennas Propag.* **2016**, *64*, 3787–3793. [CrossRef]
128. Hosseininejad, S.E.; Alarcón, E.; Komjani, N.; Abadal, S.; Lemme, M.C.; Bolívar, H.P.; Cabellos-Aparicio, A. Surveying of Pure and Hybrid Plasmonic Structures Based on Graphene for Terahertz Antenna. In Proceedings of the NANOCOM'16 3rd ACM International Conference on Nanoscale Computing and Communication, New York, NY, USA, 28–30 September 2016. [CrossRef]
129. Alu, A.; Engheta, N. Wireless at the nanoscale: Optical interconnects using matched nanoantennas. *Phys. Rev. Lett.* **2010**, *104*, 213902. [CrossRef]
130. Solis, D.M.; Taboada, J.M.; Obelleiro, F.; Landesa, L. Optimization of an optical wireless nanolink using directive nanoantennas. *Opt. Express* **2013**, *21*, 2369–2377. [CrossRef]
131. Dregely, D.; Lindfors, K.; Lippitz, M.; Engheta, M.; Totzeck, M.; Giessen, H. Imaging and steering an optical wireless nanoantenna link. *Nat. Commun.* **2014**, *5*, 4354. [CrossRef]
132. De Souza, J.L.; da Costa, K.Q. Broadband Wireless Optical Nanolink Composed by Dipole-Loop Nanoantennas. *IEEE Photonics J.* **2018**, *10*. [CrossRef]
133. Xu, Y.; Dong, T.; He, J.; Wan, Q. Large scalable and compact hybrid plasmonic nanoantenna array. *Opt. Eng.* **2018**, *57*, 087101. [CrossRef]
134. Sharma, P.; Dinesh Kumar, V. Multilayer Hybrid Plasmonic Nano Patch Antenna. *Plasmonics* **2019**, *14*, 435–440. [CrossRef]
135. Song, H.J.; Ajito, K.; Muramoto, Y.; Wakatsuki, A.; Nagatsuma, T.; Kukutsu, N. Uni-travelling-carrier photodiode module generating 300 GHz power greater than 1 mW. *IEEE Microw. Wirel. Compon. Lett.* **2012**, *22*, 363–365. [CrossRef]
136. Walthera, C.; Fischer, M.; Scalari, G.; Terazzi, R.; Hoyler, N.; Faistb, J. Quantum cascade lasers operating from 1.2 to 1.6 THz. *Appl. Phys. Lett.* **2007**, *91*, 131122. [CrossRef]
137. Pearson, J.C.; Drouin, B.J.; Maestrini, A.; Mehdi, I.; Ward, J.; Lin, R.H.; Yu, S.; Gill, J.J.; Thomas, B.; Lee, C.; et al. Demonstration of a room temperature 2.48–2.75 thz coherent spectroscopy source. *Rev. Sci. Instrum.* **2011**, *82*, 093105. [CrossRef]
138. Ojefors, E.; Grzyb, J.; Zhao, Y.; Heinemann, B.; Tillack, B.; Pfeiffer, U.R. A 820 GHz SiGe Chipset for Terahertz Active Imaging Applications. In Proceedings of the 2011 IEEE International Solid-State Circuits Conference, San Fransisco, CA, USA, 20–24 February 2011; pp. 224–226. [CrossRef]
139. Vitiello, M.S.; Scalari, G.; Williams, B.; Natale, P.D. Quantum cascade lasers: 20 Years of challenges. *Opt. Express* **2015**, *23*, 5167–5182. [CrossRef]
140. Williams, B.S.; Kumar, S.; Callebaut, H.; Hu, Q. Terahertz quantum-cascade laser at $\lambda \approx 100$ μm using metal waveguide for mode confinement. *Appl. Phys. Lett.* **2003**, *83*, 2124. [CrossRef]
141. Wu, C.; Jin, Y.; Reno, J.L.; Kumar, S. Large static tuning of narrow-beam terahertz plasmonic lasers operating at 78k. *APL Photonics* **2017**, *2*, 026101. [CrossRef]
142. Wu, C.; Khanal, S.; Reno, J.L.; Kumar, S. Terahertz plasmonic laser radiating in an ultra-narrow beam. *Optica* **2016**, *3*, 734–740. [CrossRef]
143. Nagatsuma, T.; Horiguchi, S.; Minamikata, Y.; Yoshimizu, Y.; Hisatake, S.; Kuwano, S.; Yoshimoto, N.; Terada, J.; Takahashi, H. Terahertz wireless communications based on photonics technologies. *Opt. Express* **2013**, *21*, 477–487. [CrossRef]
144. Sizov, F.; Rogalski, A. THz detectors. *Prog. Quantum Electron.* **2010**, *34*, 278–347. [CrossRef]
145. Gu, Q.; Xu, Z.; Jian, H.Y.; Pan, B.; Xu, X.; Chang, M.C.F.; Liu, W.; Fetterman, H. CMOS THz Generator with Frequency Selective Negative Resistance Tank. *IEEE Trans. Terahertz Sci. Technol.* **2012**, *2*, 193–202. [CrossRef]
146. Radisic, V.; Leong, K.M.K.H.; Mei, X.; Sarkozy, S.; Yoshida, W.; Deal, W.R. Power Amplification at 0. *65 THz using INP HEMTs. IEEE Trans. Microw. Theory Tech.* **2012**, *60*, 724–729. [CrossRef]
147. Knap, W.; Lusakowski, J. Terahertz emission by plasma waves in 60 nm gate high electron mobility transistors. *Appl. Phys. Lett.* **2004**, *84*, 2331–2333. [CrossRef]
148. Knap, W.; Teppe, F.; Dyakonova, N.; Coquillat, D.; Lusakowski, J. Plasma wave oscillations in nanometer field effect transistors for terahertz detection and emission. *J. Phys. Condens. Matter* **2008**, *20*, 384205. [CrossRef] [PubMed]
149. Otsuji, T.; Watanabe, T.; Tombet, B.; Satou, A.; Knap, W.M.; Popov, V.V.; Ryzhii, M.; Ryzhii, V. Emission and detection of terahertz radiation using two-dimensional electrons in iii-v semiconductors and graphene. *IEEE Trans. Terahertz Sci. Technol.* **2013**, *3*, 63–71. [CrossRef]

150. Zhang, X.; Chung, C.J.; Subbaraman, H.; Pan, Z.; Chen, C.T.; Chen, R.T. Design of a plasmonic-organic hybrid slot waveguide integrated with a bowtie-antenna for terahertz wave detection. In Proceedings of the SPIE 9756, Photonic and Phononic Properties of Engineered Nanostructures VI, San Francisco, CA, USA, 14 March 2016. [CrossRef]
151. Huang, K.C.; Seo, M.-K.; Sarmiento, T.; Huo, Y.; Harris, J.S.; Brongersma, M.L. Electrically driven subwavelength optical nanocircuits. *Nature Photon.* **2014**, *8*, 244–249. [CrossRef]
152. Service, R. Nanolasers. Ever-smaller lasers pave the way for data highways made of light. *Science* **2010**, *328*, 810–811. [CrossRef]
153. Ropers, C.; Neacsu, R.C.; Elsaesser, T.; Albrecht, M.; Raschke, M.B.; Lienau, C. Grating-coupling of surface plasmons onto metallic tips: A nanoconfined light source. *Nano Lett.* **2007**, *7*, 2784–2788. [CrossRef]
154. Feigenbaum, E.; Orenstein, M. Backward propagating slow light in inverted plasmonic taper. *Opt. Express* **2009**, *17*, 2465–2469. [CrossRef]
155. Andryieuski, A.; Malureanu, R.; Biagi, G.; Holmgaard, T.; Lavrinenko, A. Compact dipole nanoantenna coupler to plasmonic slot waveguide. *Opt. Lett.* **2012**, *37*, 1124–1126. [CrossRef] [PubMed]
156. Gao, Q.; Ren, F.; Wang, A.X. Plasmonic Integrated Circuits with High Efficiency Nano-Antenna Couplers. In Proceedings of the 2016 IEEE Optical Interconnects Conference (OI), San Diego, CA, USA, 9–11 May 2016. [CrossRef]
157. Harter, T.; Muehlbrandt, S.; Ummethala, S.; Schmid, A.; Nellen, S.; Hahn, L.; Freude, W.; Koos, C. Silicon-plasmonic integrated circuits for terahertz signal generation and coherent detection. *Nat. Photonics* **2018**, *12*, 625–633. [CrossRef]
158. Gorbenko, I.V.; Kachorovskii, V.Y.; Shur, M. Terahertz plasmonic detector controlled by phase asymmetry. *Opt. Express* **2019**, *27*, 4004–4013. [CrossRef] [PubMed]
159. Gomez-Diaz, J.S.; Perruisseau-Carrier, J. Graphene-based plasmonic switches at near infrared frequencies. *Opt. Express* **2013**, *21*, 15490–15504. [CrossRef] [PubMed]
160. Correas-Serrano, D.; Gomez-Diaz, J.S.; Perruisseau-Carrier, J.; Alvarez-Melcon, A.A. Graphene-based plasmonic tunable low-pass filters in the terahertz band. *IEEE Trans. Nanotechnol.* **2014**, *13*, 1145–1153. [CrossRef]
161. Chen, P.Y.; Argyropoulos, C.; Alu, A. Terahertz antenna phase shifters using integrally-gated graphene transmission-lines. *IEEE Trans. Antennas Propag.* **2013**, *61*, 1528–1537. [CrossRef]
162. Otsuji, T.; Tombet, S.B.; Satou, A.; Ryzhii, M.; Ryzhii, V. Terahertz-Wave Generation Using Graphene: Toward New Types of Terahertz Lasers. *Proc. IEEE (Early Access)* **2013**, *16*, 1–13. [CrossRef]
163. Sensale-Rodriguez, B.; Yan, R.; Kelly, M.M.; Fang, T.; Tahy, K.; Hwang, W.S.; Jena, D.; Liu, L.; Xing, H.G. Broadband graphene terahertz modulators enabled by intraband transitions. *Nat. Commun.* **2012**, *3*, 780. [CrossRef]
164. Zhang, R.; Zhang, Y.; Yu, H.; Zhang, H.; Yang, R.; Yang, B.; Liu, Z.; Wang, J. Broadband black phosphorus optical modulator in visible to mid-infrared spectral range. *arXiv* **2015**, arXiv:1505.05992. [CrossRef]
165. Li, W.; Chen, B.; Meng, C.; Fang, W.; Xiao, Y.; Li, X.; Hu, Z.; Xu, Y.; Tong, L.; Wang, H.; et al. Ultrafast all-optical graphene modulator. *Nano Lett.* **2014**, *14*, 955–959. [CrossRef]
166. Liang, G.; Hu, X.; Yu, X.; Shen, Y.; Li, L.H.; Davies, A.G.; Linfield, E.H.; Liang, H.K.; Zhang, Y.; Yu, S.F.; et al. Integrated Terahertz Graphene Modulator with 100% Modulation Depth. *ACS Photonics* **2015**, *2*, 1559–1566. [CrossRef]
167. Phare, C.T.; Lee, Y.D.; Cardenas, J.; Lipson, M. Graphene electro-optic modulator with 30 GHz bandwidth. *Nat. Photonics* **2015**, *9*, 511–514. [CrossRef]
168. Liu, M.; Yin, X.; Ulin-Avila, E.; Geng, B.; Zentgraf, T.; Ju, L.; Wang, F.; Zzhang, X. A graphene-based broadband optical modulator. *Nature* **2011**, *474*, 64–67. [CrossRef] [PubMed]
169. Singh, P.K.; Aizin, G.; Thawdar, N.; Medley, M.; Jornet, J.M. Graphene-based Plasmonic Phase Modulator for Terahertz-band Communication. In Proceedings of the 2016 10th European Conference on Antennas and Propagation (EuCAP), Davos, Switzerland, 10–15 April 2016. [CrossRef]
170. Xia, F.; Mueller, T.; Lin, Y.; Valdes-Garcia, A.; Avouris, P. Ultrafast graphene photodetector. *Nat. Nanotechnol.* **2009**, *4*, 839–843. [CrossRef] [PubMed]
171. Vicarelli, L.; Vitiello, M.S.; Coquillat, D.; Lombardo, A.; Ferrari, A.C.; Knap, W.; Polini, M.; Pellegrini, V.; Tredicucci, A. Graphene field effect transistors as room-temperature Terahertz detectors. *Nat. Mater.* **2012**, *11*, 865–871. [CrossRef]

172. Ma, Z.; Kikunage, K.; Wang, H.; Sun, S.; Amin, R.; Tahersima, M.; Maiti, R.; Miscuglio, M.; Dalir, H.; Sorger, V.J. Compact Graphene Plasmonic Slot Photodetector on Silicon-on-insulator with High Responsivity. *arXiv* **2018**, arXiv:1812.00894.
173. Viti, L.; Politano, A.; Vitiello, M.S. Black phosphorus nanodevices at terahertz frequencies: Photodetectors and future challenges. *APL Mater.* **2017**, *5*, 035602. [CrossRef]
174. Samson, Z.L.; Yen, S.C.; McDonald, K.F.; Knight, K.; Li, S.; Hewak, D.W.; Tsai, D.P.; Zheludev, N.I. Chalcogenide glasses in active plasmonics. *Phys. Status Solidi (RRL) Rapid Res. Lett.* **2010**, *4*, 274–276. [CrossRef]

Article

Real-Time Train Tracking from Distributed Acoustic Sensing Data

Christoph Wiesmeyr [1,*], Martin Litzenberger [1], Markus Waser [1], Adam Papp [1], Heinrich Garn [1], Günther Neunteufel [2] and Herbert Döller [2]

1 AIT Austrian Institute of Technology GmbH, Giefinggasse 4, 1210 Vienna, Austria; martin.litzenberger@ait.ac.at (M.L.); markus.waser@ait.ac.at (M.W.); adam.papp.pb@ait.ac.at (A.P.); heinrich.garn@ait.ac.at (H.G.)

2 NBG Fosa GmbH, Johannesgasse 15, 1010 Vienna, Austria; g.neunteufel@nbg.tech (G.N.); office.zwettl@doeller.biz (H.D.)

* Correspondence: christoph.wiesmeyr@ait.ac.at; Tel.: +43-664-8825-6065

Received: 31 October 2019; Accepted: 23 December 2019; Published: 8 January 2020

Abstract: In the context of railway safety, it is crucial to know the positions of all trains moving along the infrastructure. In this contribution, we present an algorithm that extracts the positions of moving trains for a given point in time from Distributed Acoustic Sensing (DAS) signals. These signals are obtained by injecting light pulses into an optical fiber close to the railway tracks and measuring the Rayleigh backscatter. We show that the vibrations of moving objects can be identified and tracked in real-time yielding train positions every second. To speed up the algorithm, we describe how the calculations can partly be based on graphical processing units. The tracking quality is assessed by counting the inaccurate and lost train tracks for two different types of cable installations.

Keywords: DAS; fiber optic sensing; train tracking; pattern recognition

---

1. Introduction

Railway safety is an ever-increasing issue, as traffic demand increases world wide with railways playing an important role. This is especially true in the light of the urgent need for the decarbonization of traffic. The accurate tracking of a train's real position on the track is the basis for all modern railway safety concepts. In state-of-the-art rail operations, the train's position is only known within a so-called "block" between two signals. In one block, there is only one train allowed to operate at a time, which is sufficient to prevent collisions of trains. The disadvantage of this concept is a reduced efficiency in the use of the available track infrastructure because long portions of the track stay unused. Future traffic demand calls for more flexible, thus efficient, concepts. Novel safety concepts will, therefore, be based on virtual- or moving-blocks that virtually enclose the train while in motion, providing enough safety separation between trains at all times [1,2]. An accurate, reliable, and redundant train tracking technology is an important basis for these novel concepts.

Conventional train tracking is mainly achieved by communication between train-side equipment, track-side equipment, and the interlocking system. Block-based systems need axle-counting sensors installed next to the track. They deliver a signal for each axle that passes through the detector, to ensure that a block is unoccupied before the next train enters this block. The European train safety system, ETCS, can use moving blocks in its level 3 implementation and relies on continuous positioning information of the train [3]. High accuracy differential satellite positioning systems (D-GNSS), such as GPS, Glonass, or Galileo, become available and are certainly planned to be applied in future railway systems [4], but the high safety requirements of railway operation call for highly reliable and redundant positioning systems. Furthermore, the coverage of D-GNSS and the necessary communication link back to the interlocking system is never 100% guaranteed: Tunnels, deep valleys, and tall metallic structures next to the track can degrade, or even block, radio frequency signals.

Non-coherent optical time domain reflectometry (OTDR) technology has been used for a long time for long-range monitoring of the quality and the integrity of fiber optic cable infrastructure [5], e.g., to locate break points in underwater fiber cables. Distributed acoustic sensing (DAS), a special variant of coherent (or phase sensitive) OTDR, is not only able to assess the cable quality but also to detect and locate strain and temperature changes along the cable with high sensitivity. The measurement and signal processing methods presented in this paper are based on DAS and offer an alternative or redundancy system for accurate train positioning and tracking. Other than the aforementioned technologies, it does not need equipment installed on the train nor communication between train and interlocking system. Furthermore, it does not need extra track-side equipment, with the exception of a fiber optic cable that is already in place for data connection and communication purposes. In this paper, we investigate advanced signal processing of the DAS data received from the so-called optical interrogator device for accurate train tracking. The interrogator injects a series of laser light pulses into the fiber cable and measures the back scattered (Rayleigh backscatter) light at the same end of the cable. The measured signal contains the optical path length change resulting from a refractive index change due to compression of the glass and from contraction or elongation of the fiber due to ground deformation [6]. These effects can be received from any point along the cable over a range of up to 40 km and can be measured with a positional resolution down to 0.5 m. The reader interested in the different types of DAS devices and their physical principles is referred to the survey of Bao and Chen [7].

OTDR and DAS are used, for example, for oil well monitoring [8], intrusion detection [9,10], or even for rock slide detection [11]. Vibrations of train movements near the railway track can also be monitored using a DAS device, where in most cases, fiber cables are already installed there for communication means. Several methods for train tracking using DAS signals have been proposed [12–14]. However, these algorithms were never evaluated for longer recordings and for different ground and installation conditions. All the presented methods are based on the variance of the measured DAS signals. In this contribution, we propose a method that is based on machine learning methods to detect vibrations, which is more flexible and can be more easily adapted to different recording conditions. Furthermore, the algorithm which is the basis for [15,16] has a five-second lag in time. The goal of this manuscript is to provide a detailed description and evaluation of an algorithm that can be used for real-time train position monitoring.

We will give the details on the data acquisition of DAS test data used for this investigation in the following section. The algorithmic details, as well as a discussion of the accuracy of the proposed method will be given in the later sections. We will further provide an outlook for future work and improvements.

2. Materials and Methods

In this section, we will present the signal processing methods used for train tracking, i.e., we will describe in detail how to process the raw measurement data to obtain train trajectories. The train tracking algorithm works on the basis of one-second chunks of data and is based on two main steps, i.e., vibration detection and object tracking. Before we go into details on these, we will give a short introduction on how to interpret the raw DAS signal from the Fotech Helios DAS (https://www.fotech.com/products/helios-das/ retrieved 31 October 2019)) interrogator device.

2.1. DAS Test Data Acquisition

We recorded long-term DAS data with a Fotech Helios DAS interrogator device in two different railway test sites. At one site, the track was located in a tunnel, with the single-mode fiber optic cable installed in a cable trench; in the other site, there was an open track with the single-mode fiber cable directly attached to one rails' foot with clips. For all experiments, already existing telecommunication fiber cables were used, and no special installation of cables was done for this work. At these two sites, we recorded a total of approximately 1000 h of data with 3000 trains passing the infrastructure. The total rail track length monitored during the recordings was approximately 20 km. Refer to Section 3.3 for details on the sites. Apart from these long-term recordings, two shorter recordings were available, of around one hour each, from an open track with the cable in a cable trench. The standard telecommunication fibers were used for the recordings, and the effect of fiber darkening over time was not evaluated in this study.

The Helios DAS is a coherent (C-OTDR) phase sensitive interrogator device that delivers the optical detector voltage, encoding the laser light phase change, as a 16 bit resolution signal, sampled with 150 MSamples/s. This results in a physical spacing of the DAS segments of 0.68 m, given by the light speed in the fiber. The effective spatial resolution of the measurement is limited by the laser pulse length of 100 ns, resulting in a size of the laser pulse of 20.2 m, over which the phase measurement is averaged. Table 1 summarizes the relevant measurement parameters. The cable used for test site 1 is a single-mode, stranded mini cable 60 × 5 × 12 E9/126, the cable at the second test site is a single-mode stranded fiber optic cable 5 × 12 E9/125 A-DF(ZN)2Y(BN)2Yv 5 × 12 E9/125-G652D.

Due to the nature of the optical measurement principle, neither the absolute phase nor the fiber strain is directly accessible from the signal output. Furthermore the signal is delivered with an ambiguity of 2π (equal to the "fringes" observed in a Michelson interferometer), which results in a non-linearity in very strong signals. A detailed description of the signal generation for a similar DAS setup is found in [9]. The laser pulse repetition frequency of the Helios DAS device was set to 2000 Hz for all recordings. This data stream is the input to our signal processing stages.

Table 1. Summarized parameters of measurement, optical fiber, and laser pulses used for this work. ADC = Analog-to-Digital Converter.

Parameter	Value	Unit
Laser pulse length	100	ns
Laser wavelength	1550	nm
Pulse repetition frequency	2000	Hz
Signal sampling frequency	150	kHz
Signal ADC sampling resolution	16	bit
Fiber light speed	202,020,208	m/s

2.2. Annotations

For all the different test sites, we annotated vibrations and background in the raw data. In the subsequent steps, we describe how these annotations are used to automatically detect vibrations in the DAS signal. An annotation for a test site contains 10,000 samples for background and vibration signal each, where one sample consists of the raw signal recorded in one second for one cable segment.

2.3. Vibration Detection

The vibration detection takes as input the raw measurements acquired within one second for all the cable segments along the optical fiber. Given a sampling rate of 2000 Hz and a spatial resolution of 0.68 m with a monitored fiber length of 40 km (corresponding to around 60,000 channels), the raw data consist of 2000 × 60,000 samples each second. In the following, we will denote the data matrix containing all the raw measurement data for second i as M_i. The result of the vibration detection for this second in time, v_i, is a vector of Boolean values of length 60,000, where each element of this vector indicates whether vibration is present at the second i or not for that specific cable segment. Therefore, the vibration detection step corresponds to a data reduction from M_i to v_i.

In the following, we will describe the steps we used to perform vibration detection. We use machine learning techniques to decide whether the vector $M_{i,x}$ containing the sensing samples for a given cable segment x and a given second in time i contains vibration or not. The decision is based on the spectral distribution of the raw signal $M_{i,x}$ and is done based on the following steps detailed below. Note that the decision has to be made independently for all the cable segments without any spatial averaging; in the following, therefore, we will describe the procedure for one cable channel.

1. Computation of the spectral energy in 10 frequency bins.
2. Normalization of the 10 energy values to make them sum up to 1.
3. Using Principal Component Analysis (PCA) to reduce the 10 feature values to two feature values.
4. Employ a pre-trained Support Vector Machine (SVM) to classify the feature vector of length 2 as vibration or background.

Computation of spectral energy. The energy of the frequencies of the signal $M_{i,x}$ are estimated using the Fast Fourier Transform (FFT). The squared absolute values of the Fourier transformed signal $|M_{i,x}(\xi)|^2$ are then summed up individually for ten bins

$$b_k = \sum_{\xi \in B_k} |M_{i,x}(\xi)|^2, \tag{1}$$

where B_k denotes the k-th bin with $k \in \{0, \dots, 9\}$. The frequency bins are linearly spaced and span the frequencies between 10 Hz and 990 Hz.

PCA computation. We use PCA to reduce the number of features from the frequency binning from 10 to 2. This is done by computing two linear combinations of the 10 original features optimally with respect to the variance of the data they can explain. Computationally, this PCA can be represented as a 10×2 matrix. The data used for computing the principal components are from the manually annotated test dataset described in Section 2.2.

SVM classification. SVM classification is used for each cable segment to categorize it as vibration or background for each second in time. The SVM classifier is trained on manually annotated data. We use the SVM implementation of openCV 2.4 with an RBFkernel [17].

2.4. Train Tracking

The train tracking algorithm works on the inputs from the vibration detection and works on a one-second basis. It takes the classification vector v_i and outputs the current positions of the trains present in the signal denoted by P_j, where the index j runs through the active trains at the given second i. The train tracking algorithm itself consists of three steps, which we will discuss individually subsequently:

1. Extraction of edges from the data.
2. Assigning the edges to objects and creating new objects.
3. Applying a Kalman filter to smooth the trajectories of all the objects.

Edge Detection. The edge detection is based on a K-means clustering of wavelet responses of the vibration detection vector v_i. We use Ricker wavelets of two different widths, namely 128 and 256, cable segments to compute the convolution with the detection vector v_i. From both of these convolutions, we extract 512 consecutive values as feature values for the K-means clustering algorithm. A prototypical dataset of vibration detections is used to train the algorithm with four centers which correspond to background, leading edges, trailing edges, and train signal. Note that at this point leading edges are edges that mark the start of a train viewed from the interrogator device, while trailing edges are edges that mark the end of a train. If the train moves towards the interrogator, this leading edge will also be the leading edge of the train; if the train moves away from the interrogator, the leading edge viewed from the interrogator will be the trailing edge of the train. In the following, we will call all these edges *detected edges*.

Edge Assignment and object creation. The tracking algorithm works internally with two types of objects: tracked edges and tracked trains, which we will denote by E_i, $i \in I$ and P_j, $j \in J$, respectively. Note that tracked edges are different objects from the detected edges identified in the previous step. In the following, we will describe how the detected edges are assigned to the tracked trains and the tracked edges. This is repeated every second with the corresponding detected edges. Therefore, all the steps described in this paragraph correspond to one given second in time. The algorithm starts to assign detected edges to the list of tracked trains; this is done for detected leading and trailing edges separately. We start by defining cost functions c_L and c_T for a given assignment of an edge e and a Train P by the distance between the edge and the train's leading edge if e is itself a leading edge and the distance between the edge e and the train's trailing edge if e itself is a trailing edge. Furthermore, we define a threshold T, which denotes the maximum distance between e and the corresponding edges of the trains, where we still allow an assignment to be made. The formulas for these two cost functions read as

$$c_L(e, P) = \begin{cases} |e - P_{\text{Lead}}| & \text{if } |e - P_{\text{Lead}}| < T \\ \infty & \text{if } |e - P_{\text{Lead}}| > T, \end{cases} \tag{2}$$

$$c_T(e, P) = \begin{cases} |e - P_{\text{Trail}}| & \text{if } |e - P_{\text{Trail}}| < T \\ \infty & \text{if } |e - P_{\text{Trail}}| > T. \end{cases} \tag{3}$$

Therefore, the two cost functions equal infinity if the detected edge is too far away from the corresponding edge of a given train. If the cost function $c_L(e, P) = \infty$, the edge e will not be assigned to the leading edge of the train P (this is analogous for the function c_T and the trailing edge of P). The assignment problem for a given set of detected leading edges (the assignment works analogously for trailing edges) $e_i, i \in I_e$ to a given set of trains $P_j, j \in I_j$ iteratively finds the minimum in the matrix $C_{i,j} = c_L(e_i, P_j)$ for i, j in the defined index sets. The leading edge and train with the respective indices i_0 and j_0 corresponding to this minimum, if smaller than ∞, are assigned to each other and the i_0-th row and the j_0-th column of the cost matrix C are set to ∞. This step is repeated until no more assignments can be made, i.e., the matrix C contains only ∞. The same procedure is then also repeated for the trailing edges of the trains.

After the assignment of leading and trailing edges to the trains, a subset of the originally detected edges has been assigned. The remaining detected edges are then assigned to the currently tracked edges E_i using the same iterative procedure as described for the assignment of edges to trains. We define distance functions analogous to Equation (2), (3), which leads to a definition to an assignment matrix C. With the same greedy iterative algorithm as described above, we can determine the assignments of found edges to tracked edges.

A (possibly empty) subset of the found edges cannot be assigned to a train or to a tracked edge. These edges will become tracked edges in the next time step.

The assignment process described in this section is an approximate greedy approach to the more complex quadratic assignment problem where one tries to find all the assignment index pairs (i, j) such that the corresponding total cost (i.e., the sum of the individual costs for assignments) is minimal. For better readability of the paper, we will not give a rigorous formulation of this optimization problem here since it is computationally not feasible for a real time tracking algorithm.

Kalman filtering. Each train and each tracked edge has an underlying Kalman filter. The tracked edges follow a two-dimensional state-space model consisting of position of the edge and its velocity leading to the following matrices corresponding to a constant velocity model

$$\begin{aligned} F &= \begin{pmatrix} 1 & 1 \\ 0 & 1 \end{pmatrix} \\ H &= (1,0), \end{aligned} \tag{4}$$

where F denotes the state transition model, and H denotes the observation model.

The Kalman filter for a tracked train is based on a slightly more involved state-space model. To model a moving train, we propose a four-dimensional state space consisting of leading edge position, trailing edge position, speed, and the length of the train. A measurement at a given time results in the leading and the trailing edge of the train. The resulting constant velocity model leads to the following transition and observation model

$$F = \begin{pmatrix} 1 & 0 & 1 & 0 \\ 0 & 1 & 1 & 0 \\ 0 & 0 & 1 & 0 \\ -1 & 1 & 0 & 0 \end{pmatrix}$$
$$H = \begin{pmatrix} 1 & 0 & 0 & 0 \\ 0 & 1 & 0 & 0 \end{pmatrix}. \quad (5)$$

2.5. GPU Computation for Feature Extraction

The computation of the features for the SVM classification needs to be able to process the raw data matrix M_i containing the data for second i in time and all the channels along the optical fiber. As mentioned above, for a typical setup of 40 km monitoring length and a sampling rate of 2000 Hz this leads to a data matrix of dimensions in the order of $2000 \times 60{,}000$. In the case of a Helios DAS interrogator device, the data format is unsigned int 16, which leads to a data rate of 228 MB/s. We then have to cast these data to a float datatype because the first step involves a Fourier transform that doubles the data rate as we use 32 bit float values. This Fourier transform computation has to be done for all 60,000 cable segments in parallel. The highly parallelizable nature of these computations makes general purpose GPUs (GPGPU) highly suitable for the task. The same holds true for the PCA feature reduction, which has to be applied for each cable segment individually. The SVM classifier is not available with GPU acceleration; therefore, the reduced features are downloaded to the CPU before classification. A flow chart of the complete algorithm is depicted in Figure 1.

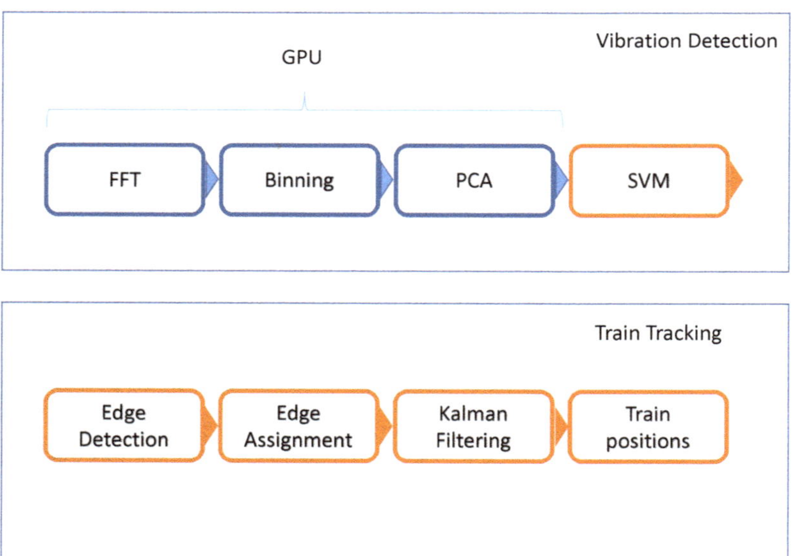

Figure 1. Flow chart of the algorithm. The top box describes the steps of the vibration detection (steps on the GPU are in blue); the bottom box describes the steps of the train tracking. FFT = Fast Fourier Transform; PCA = Principal Component Analysis; SVM = Support Vector Machine.

3. Results

This section is devoted to the results. We will start to discuss the results for the different stages of the tracking algorithm and then present the results for two long term recordings, as well as tracking accuracy results for a short-term recording where a ground truth of the train positions was available.

3.1. Train Tracking Stages

To perform the vibration detection, we first annotated vibrations in a recorded raw data file individually for the two different track conditions. These ground truth data are then used to train the PCA for feature reduction and the SVM for classification. An illustration of the vibration detection can be found in Figure 2.

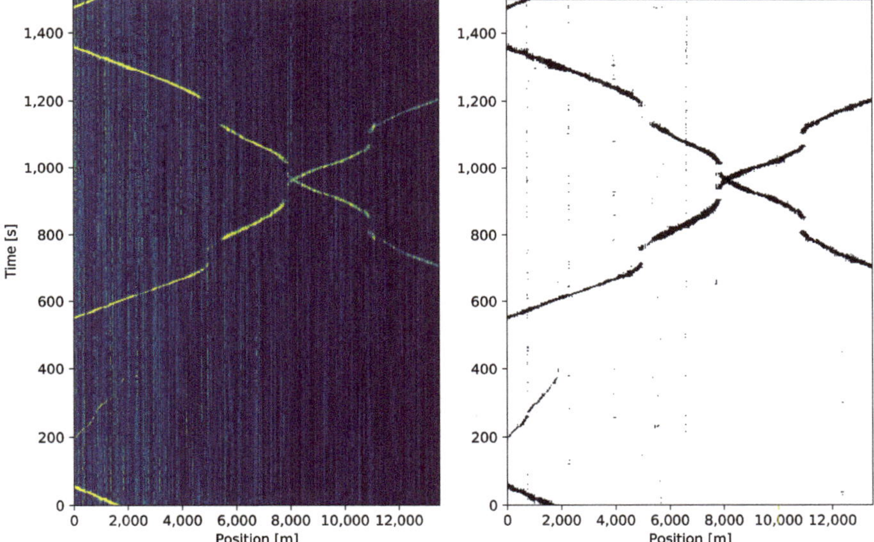

Figure 2. Left: Raw subsampled signal for visualization. **Right**: Result of the vibration detection where white pixels indicate vibration and black pixels indicate background. The gaps in the train tracks occur when a train stops since an electric train that is not moving does not emit any vibrations and is, therefore, not visible usind a Distributed Acoustic Sensing (DAS) system.

The first step of the tracking algorithm is the edge detection. In Figure 3, we give an example of the result of the edge detection from the first 35 s of the dataset shown in Figure 2. After the edge detection, the assignment and Kalman filtering are performed. For the same example dataset, we show the tracking result in Figure 4.

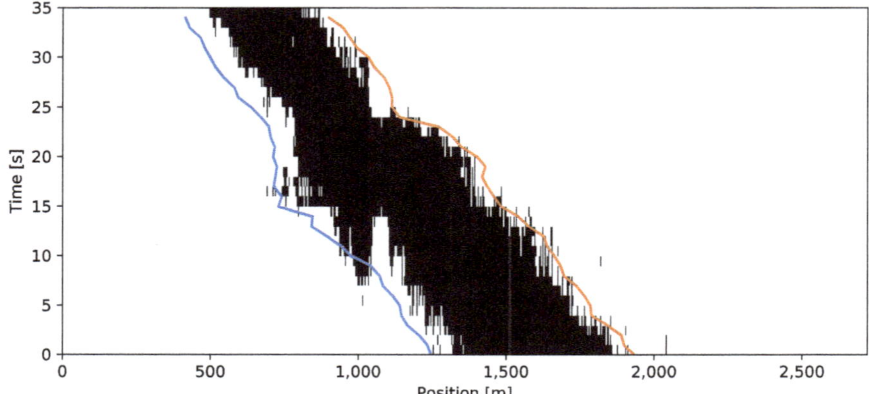

Figure 3. Leading and trailing edge of the object in blue and orange, respectively, plotted over the vibration detection results in black and white; note that the depicted signal consists of the first 35 s of the signal shown in Figure 2.

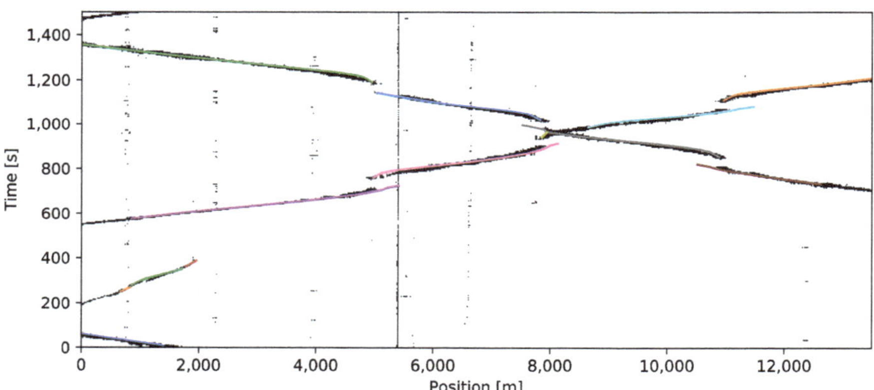

Figure 4. Train tracking result for an example dataset (colored lines), detection result in black and white; note that the depicted signal is the same as in Figure 2.

3.2. Evaluation of Tracking Accuracy

For two datasets encompassing around 3000 s of recording each, we have available ground truth positional data for two points along the track. The recording was done on an open track with the cable installed in a cable trench. The available ground truth is accurate to one second in time, which leads to an accuracy of around 30 to 40 m considering that this is the speed the objects move on the test infrastructure. The tracking errors in meters can be found in Table 2. The tracking error is in the range of 40 m, which is in the range of the available ground truth considering that both, the ground truth and the tracking algorithm, work on a one second basis that leads to a maximally two-second difference between the two systems, and the trains move at a speed of 30–40 m/s.

Table 2. Summarized performance of tracking algorithm for the two two short term recordings for single trains. Given that the ground truth and the tracking are both accurate down to one second, the evaluation is accurate down to 2 s. Considering the train speed, which is around 30–40 m/s, the errors are in the expected range.

Train ID	Error Signal 1 [m]	Error Signal 2 [m]
23468	49.64	44.20
29515	48.96	1.36
2330	51.68	38.76
73	79.56	37.40
23488	71.40	23.12
29535	44.88	17.00
23508	31.28	86.36
29555	31.28	19.72
103	64.60	57.12
90093	43.52	39.44
23528	49.64	64.60
29575	2.04	61.20
Average	47.37	40.86

3.3. *Evaluation of Tracking Reliability*

In this section, we will evaluate the tracking reliability based on two long term recording results. One is done in a tunnel where only one tube in one direction has been measured. The second test site is a standard track with the cable mounted directly on the rail. To evaluate the number of correctly tracked trains without the availability of ground truth from an alternative tracking system, the following steps were taken:

1. Identify the tracks that do not end at the end of the monitored cable section or in a station as potential tracking errors.
2. Exclude trains that stop on the open track as in that case the train ID is lost based on visual evaluation.
3. Exclude artifact tracks based on visual evaluation.
4. Count the total number of train tracks and the number of excluded tracks.

We collect the respective numbers of the correct and lost train tracks for two different test sites, where we were conducting long term measurements in Table 3. In the following subsections, we will describe the specific results for the two test sites.

Table 3. Summarized performance of tracking algorithm for the two monitored test sites.

	Test Site 1	Test Site 2
Cable stretch observed (meters)	13,600–25,900	0–8840
Cable stretch observed (segments)	20,000–38,000	0–13,000
Number of hours observed	477	538
Cable installation method	Cable trench in tunnel	Cable attached to rail on open track
Number of tracks	2174	1071
Number of correct tracks	2141	1059
% of correct trains	98%	99%

3.3.1. Test Site 1: Tunnel

The monitored section of the cable in the tunnel is from 13,600 m to 25,900 m relative distance from the interrogator device; therefore, we observe a long stretch of cable. This analysis shows that we reach 98% of correct tracks, respectively, for the two evaluated cable lengths. In this test site, no artifact tracks due to non-train-related noise was observed, meaning that the two percent incorrect tracks are attributed to lost tracks.

3.3.2. Test Site 2: Open track

This test site consists of an open track with a cable that is directly mounted on the rail, which leads to a very different scenario from the first test site. In our test period, we were not observing any lost tracks of moving trains on the open track. However, due to the cable installation method, we observed a higher amount of noise in the signal, which leads to a number of false positive detections of trains during the evaluation period. Of the 1071 total tracks, we found 12 to be artifacts, which leads to a total of 99% correct tracks. Generally, mounting the optical cable directly on the rail leads to more noise in the signal that is not caused by trains. This is not surprising since such a cable is exposed to natural elements, such as wind and rain. Furthermore, we observed that certain sections of the cable seem to be coupled, meaning that if a train enters such a section, all the cable segments in that section will vibrate. An illustration of this phenomenon, as well as of the increased noise in the signal, can be found in Figure 5. The coupling of fiber segments leads to decreased accuracy in estimating the train length because the detected leading and trailing edge do not coincide with the start and end of the train but, rather, with the start and end of the coupled segments.

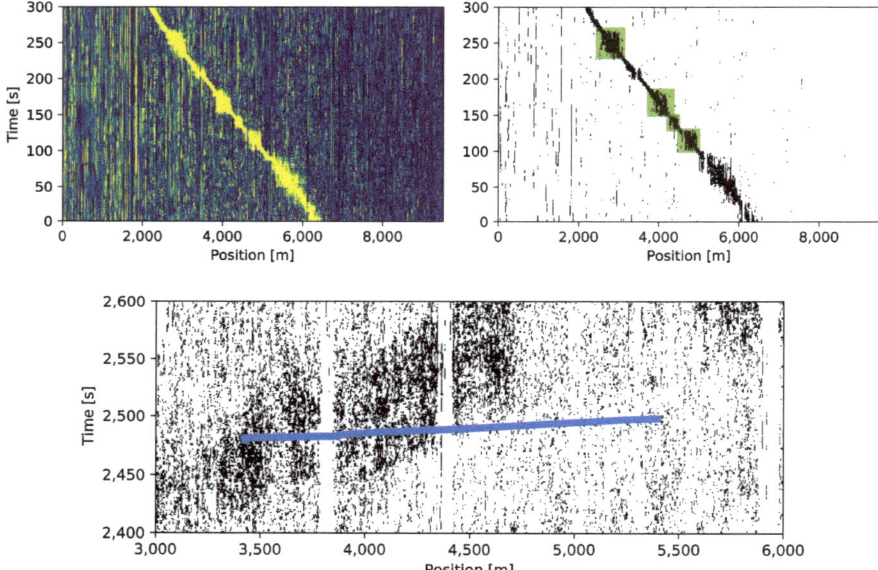

Figure 5. Top Left: Subsampled signal through band energy calculation. **Top Right**: Train Detection result for a train on test site 2; the green boxes indicate coupled cable segments; the background without train shows increased noise. **Bottom**: False positive train track due to false positive vibration detections.

3.3.3. Comparison with the Literature

The goal of this section is to discuss the presented results in comparison to the approaches discussed in the literature. The main difference between the algorithms described in the literature and our approach is the flexibility of our method through the use of machine learning.

The papers of Timofeev et al. [12,13] describe an energy based train vibration detection. According to our experience, this approach is not feasible for long stretches of monitored fiber with the recording device we used. The paper does not mention the validation data used for drawing the conclusions, nor does it mention a tracking algorithm. The accuracy of the proposed method with respect to positional accuracy is reported at 15 m. In comparison, we did not have a ground truth available with such a high accuracy.

The paper of Peng et al. [14] describes an algorithm which uses a relative variance for vibration detection. The validation data for the method are two trains recorded over 400 s. The paper does not mention any accuracy measurements for the proposed method, nor do the authors mention which tracking algorithm is used.

The earlier papers of some of the authors of this contribution [15,16] describe a similar algorithm with some important difference. The earlier versions of the algorithm used a filtering over time which introduced a delay. This has been changed in this contribution to Kalman filtering avoiding a time lag of the method. Furthermore, the edge detection has been refined to be more flexible, which is especially important for rail foot cables.

We can conclude that this is, to our knowledge, the most extensive study done on railway monitoring using DAS. In comparison to the other literature, we used flexible methods that allow for efficient re-calibration for different track conditions.

4. Discussion

In this paper, we presented a real-time train tracking algorithm that runs on the basis of one-second signals without delay. The algorithm is based on two main steps, i.e., the detection of vibrations and the tracking of trains in the signal. The performance of the algorithm was evaluated on two test sites, where one was in a tunnel with a standard cable trench and the other one was on an open track with the cable attached directly to the rail. In the tunnel, we measured a long distance which lead to occasionally lost tracks, especially on a stretch where the cable installation was sub-optimal for transferring vibrations from the train to the cable. On the open track with the cable mounted directly to the rail, we did not observe lost tracks during the test period, which is due to strong signals, because the observed cable length was considerably shorter than in the tunnel. Due to the increased noise with the cable installation method, we observed several artifact train tracks. Nevertheless, for both evaluated tracks we reached accurate tracks in more than 98% of the cases. We conclude from the results that the installation in a cable trench is advantageous for train tracking with a DAS system.

For two shorter recordings, we presented the positional error, which was around 40 m, which was also in the range of the available ground truth. For more accurate evaluations of the positional accuracy of the tracking, it would be important to get better ground truth data and perform a highly accurate calibration of the fiber cable.

For application in the railway sector, DAS shows great potential, especially when combined with other sensors for redundancy. In future research, we will evaluate methods for fusing different sensor modalities to increase the robustness of the tracking.

Author Contributions: The authors from NBG Fosa GmbH recorded the data. This involved the planning and calibration of the test sites, as well as understanding the signal behavior in different positions. The authors from AIT Austrian Institute of Technology GmbH developed the tracking algorithms and their implementation. Conceptualization, C.W. and A.P.; Data curation, G.N. and H.D.; Methodology, C.W.; Project administration, M.L. and H.G.; Software, A.P.; Supervision, M.L. and H.G.; Visualization, M.W.; Writing—original draft, C.W.; Writing—review & editing, M.L. and M.W. All authors have read and agreed to the published version of the manuscript.

Funding: This research was partially funded by the "Wiener Wirtschaftsgantur" through the research project "FOS—Real Time Methods" with the grand ID 1890985

Conflicts of Interest: The authors declare no conflict of interest.

References

1. Gao, S.; Dong, H.; Ning, B.; Zhang, Q. Cooperative Prescribed Performance Tracking Control for Multiple High-Speed Trains in Moving Block Signaling System. *IEEE Trans. Intell. Transp. Syst.* **2019**, *20*, 2740–2749. [CrossRef]

2. Hill, R.; Bond, L. Modelling moving-block railway signalling systems using discrete-event simulation. In Proceedings of the 1995 IEEE/ASME Joint Railroad Conference, Baltimore, MD, USA, 4–6 April 1995; pp. 105–111. [CrossRef]
3. Williams, C. The next ETCS Level? In Proceedings of the 2016 IEEE International Conference on Intelligent Rail Transportation (ICIRT), Birmingham, UK, 23–25 August 2016; pp. 75–79. [CrossRef]
4. Albrecht, T.; Luddecke, K.; Zimmermann, J. A precise and reliable train positioning system and its use for automation of train operation. In Proceedings of the 2013 IEEE International Conference on Intelligent Rail Transportation Proceedings, Beijing, China, 30 August–1 September 2013; pp. 134–139. [CrossRef]
5. Champavère, A. New OTDR Measurement and Monitoring Techniques. In Proceedings of the Optical Fiber Communication Conference (OFC 2014), San Francisco, CA, USA, 9–13 March 2014; p. W3D.1. [CrossRef]
6. Dean, T.; Hartog, A.; Papp, B.; Frignet, B. Fibre optic based vibration sensing: Nature of the measurement. In Proceedings of the 3rd EAGE Workshop on Borehole Geophysics, Athens, Greece, 19–22 April 2015.
7. Bao, X.; Chen, L. Recent Progress in Distributed Fiber Optic Sensors. *Sensors* **2012**, *12*, 8601–8639. [CrossRef] [PubMed]
8. Baldwin, C.S. Brief history of fiber optic sensing in the oil field industry. In Proceedings of the Fiber Optic Sensors and Applications XI, Baltimore, MD, USA, 5–9 May 2014; p. 909803. [CrossRef]
9. Juarez, J.C.; Maier, E.W.; Choi, K.N.; Taylor, H.F. Distributed fiber-optic intrusion sensor system. *J. Lightwave Technol.* **2005**, *23*, 2081. [CrossRef]
10. Peng, F.; Wu, H.; Jia, X.H.; Rao, Y.J.; Wang, Z.N.; Peng, Z.P. Ultra-long high-sensitivity Φ-OTDR for high spatial resolution intrusion detection of pipelines. *Opt. Exp.* **2014**, *22*, 13804–13810. [CrossRef] [PubMed]
11. Chen, C.; Chen, R.; Wei, F.; Wu, D. Experimental and application of spiral distributed optical fiber sensors based on OTDR. In Proceedings of the 2011 International Conference on Electric Information and Control Engineering, Wuhan, China, 15–17 April 2011; pp. 5905–5909. [CrossRef]
12. Timofeev, A.V.; Egorov, D.V.; Denisov, V.M. The Rail Traffic Management with Usage of C-OTDR Monitoring Systems. *World Acad. Sci. Eng. Technol. Int. J. Comput. Electr. Autom. Control Inf. Eng.* **2015**, *9*, 1492–1495.
13. Timofeev, A.V. Monitoring the Railways by Means of C-OTDR Technology. *Int. J. Mech. Aerosp. Ind. Mech. Eng.* **2015**, *9*, 634–637.
14. Peng, F.; Duan, N.; Rao, Y.J.; Li, J. Real-Time Position and Speed Monitoring of Trains Using Phase-Sensitive OTDR. *Photonics Technol. Lett.* **2014**, *26*, 2055–2057. [CrossRef]
15. Papp, A.; Wiesmeyr, C.; Litzenberger, M.; Garn, H.; Kropatsch, W. A real-time algorithm for train position monitoring using optical time-domain reflectometry. In Proceedings of the 2016 IEEE International Conference on Intelligent Rail Transportation (ICIRT), Birmingham, UK, 23–25 August 2016; pp. 89–93. [CrossRef]
16. Papp, A.; Wiesmeyr, C.; Litzenberger, M.; Garn, H.; Kropatsch, W. Train Detection and Tracking in Optical Time Domain Reflectometry (OTDR) Signals. In Proceedings of the German Conference on Pattern Recognition, Hannover, Germany, 12–15 September 2016; pp. 320–331.
17. Bradski, G. The OpenCV Library. *Dr. Dobb's J. Softw. Tools* **2000**. Available online: https://github.com/opencv/opencv/wiki/CiteOpenCV (accessed on 7 January 2020).

Article

Towards the Development of Rapid and Low-Cost Pathogen Detection Systems Using Microfluidic Technology and Optical Image Processing

Abdelfateh Kerrouche [1,*], Jordan Lithgow [1], Ilyas Muhammad [1] and Imed Romdhani [2]

[1] School of Engineering and the Built Environment, Edinburgh Napier University, Edinburgh EH10 5DT, UK; 40207656@live.napier.ac.uk (J.L.); muhammad.ilyas@napier.ac.uk (I.M.)

[2] School of Computing, Edinburgh Napier University, Edinburgh EH10 5DT, UK; i.romdhani@napier.ac.uk

* Correspondence: a.kerrouche@napier.ac.uk; Tel.: +44-(0)131-455-2586

Received: 29 February 2020; Accepted: 1 April 2020; Published: 7 April 2020

Abstract: Waterborne pathogens affect all waters globally and proceed to be an ongoing concern. Previous methods for detection of pathogens consist of a high test time and a high sample consumption, but they are very expensive and require specialist operators. This study aims to develop a monitoring system capable of identifying waterborne pathogens with particular characteristics using a microfluidic device, optical imaging and a classification algorithm to provide low-cost and portable solutions. This paper investigates the detection of small size microbeads (1–5 µm) from a measured water sample by using a cost-effective microscopic camera and computational algorithms. Results provide areas of opportunities to decrease sample consumption, reduce testing time and minimize the use of expensive equipment.

Keywords: pathogen detection; microfluidics; image processing; computational algorithms

1. Introduction

1.1. Water Borne Pathogen

Due to the increasing concern for water scarcity, it is expected that by 2025, half of the world's population will endure 'water-stressed' areas [1]. This is due to climate change affecting both the developing and developed world. Already, 1.1 billion people lack access to clean water and 2.7 billion have a water shortage for at least one month of the year, exposing waterborne diseases such as cholera, typhoid fever and hepatitis A [2]. As a result, approximately 525,000 children under the age of five die annually from diarrhea diseases alone. According to the World Health Organization (WHO), lower respiratory infections and diarrheal diseases are the top two causes of deaths in low-income countries in 2016. From this, 11–20 million people get an illness due to typhoid fever, and 128,000–161,000 people die from this annually [2]. Typhoid fever is caused by the Enterobacteriaceae Salmonella Typhimurium, which typically is around 2–5 µm in size. In addition to this, the developed world suffers from pathogens such as Cryptosporidium, which can survive months within waters [3,4]. This particular pathogen causes gastrointestinal and respiratory illnesses, although the death rate is much lower than typhoid fever. Detection protocols such as the U.S. Environmental Protection Agency (EPA) method 1623.1 [5] stipulate a procedure for *Cryptosporidium* detection. This method consists of several steps involving filtration, elution, centrifugation, immuno-magnetic separation (IMS) and staining with fluorescent dyes, followed by microscopic examination for the manual identification and enumeration of oocysts.

Bridle [6] has previously established that research has been conducted to identify advances in obtaining results of waterborne pathogen movements. In compliance with WHO, waterborne pathogens have to be carefully monitored and analyzed to implement the safety of water intake. Considering this, microfluidic devices (a set of micro-channels etched into a material e.g., glass [7]) can take a very small sample and conduct a process to analyses the safety of water through an experimental monitoring method.

1.2. Microfluidic Technology

Newly introduced technology is continuously being miniaturized, hence the attractiveness of microfluidic devices. Microfluidic technology was introduced around 20 years ago and is maturing as a technology, despite the initial lack of success on the market [8,9]. The device was influenced from microelectronic chips; however, the use and analysis of fluid movement was changed from the use of pathways for electrons. Microfluidic devices accommodate more mixers and separators rather than transistors and other microelectronic hardware [10]. This increases microfluidics applications to be developed to perform in applications, such as drug testing in nanosystems, biochemistry immunoassay and gene sequencing etc., to proceed at a faster, cheaper rate, whilst decreasing the sample and reagent consumptions [11]. From these benefits, other applications use the devices to provide aid in their success, such as cosmetic formulations, biology cell culture and energy plasma confinement [12]. The science behind microfluidic technology is to manipulate and control small levels of fluid from microliters (μL) to picoliters (pL) within small micrometer channels, allowing for the precision of experiments to be maximized and a higher probability of detection to be available. For the purpose of this paper, the main focus will be to detect and monitor 'acting' pathogens via a smart designed system. In the past few years, microfluidic systems have been intensely researched to advance applications in the fields of chemistry, biology, genomics, proteomics, pharmaceuticals, biodefense [13] emerging as a powerful useful technology. Large processes that once were carried out in bulky, slow, expensive equipment are now processed on a single miniaturized chip. Microfluidic chips are purposely designed to monitor the movement of cells, pathogens or nanoparticles within the microchannels, allowing easy detection and removal methods under controlled conditions. In addition, microchannels typically have a width of 100–800 μm which allows advantages of sample consumption to be reduced, whilst also increasing the speed of results. This allows increasing efficiency and an inexpensive process [12,13].

1.3. Microfluidic Flow Cytometry

Microfluidic flow cytometers techniques are one of the most powerful approaches for a high resolution of cell analysis. This technique was developed from a traditional flow cytometer, which aids in biological applications and clinic research [14]. Flow cytometers are popular due to the ability to sort, count and detect individual cells, while also measuring scientific characteristics and the handling of cell populations [14,15]. However, flow cytometers do not effectively detect particles and cells smaller than 0.5 μm in diameter via light scattering [16]. Light scattering manipulates singles out pathogens based on size and shape. Microfluidic flow cytometry technique aims to process a sample by 3 stages: external pumping force of the sample, focusing of particles within the sample and sorting of particles by separation. This allows particles to be manipulated and analyzed faster, whilst being less expensive than a conventional flow cytometer [17,18]. The limitations of a microfluidic flow cytometer are that, due to a small flow rate of the external pump force, it will consist of a non-steady flow, which contributes to installing extra fabrication steps and limited range of flow rate, increasing the production cost [19].

In addition, optical detection within activated flow cytometry for particle detection is well advanced for microfluidic techniques. For instance, fluorescent dye is the most common optical detection technique due to its high sensitivity, which illuminates targets within samples under a specific wavelength of a light source [6]. This allows reflection of the particle to be transmitted and detected from the optical point of a microscope. The development and integration of different controlling units (e.g., pumps, valves and switches) into microfluidic devices are challenging for an automated process. This is due to the response time needed to redirect the particles into the desired output channels [18]. New advances for optical detection and the separation of cells or particles in the modern flow cytometer defeat these drawbacks and reduce the bulkiness of the system. Integrating optical fibers and waveguides to within the device around the microfluidic channels allows results to be optimized and analyzed without the use of a microscope or camera, as Lin et al. [19] proved with the detection of Phalaenopsis pathogens using optical fibres.

2. Sample Preparation

2.1. Methodology

To monitor the movement of a waterborne pathogen, a micro-pump (The Aladdin single-syringe infusion) drives a sample of 3 mL through a microfluidic device. To simulate the movement of a pathogen inside the microfluidic channel, a polymer fluorescent green microsphere beads are used with a range size of 1–5 µm.

A USB microscopic camera (The Dino-Lite AM4515T8 Edge 700x~900) was set up and positioned at a perspective angle. The camera emits an ultraviolet light source, causing the fluorescent microspheres to reflect to the camera. As shown in Figure 1a, the camera was secured in a position using a clamp to obtain a perspective view of the microfluidic device. The camera's focus point was altered to output a clear view of the microfluidic channels on the software. Five hundred micrograms of polymer fluorescent green microspheres were diluted and mixed into a 50 mL test tube containing clean fluid (Propane-2-Ol, "IPA"). This allowed the sample to be withdrawn easily with the use of a BD Plastipak 3 mL max syringe. The tubing was connected to the input port of the microfluidic device. Tubes were connected to the output ports of the microfluidic device to release the wastage processed. These green polymer beads are used in this paper because they have the same size and the same color as the waterborne pathogen: Cryptosporidium. Figure 1b shows an image of a microbead inside the microfluidic channel.

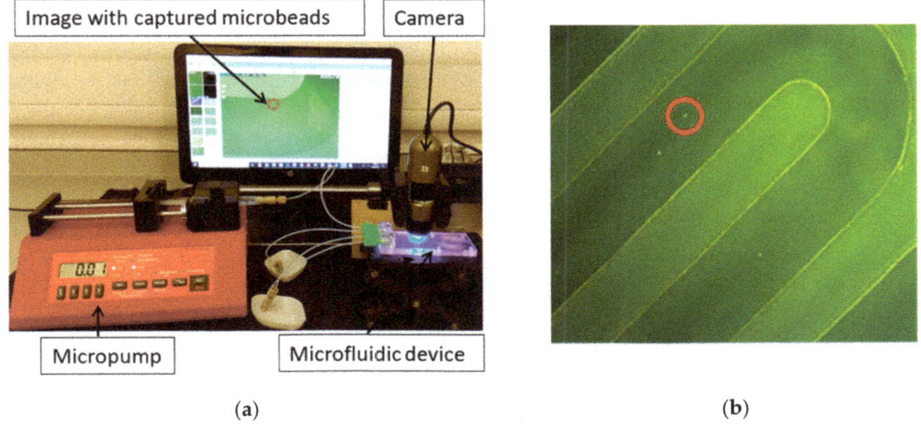

(a) (b)

Figure 1. (a) The system setup includes a syringe pump and a USB camera with 500–900 times zoom; (b) image detection of the microbeads of 1–5 µm within the microfluidic channel.

The tests were performed in the lab with controlled ventilation. The temperature was set to an ambient room temperature of 22 degrees Celsius, and the humidity was measured for an indoor level of 43%. These two parameters of ambient temperature and relative humidity can affect optical image quality taken by the camera and can also affect the fluid movement inside the channel, as the microfluidic is made from a clear plastic.

2.2. Velocity Tests

Velocity tests of microbeads have been carried out in this research to study the movement of beads inside the microfluidic channel. The micro-pump injected the sample at a controlled flow rate through the microchannel of a custom designed microfluidic chip via capillary tubes. Velocity measurements will help to identify the limitation of this type of microscopic camera in terms of the minimum frames per second needed for this application with a selective flow rate. This is important, as lower frame rates can result in choppy or a broken movement.

Flow rates were tested from 5–235 μL/min (in increments of 10 μL/min) to find the most probable flow paths for a pathogen cell. Videos from the camera stored the movement of the fluorescent bead within the microchannel at a rated flow rate in increments of 30 μL/min. A highlighted bead expresses the motion of which it travels at a specific flow rate. Flow rates between 5–35 μL/min resulted in no movement of particles therefore were not included. The movement is processed from the images using Matlab, allowing analysis of where most beads flow in terms of the 0.8 mm diameter channel, as shown in Figure 2.

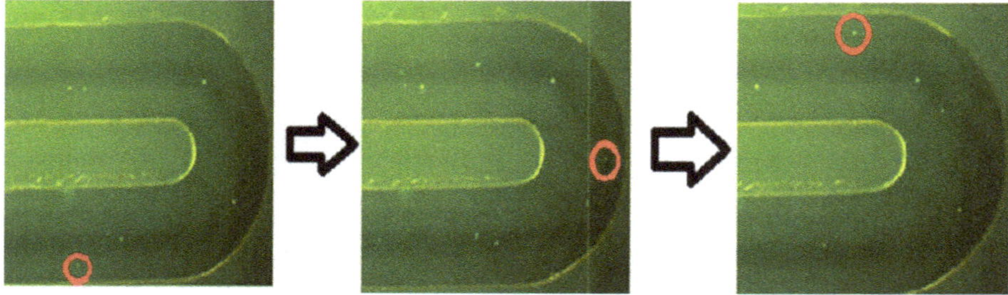

Figure 2. Detection of microsphere movement at 45 μL/min.

In order to find the velocity of the fluorescent beads at different flow rates, a distance is fixed at 250 mm for all flow rates tested using the software provided by the camera, as shown in Figure 3, The general equation for finding speed is Equation (1), where v is the velocity, d is the distance and t is the time. The time is found using the video recorded by the Dino Lite camera—this is also highlighted in Figure 3.

$$v = d/t \quad (1)$$

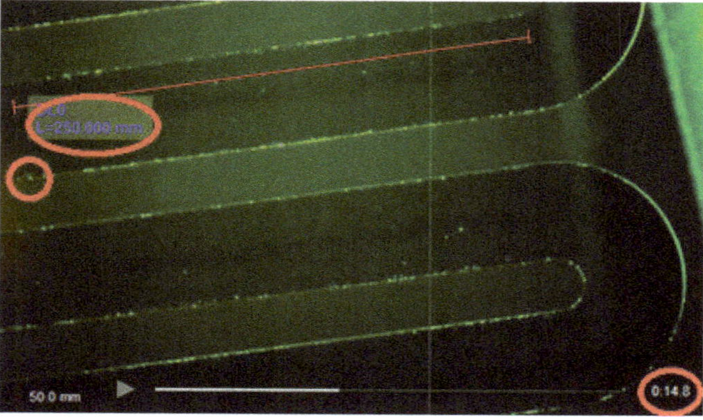

Figure 3. Dino Lite 2.0 software with evidence of fixed distance of 250 mm and a time interval of when microsphere travels the distance.

An average time is recorded from multiple tests for microspheres travelling in 3 sections and these are inside, middle and outside, as described in Figure 4.

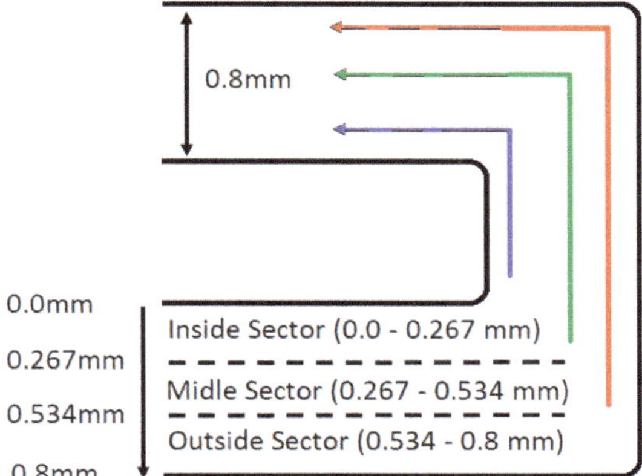

Figure 4. Drawing illustrates the three sections of the microfluidic channel.

3. Results

3.1. Fluorescent Microspheres Detection

The main objective of this study is to monitor the movement of waterborne pathogen cells using a microscopic camera and a light source integrated into a microfluidic device. The micro-pump is configured at a very low flow rate of 45 µL/min, to investigate the possibility of detection. After the operating equipment is correctly configured, 500 µg of a solid bead powder is diluted into a 50 mL test tube containing clean fluid (Propane-2-Ol, "IPA"). This master sample is mixed using ultrasonic and steel ball merging to gain a homogeneous bead distribution. This allowed samples to be withdrawn easily, by extracting 5 µL from the master sample and added it into 10 mL and 100 mL of IPA. The first images are obtained using the method in Section 2.1 to produce a high reflection rate of fluorescent microspheres. Therefore, a diluting process is performed to receive more realistic images.

Figure 5a demonstrates the mass of fluorescent green beads within the master sample. Figure 5b shows image with diluted sample, which contains a less dense concentration of fluorescent beads. Therefore, a further dilution process is performed to result the image in Figure 5c, where only a very small concentration of beads are detected, expressing a more realistic result of pathogen movement.

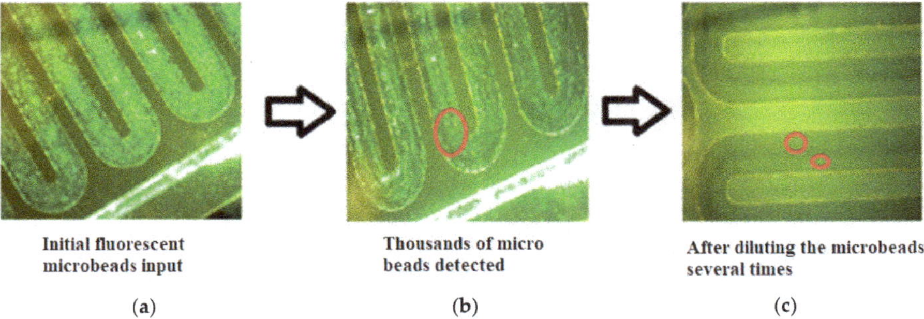

Figure 5. Diluting process images illustrated by (**a**–**c**).

3.2. Image Processing Using Computational Algorithm

The microsphere beads used in this research are inert, nontoxic and do not contaminate the device, so there is no requirement to move it to a dedicated biology laboratory. However, the manufacturer (Cospheric) does not specify the size distribution of the beads, hence the number of beads per milligram can vary (calculated with the given diameter range) between 1.2×10^7 and 1.5×10^9 (using ideal spheres and the material density 1.3 g/cc). This makes a measurement by volume and weight not suitable.

Different computational methods and algorithms for biological microscopic image analysis have demonstrated a comprehensive contribution for efficient and quantitative study of microorganism cells [20,21]. Experiments have been conducted in this paper to detect and quantify fluorescent microbeads (1–5 μm) in a measured sample of a clean water. Detailed steps are outlined as: (i) taking 2 μL of sample via micropipette; (ii) placing one drop on top of a microscopic glass slide; (iii) captureing an image of the full drop under microscopic camera; (iv) capturing zoomed images of the same drop frame by frame and finally (v) analyzing captured images by using computational algorithms using Matlab. The Circular Hough Transform (CHT) algorithm [22,23] based code in Matlab, is used to find circular/spherical shapes in the captured image as one frame of a drop, as discussed above. The CHT algorithm is used to mark/circle specific sized microbeads for quantification. CHT algorithm accuracy rate is affected by different parameters/factors, such as minimum and maximum radius of the target circular shape, object polarity in the sense of dark or bright, and computation method. The results using CHT algorithm are approximately 80%, as reported in Table 1.

Table 1. Values of different parameters used during different iterations.

Iteration	minRadius	maxRadius	Sensitivity	Total Circles	Total Beads	Accuracy
1	1	10	0.992	127	238	53.36%
2	1	30	0.992	153	238	64.29%
3	1	50	0.992	169	238	71.01%
4	1	100	0.992	170	238	71.43%
5	1	600	0.992	190	238	79.83%

Figure 6 shows resulting images of different iterations, which refer to the number of runs of the CHT algorithm applied in each image capture. First, the image is converted to grayscale, as color images are not giving a good segmentation. Then, minimum and maximum radius are defined as 1 and 600 pixels respectively. Finally, the object polarity is set to dark.

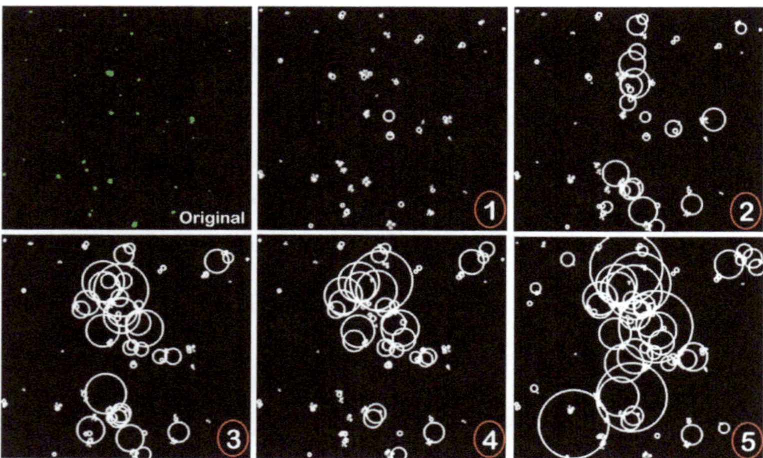

Figure 6. Original image and resulting images of different iterations, as shown in Table 1.

3.3. Microsphere Detection with Change of Flow Rate

Table 2 illustrates a sample of data collected in terms of number of beads over a fixed period of time (30 s). This time is fixed by Matlab to count the number of the total beads passed within the microfluidic channel, as shown in Figure 3. The data shown in Table 2 demonstrates an overview of the highest probable chances microspheres are going to travel inside the channel at specific flow rates. The evidence of the movement detection allows an analysis of where most beads flow in terms of the 0.8 mm diameter channel displayed in Table 2. The microfluidic channel diameter is then divided into two sections on Table 2 (outside to middle, middle to inside) to demonstrate where the bead is more likely to travel.

Table 2. Number of beads counted after 30 s. Terms of specific flow rates against specific areas of microfluidic channels at the same time for each test.

Diameter (mm)	45 µL/min	75 µL/min	105 µL/min	135 µL/min	165 µL/min	195 µL/min	225 µL/min
0.7–0.8	2	1	1	1	0	1	0
0.6–0.7	0	1	4	0	1	0	1
0.5–0.6	1	0	4	0	0	0	0
0.4–0.5	2	2	2	0	0	0	0
0.3–0.4	1	0	2	1	2	1	1
0.2–0.3	0	2	0	3	0	0	1
0.1–0.2	0	2	1	3	1	1	0
0.0–0.1	0	0	1	1	2	4	3

As it could be seen in Table 2, between the outside of the meander 0.8 mm to the middle sector at 0.4 mm, there is a higher probability that the bead will travel between the flow rates of 45–105 µL/min.

Figure 7 shows results from the data collected over three tests for each flow rate to determine the recovery rate, which is the number of beads detected in the region in question over the first 100 beads detected within the channel. As demonstrated in Figure 4, 69% of beads can be recovered from the sample between the outside of the meander 0.8 mm to the middle sector at 0.4 mm. On the other side, from the middle of the channel at 0.4 mm to the inside 0.0 mm the probability that the bead will flow between these points is at the flow rate between 135–225 µL/min. This provides a recovery rate of 88%, conveying that the increase of flow rate increases the consistency of where the microsphere will travel. However, this will not allow an easy detection, as the bead's velocity will increase too, as explained in Section 3.4.

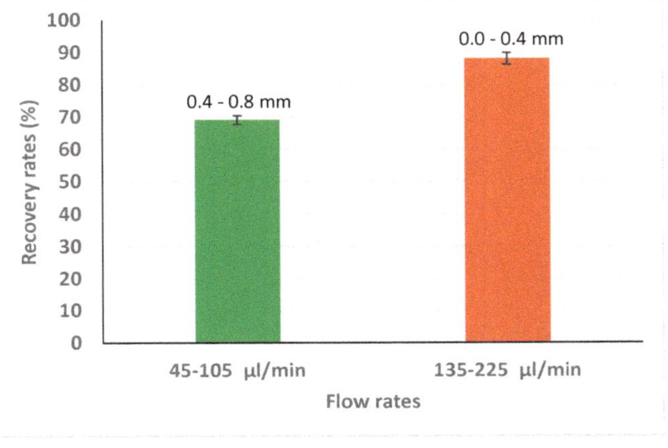

Figure 7. Recovery rates within the microfluidic channel.

From these results, fabrication of a microchannel can be inserted for the pathogen to be removed from the sample. Being able to do this will allow microfluidic technology to essentially replace most filtration methods used to remove waterborne pathogen such as EPA *Cryptosporidium* detection method [5].

3.4. Velocity Measurements

As this method detects the movement of a fluorescent bead, it allowed further testing to find a single flow rate, where a pathogen cell will consistently flow on the same path. This provides an opportunity to locate the pathogen within the sample, which helps with placing the camera at the right position for detection. The microfluidic channel diameter is divided into three sections to demonstrate where the bead is more likely to travel, as shown in Figure 4.

Figure 8 illustrates the average velocity of the microsphere travelling at the representing flow rate within the microchannel. Results demonstrate that the gradual increase of flow rate up to 145 μL/min slightly increases the velocity of the microsphere. However, after this flow rate, the velocity of the beads increased significantly in all three parts of the microfluidic channel. This is due to the extremely high velocity of the microsphere that is unable to be detected by the camera—and only some 'slower' microspheres could be detected as altering the average velocity.

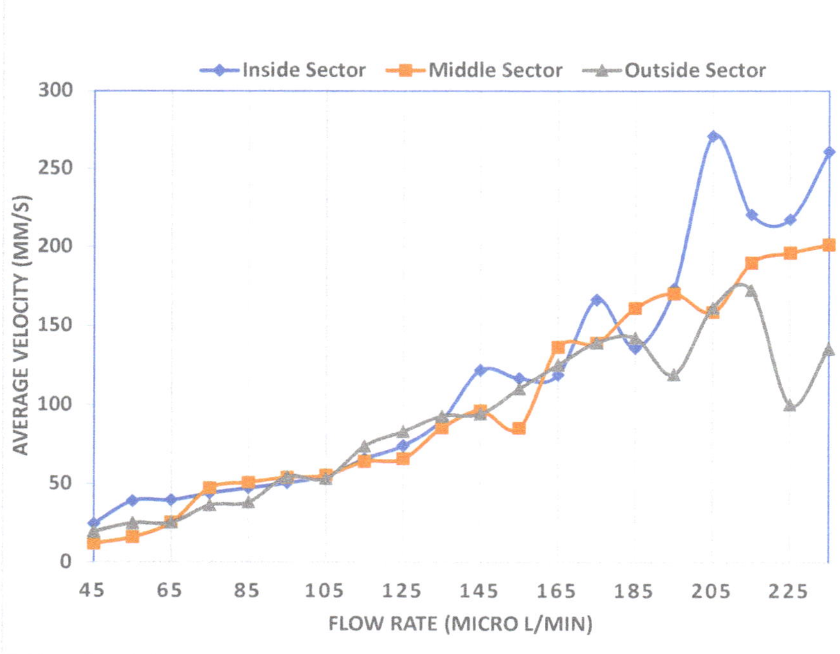

Figure 8. Gradual increase of velocity for the microsphere travelling within the microfluidic channel with respect of flow rate increase.

4. Discussion and Conclusions

This research aims to design a smart monitoring system to detect a waterborne pathogen cell using microfluidic technology and a cost-effective microscopic camera. The sample was driven through the designed microfluidic channel at a selected flow rate on a micro-pump. The pathogen movement was tracked by a high-quality camera at a perspective angle. With repeated experiments and alterations of the input flow rate, the output is hypothesized to find a consistent 'track line' that the pathogen travels within the channel, so a separation channel can be fabricated to separate the pathogen from the sample.

The prominent challenges for the pathogen detection methods in practice are: (i) physical characteristics of different groups of pathogens, (ii) large volume of water containing low concentration of target pathogens and (iii) sample preparation protocols. The key advantages for the results delivered by this new method allows sample consumption to be decreased, the reduction of testing time and reducing the use of expensive equipment, providing an attractive cost. This will benefit existing methods such as Flow Cytometry and EPA Cryptosporidium detection approach by replacing procedures with a less complicated process that does not need specialized persons to operate. This method provides new possibilities for further analysis and research to improve results on finding precise intersection points for high recovery rates of waterborne pathogens with different sizes such as Giardia or *Escherichia coli* (*E. coli*). This provides vital intersection points for a microchannel to be fabricated to remove the waterborne pathogen.

A key objective for this paper is to investigate the design of an intersection point on the microchannel to separate the waterborne pathogen. To achieve this, the micro pump flow rates were changed slightly to find a consistent path of which the microsphere will travel. Figure 3 displays the movement of microspheres around a meander. This gives a general observation that the slower the flow rate is (45 µL/min), the microsphere movement is towards the outside of the microchannel. Comparing this to 225 µL/min flow rate, it portrays that the movement of the microsphere is on the inside of the microchannel. From the videos taken by the camera, Table 2 gathered data from the movement of the microsphere. This shows that, at slower flow rates of 45–105 µL/min, the microsphere has a higher probability of 69% to be recovered at 0.4–0.8 mm of the microchannel. Whereas at faster flow rates of 135– 225 µL/min, the recovery rate has a probability of 88% at 0.0–0.4 mm of the microchannel. However, as stated in Section 3.4, due to the fast velocities of microspheres at these flow rates, the detection became more difficult. This could slightly alter the probability rate of recovery, due to the possibility of some microspheres not been detected.

The detection accuracy of the beads with the CHT algorithm used in this paper can be improved using different platforms such as ImageJ or Python. The processed image of a full drop can be performed using sequential steps to study different parameters such as target object size, the distance between two objects and threshold.

Microfluidic chips can be manufactured using different types of materials such as PDMS (PolyDiMethylSiloxane), Silicon, Glass or PMMA (PolyMethylMethAcrylate). These materials can be investigated to improve the images taken by the camera in order to improve recovery rates. Alongside this, the system described in this paper can be reduced even more into a single automated unit, to be able to optimize results that requires no manual intervention.

From all factors discussed in this paper, the idea of detecting movement of a waterborne pathogen using a cost-effective camera is very promising. This benefits engineering problems by minimizing the need for expensive laboratories and equipment to commence procedures on tackling waterborne pathogens.

Author Contributions: Conceptualization, A.K. and J.L.; methodology, A.K.; software, A.K. and I.M.; validation, A.K. and I.R.; formal analysis, A.K.; investigation, A.K; resources, A.K.; data curation, A.K.; writing—original draft preparation, A.K.; writing—review and editing, A.K.; visualization, A.K.; supervision, A.K.; project administration, A.K.; funding acquisition, A.K. and J.L. Please turn to the CRediT taxonomy for the term explanation. Authorship must be limited to those who have contributed substantially to the work reported. All authors have read and agreed to the published version of the manuscript.

Funding: This research received no external funding.

Acknowledgments: The authors would like to acknowledge the support from Helen Bridle, Heriot Watt University.

Conflicts of Interest: The authors declare no conflict of interest.

References

1. World Health Organization (WHO). "Drinking Water". 2018. Available online: https://www.who.int/topics/drinking_water/en/ (accessed on 29 February 2020).
2. World Health Organization (WHO). "Top 10 Causes of Death". 2018. Available online: https://www.who.int/news-room/fact-sheets/detail/the-top-10-causes-of-death (accessed on 29 February 2020).
3. Cryptosporidium: Drinking Water Health Advisory. EPA-822-R-01-009. Available online: https://www.epa.gov/sites/production/files/2015-10/documents/cryptosporidium-report.pdf (accessed on 29 February 2020).
4. Kerrouche, A. Megasonic sonication for cost-effective and automatable elution of cryptosporidium from filters and membranes. *J. Microbiol. Methods* **2015**, *118*, 123–127. [CrossRef] [PubMed]
5. Method 1623.1, 2012. *Method 1623.1: Cryptosporidium and Giardia in Water by Filtration/IMS/FA*; Environmental Protection Agency: Washington, DC, USA, 2012.
6. Bridle, H. *Overview of Waterborne Pathogens, Waterborne Pathogens: Detection Methods and Applications*, 1st ed.; Academic Press: Cambridge, MA, USA, 2014.
7. Whitesides, G.M. The origins and the future of microfluidics. *Nature* **2006**, *442*, 368–373. [CrossRef] [PubMed]
8. Tian, W.C.; Finehout, E. *Microfluidic Diagnostic Systems for the Rapid Detection and Quantification of Pathogens, Microfluidics for Biological Applications*; Springer: Boston, MA, USA, 2008.
9. Ghosal, S. Microfluidics. In *Encyclopaedia of Complexity and Systems Science*; Meyers, R., Ed.; Springer: New York, NY, USA, 2009.
10. Hansen, C.B.; Kerrouche, A.; Tatari, K.; Rasmussen, A.; Ryan, T.; Summersgill, P.; Desmulliez, M.P.Y.; Bridle, H.; Albrechtsen, H.J. Monitoring of drinking water quality using automated ATP quantification. *J. Microbiol. Methods* **2019**, *165*, 105713. [CrossRef] [PubMed]
11. Chen, Q.L.; Cheung, K.L.; Kwan, Y.W.; Kong, S.K.; Ho, H.P. A centrifugal microfluidics platform for potential application on immobilization-free bead-based immunoassays. *Appl. Mech. Mater.* **2013**, *289*, 39–44. [CrossRef]
12. Bridle, H.; Miller, B.; Desmulliez, M.P.Y. Application of microfluidics in waterborne pathogen monitoring: A review. *Water Res.* **2014**, *55*, 256–271. [CrossRef] [PubMed]
13. Maxine, Y. A review of state-of-the-art microfluidic technologies for environmental applications: Detection and remediation. *Glob. Chall.* **2019**, *3*. [CrossRef]
14. Gone, Y.; Na Fan, X.Y.; Bei, P.; Hai, J. New advances in microfluidic flow cytometry. *Electrophoresis* **2019**, *40*, 1212–1229.
15. Lukes, J.R.; Hong, J. *Flow Cytometer Lab-on-a-Chip Devices, Encyclopaedia of Microfluidics and Nanofluidics*; Springer: New York, NY, USA, 2015.
16. Wu, M.; Piccini, M.; Koh, C.Y.; Lam, K.S.; Singh, A.K. Single cell microRNA analysis using microfluidic flow cytometry. *PLoS ONE* **2013**, *8*. [CrossRef] [PubMed]
17. Piyasena, M.E.; Graves, S.W. The intersection of flow cytometry with microfluidics and microfabrication. *Lab Chip* **2014**, *14*, 1044. [CrossRef] [PubMed]
18. Ditterich, P.S.; Malik, F.; Schwille, P. *Cytometry on Microfluidic Chips, Microfluidic Technologies for Miniaturized Analysis Systems*; Springer: Dortmund, Germany, 2007; Chapter 14.
19. Lin, C.L.; Chang, W.H.; Wang, C.H.; Lee, C.H.; Chen, T.Y.; Jan, F.J.; Lee, G.B. A microfluidic system integrated with buried optical fibers for detection of Phalaenopsis orchid pathogens. *Biosens. Bioelectron.* **2015**. [CrossRef] [PubMed]
20. Meijering, E.; Dzyubachyk, O.; Smal, I. Chapter nine—Methods for cell and particle tracking. *Methods Enzymol.* **2012**, *504*, 183–200. [CrossRef] [PubMed]
21. Li, C.; Wang, K.; Xu, N. A survey for the applications of content-based microscopic image analysis in microorganism classification domains. *Artif. Intell. Rev.* **2019**, *51*, 577–646. [CrossRef]
22. Kis, B. Counting Bacteria Colonies Based on Image Processing Methods. In Proceedings of the Medical Technologies Congress (TIPTEKNO), Selçuk, Turkey, 3–5 October 2019; pp. 1–4. [CrossRef]
23. Bonteanu, P. A new pupil detection algorithm based on circular Hough transform approaches. In Proceedings of the IEEE 25th International Symposium for Design and Technology in Electronic Packaging, Cluj-Napoca, Romania, 23–26 October 2019; pp. 260–263. [CrossRef]

Article

Dependence-Analysis-Based Data-Refinement in Optical Scatterometry for Fast Nanostructure Reconstruction

Zhengqiong Dong [1], Xiuguo Chen [2,*], Xuanze Wang [1], Yating Shi [2,*], Hao Jiang [2] and Shiyuan Liu [2]

1 Hubei Key Laboratory of Manufacture Quality Engineering, Hubei University of Technology, Wuhan 430068, Hubei, China; dongzhq@mail.hbut.edu.cn (Z.D.); wangxz@mail.hbut.edu.cn (X.W.)
2 State Key Laboratory for Digital Manufacturing Equipment and Technology, Huazhong University of Science and Technology, Wuhan 430074, Hubei, China; hjiang@hust.edu.cn (H.J.); shyliu@hust.edu.cn (S.L.)
* Correspondence: xiuguochen@hust.edu.cn (X.C.); yatingshi@hust.edu.cn (Y.S.)

Received: 16 August 2019; Accepted: 27 September 2019; Published: 30 September 2019

Abstract: Optical scatterometry is known as a powerful tool for nanostructure reconstruction due to its advantages of being non-contact, non-destructive, low cost, and easy to integrate. As a typical model-based method, it usually makes use of abundant measured data for structural profile reconstruction, on the other hand, too much redundant information significantly degrades the efficiency in profile reconstruction. We propose a method based on dependence analysis to identify and then eliminate the measurement configurations with redundant information. Our experiments demonstrated the capability of the proposed method in an optimized selection of a subset of measurement wavelengths that contained sufficient information for profile reconstruction and strikingly improved the profile reconstruction efficiency without sacrificing accuracy, compared with the primitive approach, by making use of the whole spectrum.

Keywords: optical scatterometry; inverse problem; profile reconstruction; dependence analysis; data refinement

---

1. Introduction

Nano-metrology is the only effective method that ensures the reliability and consistency of nano-manufacturing [1–3]. Compared with other techniques such as scanning electron microscopy (SEM), atomic force microscopy (AFM) [4], and near-field scanning optical microscope (NSOM) [5], optical scatterometry [6,7], also known as optical critical dimension metrology or optical critical dimension (OCD) metrology, is more suitable for monitoring, assessing, and optimizing the nano-manufacturing processes due to its advantages of being non-contact, non-destructive, low in cost, and easy to integrate, etc. Recently, optical scatterometry has been applied in many fields with great success, such as the process control for back-end-of-the-line (BEOL) [8], the in-chip critical dimension (CD), overlay metrology [9], and in-situ measurement of pattern reflow in nanoimprinting [10].

In general, optical scatterometry involves two procedures: the forward optical modeling of sub-wavelength structures and the reconstruction of structural profiles from the measured signatures [11]. Here, the general term signatures means the scattered light information from the diffractive grating structure, which can be in the form of reflectance, ellipsometric angles, Stokes vector elements, or Mueller matrix elements. The forward optical model describes the light-nanostructure interaction by solving the complex Maxwell's equations. There are many reliable forward-modeling techniques such as the rigorous coupled-wave analysis (RCWA), the finite element method (FEM), the boundary element method (BEM), or the finite-difference time-domain (FDTD) method. The profile reconstruction process conducted under a fixed measurement configuration is an inverse problem with the objective of optimizing a set of floating profile parameters (e.g., CD, sidewall angle, and height) whose theoretical signatures can best match the measured ones through regression analysis or library search [12–14]. The measurement configuration is defined as a combination of specially selected wavelengths and incident and azimuthal angles.

As a typical model-based method, abundant measured data is usually collected under the measurement configurations with fixed incidence and azimuthal angles, but a wide waveband for spectroscopic scatterometry, or with fixed measurement wavelength and azimuthal angle, but incidence angular range for angle-resolved scatterometry. Taking the inverse structural profile reconstruction in spectroscopic scatterometry as an example, the measurement sensitivities of unknown profile parameters are affected by the fixed incident and azimuthal angles, which can be improved by traversing their possible values; the size of fitting data usually directly depends on the number of selected wavelength points within the waveband. A large sized measured data set should be used to guarantee the fidelity of the fitting results, but on the other hand, too much redundant information will significantly degrade the efficiency during the inverse problem solving process due to the frequent iteration of the forward model. Moreover, the employment of the redundant information, which is insensitive to the measurands in solving inverse problems, leads to the loss in precision as well. Thus, an appropriate measurement configuration with high sensitivity and a small amount of fitting data will benefit the measurement with both precision and speed.

Many instrument optimization and spectrum denoising methods have been proposed with the objective of improving the quality of measured data [15–17]. As the measurement precision mainly depends on the sensitivity of model parameters, several approaches based on local sensitivity analysis (LSA) also have been developed to determine optimal incident and azimuthal angles for spectroscopic scatterometry [18–21]. We have previously provided an optimization method based on global sensitivity analysis (GSA) to determine an optimal combination of the fixed incident and azimuthal angles corresponding to the best measurement precision [22]. However, the present measurement-configuration optimization methods only focus on the improvement of measurement precision, the speed of the reconstruction process has not been considered. Usually, a vector consisting of several data points at multiple measurement wavelengths or incidence angles is calculated by a forward model to match the measured spectrum. The forward model's computing time linearly depends on the number of selected data points and it will be called frequently, therefore, it is highly desirable to find an appropriate strategy that can select the measurement configurations containing sufficient information for structural parameter reconstruction to enhance the measurement speed without sacrificing the accuracies, especially for the nanostructures whose forward model is very complex and time-consuming.

In this paper, we propose to refine the measured signatures by identifying and removing the measurement configurations with redundant information based on dependence analysis before the reconstruction process. As for the mathematical inverse problem, Twomey's pioneer work on atmospheric measurements has demonstrated that in a chosen set of data the most linearly independent measurements contain all the information needed for an inverse problem solution [23–25]. Then, Assaad estimated additional reflectance values using the several acquired independent measurements to obtain a larger set of measurement data in optical scatterometry, which is desired to reduce the uncertainty range of the parameter estimates [26]. Inspired by these analyses, we conducted information content analysis of the measured signatures based on dependence-analysis theory for optical scatterometry. The analysis was conducted by the eigen-decomposition of the Jacobian matrix $\mathbf{JJ}^{\mathrm{T}}$ in the linear model to allocate the most independent measurements in the acquired data set. The few remaining independent measurements were then adopted to reduce the reconstruction time, which is crucial for real-time and in-process applications.

It should be noted that this paper does not intend to discuss the existence and uniqueness of the inverse problem in optical scatterometry with respect to measurement conditions, which is a tough and still open issue in mathematics [27–29]. The main purpose of this paper is to provide an effective approach to choose a subset of measurement configurations from the whole optical wavelength spectrum or angular range that contains sufficient information for efficient structural profile reconstruction without sacrificing accuracy. The selected subset of measurement configurations may not be unique; its size may not be the smallest, but it should provide a higher reconstruction accuracy than its randomly selected counterparts and a higher reconstruction efficiency than the primitive approach that analyzes the whole optical wavelength spectrum or angular range.

2. Method

The inverse problem in optical scatterometry is described as an objective to minimize a least-square function, which can be generally expressed as:

$$\chi^2 = \sum_{i=1}^{N}\left\{[\mathbf{y}_i - \mathbf{f}(\mathbf{x},\ \mathbf{a}_i)]^{\mathrm{T}}\mathbf{w}_i[\mathbf{y}_i - \mathbf{f}(\mathbf{x},\ \mathbf{a}_i)]\right\}, \tag{1}$$

where, $\mathbf{x} = [x_1, x_2, \ldots, x_m]^{\mathrm{T}}$ represents a column vector of m profile parameters (e.g., CD, sidewall angle, and height) and $\mathbf{f}(\mathbf{x},\ \mathbf{a}_i)$ is a column vector containing the calculated signature according to the forward optical model under the ith measurement configuration $\mathbf{a}_i$. The measurement configuration $\mathbf{a}_i$ is a combination of fixed azimuthal angle φ, fixed incident angle θ, and the ith measurement wavelength λ_i for spectroscopic scatterometry, namely $\mathbf{a}_i = [\varphi, \theta, \lambda_i]^{\mathrm{T}}$, and is a combination of fixed azimuthal angle φ, fixed measurement wavelength λ, and the ith incident angle θ_i for angle-resolved scatterometry, namely, $\mathbf{a}_i = [\varphi, \theta_i, \lambda]^{\mathrm{T}}$. Correspondingly, the N in Equation (1) denotes the number of measurement configurations, which can be the number of measurement wavelengths for spectroscopic scatterometry or the number of incidence angles for angle-resolved scatterometry. The column vector $\mathbf{y}_i = [y_{i1}, y_{i2}, \ldots, y_{il}]^{\mathrm{T}}$ consists of the corresponding measured signature under the same measurement configuration $\mathbf{a}_i$. Here, the value of l depends on the specific types of the measured signature. If the measured signature is reflectance, $l = 1$, while if the measured signature is a Mueller matrix, $l = 16$ (or $l = 15$ for the normalized Mueller matrix). The weighting matrix $\mathbf{w}_i$ is a $l \times l$ positive definite matrix, which is usually chosen to be the inverse of the covariance matrix of the measured signature under the ith measurement configuration $\mathbf{a}_i$. In this case, Equation (1) relates to the commonly used multivariate chi-square statistics χ^2. Writing the right side of Equation (1) in a matrix expression, the solution $\hat{\mathbf{x}}$ of the inverse problem can be obtained by:

$$\hat{\mathbf{x}} = \arg\min_{\mathbf{x}\in\Omega}\left\{[\mathbf{y} - \mathbf{f}(\mathbf{x},\ \mathbf{a})]^{\mathrm{T}}\mathbf{w}[\mathbf{y} - \mathbf{f}(\mathbf{x},\ \mathbf{a})]\right\}, \tag{2}$$

here, Ω is the associated profile parameter domain.

The use of χ^2 is built on top of the belief that the measurement errors are normally distributed with zero mean, namely, the measured signature $\mathbf{y}$ can be expressed as:

$$\mathbf{y} = \mathbf{f}(\hat{\mathbf{x}},\ \mathbf{a}) + \varepsilon, \tag{3}$$

where ε is a vector of multiple random and independent variables with each element subjected to a normal distribution with zero mean. Assuming a profile parameter vector $\mathbf{x}$ is close enough to the true or nominal parameter vector $\mathbf{x}^*$ under a measurement configuration $\mathbf{a}_i$, and the function $\mathbf{f}(\mathbf{x}, \mathbf{a}_i)$ is sufficiently smooth, then the function value $\mathbf{f}(\mathbf{x}, \mathbf{a}_i)$ can be expanded in the vicinity of $\mathbf{x}^*$ using the first-order Taylor expansion formulation [30]:

$$\mathbf{f}(\mathbf{x},\ \mathbf{a}_i) \approx \mathbf{f}(\mathbf{x}*,\ \mathbf{a}_i) + \mathbf{J}(\mathbf{x}*,\ \mathbf{a}_i)\cdot\Delta\mathbf{x}, \tag{4}$$

where $\mathbf{J}(\mathbf{x}^*, \mathbf{a}_i)$ is the Jacobian matrix with respect to $\mathbf{x}$ at $\mathbf{x} = \mathbf{x}^*$ and $\Delta\mathbf{x}$ represents the error in $\mathbf{x}$ and is given by $\Delta\mathbf{x} = \mathbf{x} - \mathbf{x}^*$. The differences between the measured signatures and the expected values can be expressed as:

$$\Delta\mathbf{y}_i \approx \mathbf{J}(\mathbf{x}*, \mathbf{a}_i) \cdot \Delta\mathbf{x} + \varepsilon_i. \tag{5}$$

To determine the degree of independence among N measurements, Equation (5) is multiplied by an arbitrary factor η_i and is summed over all i, to give:

$$\sum_{i=1}^{N} \eta_i \Delta\mathbf{y}_i = \sum_{i=1}^{N} \eta_i \mathbf{J}(\mathbf{x}*, \mathbf{a}_i) \cdot \Delta\mathbf{x} + \sum_{i=1}^{N} \eta_i \varepsilon_i, \tag{6}$$

where, $\sum\limits_{i=1}^{N} \eta_i^2 = 1$. Then, the difference $\Delta\mathbf{y}_j$ for a certain measurement configuration j can be expressed as [23]:

$$\Delta\mathbf{y}_j = -\eta_j^{-1} \sum_{i \neq j} \eta_i \Delta\mathbf{y}_i + \eta_j^{-1} \left\{ \sum_{i=1}^{N} \eta_i \mathbf{J}(\mathbf{x}*, \mathbf{a}_i) \cdot \Delta\mathbf{x} + \sum_{i=1}^{N} \eta_i \varepsilon_i \right\}. \tag{7}$$

The first term of Equation (7) on the right side is completely dependent on the other differences and represents the predictable part of $\Delta\mathbf{y}_j$; the bracketed expression on the right side is unpredictable and independent of the other measurements. When the first term in the bracketed expression does not exceed the second random measurement error part, Equation (7) provides a way for predicting $\Delta\mathbf{y}_j$ from other measurements, in other words, the $\Delta\mathbf{y}_j$ is redundant information and can be obtained by others; this is based on:

$$\left| \sum_{i=1}^{N} \eta_i \mathbf{J}(\mathbf{x}*, \mathbf{a}_i) \cdot \Delta\mathbf{x} \right| \leq \left| \sum_{i=1}^{N} \eta_i \varepsilon_i \right|. \tag{8}$$

It follows that the smaller the left term of Equation (8), the better the prediction. The question of mutual independence can thereby be examined by considering the minimal value of $\left|\boldsymbol{\eta}^{\mathrm{T}} \cdot \mathbf{J}(\mathbf{x}*, \mathbf{a})\right|^2$, which can be achieved by a suitable selection of η_i. Here, $\boldsymbol{\eta}$ is a N dimensional vector consisting of elements η_i (i = 1, 2, ..., N).

According to the Schwartz inequality, the minimum value of $\left|\boldsymbol{\eta}^{\mathrm{T}} \cdot \mathbf{J}(\mathbf{x}*, \mathbf{a})\right|^2$ is equal to the smallest $\zeta_{\min}$ of the eigenvalues of the matrix $\mathbf{J}(\mathbf{x}*, \mathbf{a}) \cdot \mathbf{J}^{\mathrm{T}}(\mathbf{x}*, \mathbf{a})$ when the η_i values are equal to the N components of the corresponding eigen-vector. If the condition by the Equation (8) holds, the difference $\Delta\mathbf{y}_i$ that corresponds to the smallest η_i of the eigen-vector for $\zeta_{\min}$ is the most linearly dependent. Consequently, the value of $\Delta\mathbf{y}_i$ can be calculated by the first term on the right side of Equation (7) from the other measurement differences with no loss of information. Therefore, we can identify the redundant measurement configuration by:

$$\mathbf{a}_i^{\text{redundant}} = \underset{\mathbf{a} \in \mathrm{P}}{\operatorname{argmin}} [\eta(\mathbf{a}_i)], \tag{9}$$

here, P is the associated configuration domain.

The optimization procedure can be summarized as follows:

Firstly, give a slight deviation from the nominal dimensions of the investigated sample by calculating the signature difference $\Delta\mathbf{y}_i = [\Delta y_{i1}, \Delta y_{i2}, \ldots, \Delta y_{il}]^{\mathrm{T}}$ (i = 1, 2, ..., N) under N measurement configurations according to Equation (5);

Secondly, for each element Δy_{ij} (j = 1, 2, ..., l) in $\Delta\mathbf{y}_i$, carry out the eigen-analysis according to Equations (6)–(9) to identify and eliminate the measurement configurations with redundant information one-by-one; since at least m measured data points are required to determine the m unknown parameters for any mathematical inverse problem in theory [21–24], the achieved set S_j will contain m non-redundant measurement configurations;

Thirdly, considering that different S_j may contain identical measurement configurations, the union of S_j for all $j = 1, 2, \ldots, l$ is the final set of non-redundant measurement configurations, namely,

$$S_{opt} = \bigcup_{j=1}^{l} S_j. \quad (10)$$

After that, the inverse problem described in Equation (2) should be solved under the above achieved measurement configurations, namely, $\mathbf{a} \in S_{opt}$.

It is noted that the above refinement of measurement configurations involves the calculation of the Jacobian matrix $\mathbf{J}(\mathbf{x}^*, \mathbf{a}_i)$, which is typically only valid at the vicinity of $\mathbf{x}$. To ensure that the refinement results of measurement configurations are valid in the changes of profile parameters, it is necessary to repeat the above refinement process of measurement configurations at k different values of $\mathbf{x}$. In this case, the final set of non-redundant measurement configurations will be the union of the achieved set at each value of $\mathbf{x}$. Here, the value of k depends on the types of measured signatures as well as the complexity of the nanostructure under measurement. As a rule of thumb, k should take a relatively large value for a complicated nanostructure with a simple type of measured signature, such as reflectance.

3. Experiments

3.1. Experimental Setup

We took spectroscopic scatterometry as an example to demonstrate the capability of the proposed method in identifying and eliminating the measurement configurations with redundant information. The experiments were carried out by a commercial Mueller matrix ellipsometer (ME-L ellipsometer, Wuhan Eoptics Technology Co., China). As schematically shown in Figure 1, the system layout of the dual rotating-compensator ellipsometer in order of light propagation is $PC_{r1}(\omega_1)SC_{r2}(\omega_2)A$, where P and A stand for the fixed polarizer and analyzer, respectively, C_{r1} and C_{r2} refer to the 1st and 2nd frequency-coupled rotating compensators, respectively, and S stands for the sample [31,32]. The 1st and 2nd compensators rotate synchronously at $\omega_1 = 5\omega$ and $\omega_2 = 3\omega$, respectively, where ω is the fundamental mechanical frequency. With the light source used in this ellipsometer, the spectral range is from 200 to 1000 nm, covering the spectral range of 300–900 nm used in this work. The beam diameter can be changed from the nominal values of ~3 mm to a value less than 200 μm with the focusing lens. The two arms of the ellipsometer and the sample stage can be rotated to change the incidence and azimuthal angles in experiments. Except for the reflectance and ellipsometric angles of the sample under measurement, the 16 Mueller matrix elements also can be obtained with the dual-rotating compensator setting. An in-house developed MATLAB® (version R2017a, The MathWorks, Inc., Natick, MA, USA) program for analyzing the measured signatures ran on a workstation equipped with double 2.0 GHz Intel Xeon CPUs. The forward optical model in this program was developed based on rigorous coupled-wave analysis (RCWA) [33–35], and the inverse optical scattering problem was solved through the commonly used Levenberg-Marquardt (LM) algorithm [36].

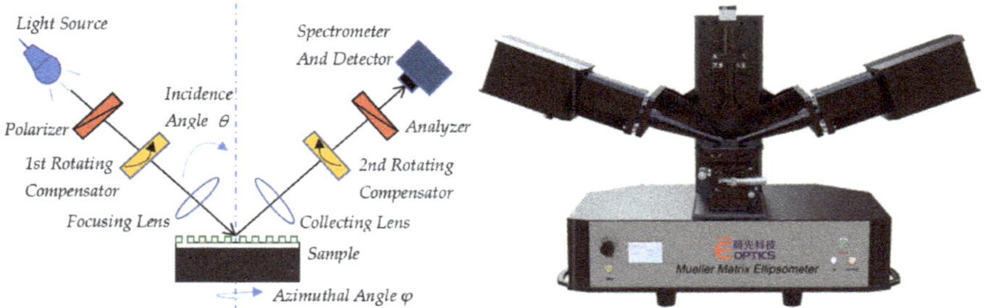

Figure 1. Principle and instrument of the dual rotating-compensator Mueller matrix ellipsometer.

3.2. Experimental Results on 2D Grating

The first investigated sample was a 2D Si grating whose SEM cross-section image is shown in Figure 2. The optical properties of Si were taken from Reference [37]. As depicted in Figure 2, the cross section of the Si grating could be characterized by a symmetrical trapezoidal model with top width (W), grating height (H), sidewall angle (SWA), and period (P), whose dimensions obtained from Figure 2 were W = 350 nm, H = 472 nm, SWA = 87.63°, and P = 800 nm.

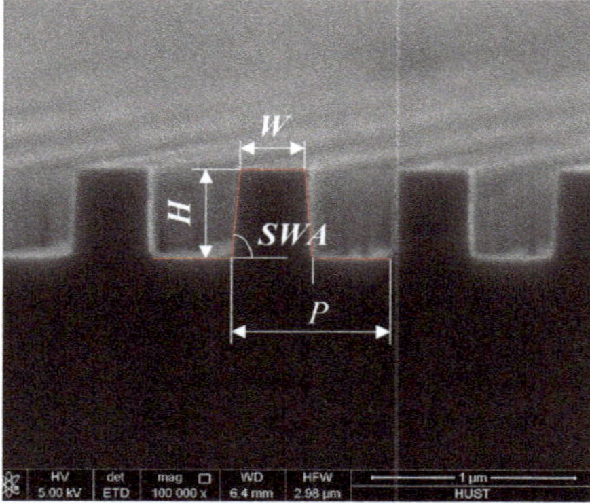

Figure 2. Scanning electron microscopy (SEM) cross-section image of the investigated 2D Si grating sample.

In this case, the profile parameters under study were W, H, and SWA, while the grating period P was fixed at its nominal dimension, namely, $m = 3$. The measured signature contained 15 Mueller matrix elements normalized to the first element, namely, $l = 15$. As mentioned above, the measurement configuration $\mathbf{a} = [\varphi, \theta, \lambda]$ was defined as a combination of fixed azimuthal angle φ, fixed incident angle θ, but a wide waveband λ for spectroscopic scatterometry. As a simple example to demonstrate our data refinement method, here, the incident and azimuthal angles were fixed at $\theta = 65°$ and $\varphi = 30°$, the wavelength covered a fixed range of 300~900 nm with a step of 5 nm. We assumed that the dimensions of the three profile parameters under investigation had a deviation of about 1% from their nominal values. The differences of the 15 Mueller matrix elements between the actual and nominal profiles were calculated by our forward RCWA model under each wavelength point according to Equation (5). After considering all the 15 Mueller matrix elements and the wavelengths used, the enanalysis and eigen-analysis procedures described above were performed on the differences. According to Equations (6)–(9), the wavelength points with redundant information were identified and eliminated from the set of measured signatures in a repetitive manner. Since the investigated sample shown in Figure 2 is a simple structure, and the model output Mueller matrix had 15 elements, we achieved the 3 × 15 optimized data sets by traversing the three profile parameters and the 15 Mueller matrix elements. Through merging the same wavelengths into one for the 3 × 15 optimized data sets, the remaining unions S_{opt}, containing 19 points, were obtained and shown by the black circles in Figure 3, which are supposed to contain enough information needed for the profile reconstruction of the 2D Si grating.

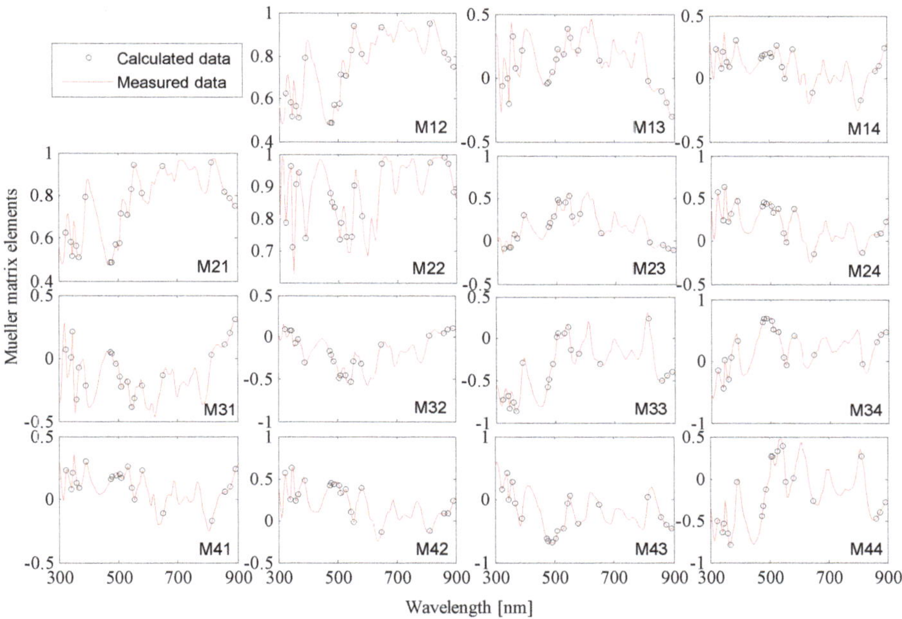

Figure 3. Fitting results of the calculated and the ellipsometer-measured Mueller matrix elements with the incidence and azimuthal angles fixed at $\theta = 65°$ and $\varphi = 30°$, respectively.

To examine the validity of the remaining information, regression analysis was conducted to reconstruct the three profile parameters from the full spectral range of 300–900 nm, the optimal spectrum containing 19 points, and the spectrum of 19 random points. The non-linear LM algorithm was applied to fit the measured 15 Mueller matrix elements with the modeled ones, which converged very quickly to a minimum when a suitable initial condition was chosen. During the fitting procedure, the root-mean-squared error (RMSE) was used to quantify the goodness of fit, which summed over all the measurement wavelengths the differences between the measured data and model-generated data. The lower the RMSE value, the better the fit or agreement between the measured and model-generated data. Figure 3 presents the fitting result of the model-calculated best-fit Mueller matrix elements (black circles) and the measured spectrum including (red lines) 19 optimal points.

Table 1 shows the extracted profile parameter values for the fits and the SEM-measured values. The uncertainties associated with the SEM-measured values were estimated by manually measuring the SEM micrographs. The uncertainties appended to the scatterometry-determined values were estimated from 30 repeated measurements. All uncertainties correspond to a 95% confidence level. As shown in Table 1, the parameters *W*, *H*, and *SWA* of the 2D grating, extracted by fitting the optimized spectrum, approximately agree with the ones obtained by using the full spectrum, while the former method is over three times faster than the later one; the time spent between the optimal and random measuring modes is slightly different, and the reason may be that the former iterations are a little smaller than the latter during the profile reconstruction procedure as that latter has more sufficient information. Figure 3 shows that the model-calculated 15 Mueller matrix elements match well with the measured spectrum. The experiments performed on the 2D Si grating provide preliminary evidence that the advantages of the proposed method are in speeding up the reconstruction procedure and effectively improving the precision of the extracted parameters.

Table 1. Comparison of fitting results of the 2D grating extracted from different spectrum and SEM measurements.

Measuring Mode	Dimensions (nm/°)			RMSE	Time (s)
	W /nm	H /nm	SWA /°		
SEM	350.3 ± 4.74	472.1 ± 4.87	87.63 ± 0.611	—	—
Full spectrum	347.3 ± 0.17	468.9 ± 0.20	86.89 ± 0.019	9.44	548
Optimal (19)	**347.4 ± 0.15**	**476.9 ± 0.12**	**87.27 ± 0.008**	**8.27**	**143**
Random (19)	345.2 ± 0.74	486.4 ± 0.86	86.31 ± 0.103	13.59	225

3.3. Experimental Results on 3D Grating

The second sample was a photoresistive grating consisting of a cylinder array deposited on silicon with a bottom anti-reflective coating (BARC), which is a 3D grating whose SEM cross-section image is shown in Figure 4. The 3D grating of the cylinder array was chosen for this study due to the high computation effort of its forward optical model. As depicted in Figure 4, the cross section of the 3D grating is characterized by a model with cylinder diameter (D) and height (H_1), BARC height (H_2), and periods of arrays in two directions (P_x and P_y). The dimensions obtained from Figure 4 are D = 226.7 nm, H_1 = 355.1 nm, H_2 = 104.3 nm, and $P_x = P_y$ = 835 nm. In our experiments, optical properties of the photoresist were modelled by a two-term Forouhi–Bloomer model [38], whose eight undetermined parameters were determined by measuring a photoresist film deposited on the silicon substrate using the above ellipsometer and were A_1 = 0.006029, A_2 = 0.020598, B_1 = 14.195303 eV, B_2 = 14.196431 eV, C_1 = 50.523937 eV^2, C_2 = = 50.537878 eV^2, n (∞) = 1.436087, and E_g = 4.774050 eV. Similarly, optical properties of the BARC were modelled using the Tauc–Lorentz model [39], whose five undetermined model parameters were obtained as ε_1 (∞) = 1.42680, E_g = 3.45971 eV, A = 21.14955, E_0 = 9.94921, C = 0.98767. For more details about physical meanings of the above parameters, one can consult References [38,39], which are omitted here for brevity.

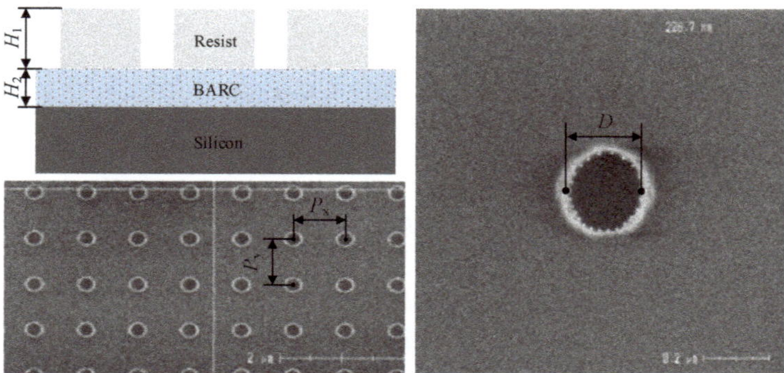

Figure 4. SEM cross-section image of the investigated 3D grating sample.

Here, the investigated profile parameters of the 3D grating included the D, the H_1, and the H_2, namely, m = 3, while the two grating periods were fixed at their nominal dimensions. In contrast to the experiment settings above, the incidence and azimuthal angle were fixed at θ = 65° and φ = 0°, and the terms of N, C, and S parameters versus wavelength were used as measured signatures, which were derived from ellipsometric Psi (Ψ) and Delta (Δ) and taken as $N = \cos(2\Psi)$, $C = \sin(2\Psi)\cos(\Delta)$, and $S = \sin(2\Psi)\sin(\Delta)$. Considering that the investigated 3D sample deviated largely from its nominal values, its forward model was highly complex and nonlinear, and the model output was N, C, and S, namely, $l = 3$, we assumed that the dimensions of the three profile parameters changed $k = 6$ times to ensure that the optimization procedure was stable, and the deviations were about ±1%, ±5%, or ±10% from their nominal dimensions, respectively. According to our proposed method, 3 × 3 × 6 optimized data sets remained, then, the union S_{opt} of the remaining data sets, consisting of the 36 wavelength points shown in Figure 5, were achieved for the inverse problem solution.

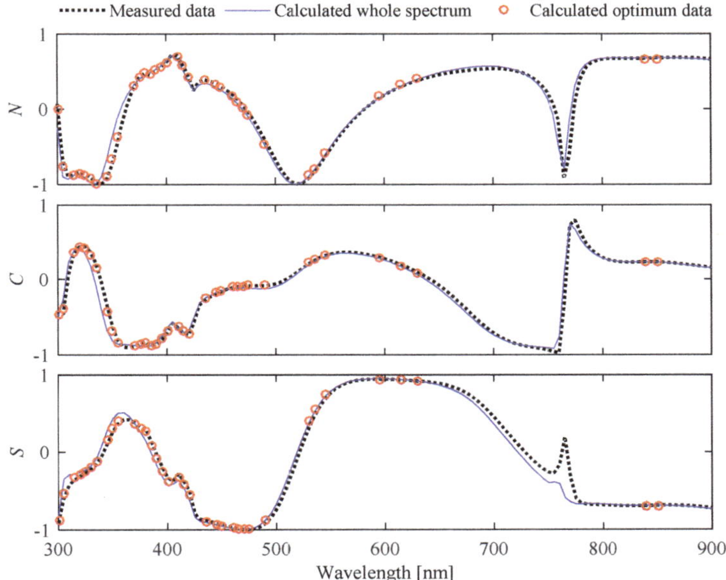

Figure 5. Fitting results of the calculated and the ellipsometer-measured N, C, and S parameters at the incidence angle $\theta = 65°$ and azimuthal angle $\varphi = 0°$.

In the same way, the regression analysis was performed to extract the three unknown profile parameters, and the non-linear LM algorithm was also applied to fit the measured signatures with the modeled ones. The extracted profile parameter values for the fits and the SEM-measured values are shown in Table 2. The uncertainties appended to the SEM-measured and scatterometry-measured values in Table 2 were estimated in a similar manner to those in Table 1. Figure 5 presents the fitting result of the model-calculated best-fit N, C, and S parameters and the measured spectrum (black dots) containing whole range (blue lines) and only 36 optimal points (red circles), respectively.

Table 2. Comparison of fitting results of the 3D photoresist grating extracted from different spectrum and SEM measurements.

Measuring Mode	Dimensions (nm)			RMSE	Time (hour)
	D /nm	H_1 /nm	H_2 /nm		
SEM	226.7 ± 28.32	355.1 ± 2.05	104.3 ± 1.30	—	—
Full spectrum	249.8 ± 2.56	344.3 ± 1.44	107.6 ± 0.62	28.77	14
Optimal (36)	**250.4 ± 2.21**	**343.9 ± 1.03**	**106.4 ± 0.51**	**17.45**	**4.5**
Random (36)	262.4 ± 2.38	338.7 ± 2.42	108.1 ± 0.89	26.50	6

From Table 2, the three average values of the 3D grating dimensions extracted using 36 optimal points are very close to the values obtained using the full range spectrum, but the extracted parameter D has about a 25 nm deviation from the SEM-measured value, which is probably caused by the obvious non-uniformity of the grating surface depicted in Figure 4, and the SEM-measured values include estimation errors in manually measuring the SEM micrographs. Table 2 also shows that the proposed method results in the lowest value of RMSE, the highest measurement precision, and the least time cost. It is observed from Figure 5 that the best-fit N, C, and S parameters using the optimized spectrum are in better agreement with the measured signatures compared to those of the full range. Consequently, we may conclude that the proposed dependence-analysis-based data-refinement method can be applied to determine several optimal measurement points without any loss in the accuracies for profile reconstruction in optical scatterometry.

4. Conclusions

In summary, we proposed a method to identify and eliminate the measurement configurations with redundant information based on dependence analysis in optical scatterometry. By assuming that the dimensions of profile parameters under investigation have some deviation from their nominal values, the differences between the actual and nominal dimensions were calculated by the forward optical model under each measurement point. A formulation was derived to identify the measurement configurations with redundant information through performance of a dependence analysis followed by an eigen-analysis. By eliminating redundant information from the measured data in a repetitive manner, a few optimal points remained and were used for the reconstruction process.

Experiments performed on a 2D Si grating and a 3D photoresist grating revealed that the reconstructed grating profiles from the optimally selected subset of measurement wavelengths according to the proposed method have a higher accuracy than those from the randomly selected counterparts with a higher efficiency than the analysis making use of the whole spectrum. This suggests that the proposed dependence-analysis-based data-refinement method can be a powerful tool to enhance the reconstruction speed of nanostructure metrology using scatterometry without sacrificing the accuracies, especially for the nanostructures whose forward model is very complex and time-consuming.

Author Contributions: Conceptualization, X.C. and Z.D.; methodology and formal analysis, Z.D. and Y.S.; writing—original draft preparation, Z.D.; writing—review and editing, X.C., Z.D., X.W., Y.S., H.J., and S.L.; supervision, Z.D., Y.S., and X.C.; project administration, Y.S. and X.C.; funding acquisition, Z.D., Y.S., X.C., and S.L.

Funding: This research was funded by the National Natural Science Foundation of China (Grant Nos. 51775217, 51727809, and 51525502), the Natural Science Foundation of Hubei Province of China (Grant Nos. 2018CFB290 and 2018CFB559), the China Postdoctoral Science Foundation (Grant Nos. 2016M602269 and 2019M652633), the National key research and development program of China (Grant No. 2017YFF0204705), and the National Science and Technology Major Project of China (Grant No. 2017ZX02101006-004).

Acknowledgments: The authors would like to thank Zhimou Xu from the School of Optical and Electronic Information of Huazhong University of Science and Technology, and Shanghai Micro Electronics Equipment Co., Ltd. (Shanghai, China) for preparing the samples.

Conflicts of Interest: The authors declare no conflict of interest.

References

1. Fang, F.Z.; Zhang, X.D.; Gao, W.; Guo, Y.B.; Byrne, G.; Hansen, H.N. Nanomanufacturing – Perspective and applications. *CIRP Ann. Manuf. Technol.* **2017**, *66*, 683–705. [CrossRef]
2. Bundary, B.; Solecky, E.; Vaid, A.; Bello, A.F.; Dai, X. Metrology capabilities and needs for 7nm and 5nm logic nodes. *Proc. SPIE* **2017**, *10145*, 101450G.
3. Sunkoju, S.; Schujman, S.; Dixit, D.; Diebold, A.; Li, J.; Collins, R.; Haldar, P. Spectroscopic ellipsometry studies of 3-stage deposition of $CuIn_{1-X}Ga_XSe_2$ on Mo-coated glass and stainless steel substrates. *Thin Solid Films* **2016**, *606*, 113–119. [CrossRef]
4. Lereu, A.L.; Passian, A.; Farahi, R.H.; Abel-Tiberini, L.; Tetard, L.; Thundat, T. Spectroscopy and imaging of arrays of nanorods toward nanopolarimetry. *Nanotechnology* **2012**, *23*, 045701. [CrossRef]

5. Hansen, H.N.; Carneiro, K.; Haitjema, H.; Chiffre, L.D. Dimensional micro and nano metrology. *CIRP Ann. Manuf. Technol.* **2006**, *55*, 721–743. [CrossRef]
6. Huang, H.T.; Kong, W.; Terry, F.L., Jr. Normal-incidence spectroscopic ellipsometry for critical dimension monitoring. *Appl. Phys. Lett.* **2001**, *78*, 3983–3985. [CrossRef]
7. Matthias, W.; Johannes, E.; Jürgen, P.; Max, S.; Alexander, D.; Bernd, B. Metrology of nanoscale grating structures by UV scatterometry. *Opt. Express* **2017**, *25*, 2460–2468.
8. Faruk, M.G.; Zangooie, S.; Angyal, M.; Watts, D.K.; Sendelbach, M.; Economikos, L.; Herrera, P.; Wilkins, R. Enabling scatterometry as an in-line measurement technique for 32nm BEOL application. *IEEE Trans. Semicond. Manuf.* **2011**, *24*, 499–512. [CrossRef]
9. Kim, Y.N.; Paek, J.S.; Rabello, S.; Lee, S.; Hu, J.; Liu, Z.; Hao, Y.; McGahan, W. Device based in-chip critical dimension and overlay metrology. *Opt. Express* **2009**, *17*, 21336–21343. [CrossRef] [PubMed]
10. Patrick, H.J.; Gemer, T.A.; Ding, Y.F.; Ro, H.W.; Richter, L.J.; Soles, C.L. Scatterometry for in situ measurement of pattern flow in nanoimprinted polymers. *Appl. Phys. Lett.* **2008**, *93*, 233105. [CrossRef]
11. Paz, V.F.; Peterhänsel, S.; Frenner, K.; Osten, W. Solving the inverse grating problem by white light interference Fourier scatterometry. *Light Sci. Appl.* **2012**, *1*, e36.
12. Chen, X.; Liu, S.; Zhang, C.; Jiang, H. Measurement configuration optimization for accurate grating reconstruction by Mueller matrix polarimetry. *J. Micro/Nanolith. MEMS MOEMS* **2013**, *12*, 033013. [CrossRef]
13. Zhu, J.; Liu, S.; Zhang, C.; Chen, X.; Dong, Z. Identification and reconstruction of diffraction structures in optical scatterometry using support vector machine method. *J. Micro/Nanolith. MEMS MOEMS* **2013**, *12*, 013004. [CrossRef]
14. Zhang, C.; Liu, S.; Shi, T.; Tang, Z. Fitting-determined formulation of effective medium approximation for 3D trench structures in model-based infrared reflectrometry. *J. Opt. Soc. Am. A* **2011**, *28*, 263–271. [CrossRef] [PubMed]
15. Zallat, J.; Aïnouz, S.; Stoll, M.P. Optimal configurations for imaging polarimeters: Impact of image noise and systematic errors. *J. Opt. A Pure Appl. Opt.* **2006**, *8*, 807–814. [CrossRef]
16. Zaharov, V.V.; Farahi, R.H.; Snyder, P.J.; Davison, B.H.; Passian, A. Karhunen-Loève treatment to remove noise and facilitate data analysis in sensing, spectroscopy and other applications. *Analyst* **2014**, *139*, 5927–5935. [CrossRef] [PubMed]
17. Mu, T.; Chen, Z.; Zhang, C.; Liang, R. Optimal configurations of full-Stokes polarimeter with immunity to both Poisson and Gaussian noise. *J. Opt.* **2016**, *18*, 055702. [CrossRef]
18. Logofătu, P.C. Sensitivity analysis of grating parameter estimation. *Appl. Opt.* **2002**, *41*, 7179–7186. [CrossRef]
19. Ku, Y.; Wang, S.; Shyu, D.; Smith, N. Scatterometry-based metrology with feature region signature matching. *Opt. Express* **2006**, *14*, 8482–8491. [CrossRef]
20. Vagos, P.; Hu, J.; Liu, Z.; Rabello, S. Uncertainty and sensitivity analysis and its applications in OCD measurement. *Proc. SPIE* **2009**, *7272*, 72721N.
21. Dong, Z.; Liu, S.; Chen, X.; Shi, Y.; Zhang, C.; Jiang, H. Optimization of measurement configuration in optical scatterometry for one-dimensional nanostructures based on local sensitivity analysis. *J Infrared Millim Waves* **2016**, *1*, 116–122.
22. Dong, Z.; Liu, S.; Chen, X.; Zhang, C. Determination of an optimal measurement configuration in optical scatterometry using global sensitivity analysis. *Thin Solid Films* **2014**, *562*, 16–23. [CrossRef]
23. Twomey, S. Indirect measurements of atmospheric temperature profiles from satellites: II. Mathematical aspects of the inverse problem. *Mon Weather Rev* **1966**, *96*, 363–366. [CrossRef]
24. Twomey, S.; Howell, H.B. Some Aspects of the Optical Estimation of Microstructure in Fog and Cloud. *Appl. Opt.* **1967**, *6*, 2125–2131. [CrossRef] [PubMed]
25. Twomey, S. Information content in remote sensing. *Appl. Opt.* **1973**, *13*, 942–945. [CrossRef] [PubMed]
26. Al-Assaad, R.A. Scatterometry for Semiconductor Sub-Micrometer and Nanometer Critical Dimension Metrology. Ph. D. Thesis, University of Texas, Austin, TX, USA, 2006.
27. Kirsch, A. Uniqueness theorems in inverse scattering theory for periodic structures. *Inverse Probl.* **1994**, *10*, 145–152. [CrossRef]
28. Kirsch, A. *An Introduction to the Mathematical Theory of Inverse Problems*, 2nd ed.; Springer: New York, NY, USA, 2011.
29. Yang, J.; Zhang, B. Uniqueness results in the inverse scattering problem for periodic structures. *Math. Method. Appl. Sci.* **2012**, *35*, 828–838. [CrossRef]
30. Chen, X.; Liu, S.; Gu, H.; Zhang, C. Formulation of error propagation and estimation in grating reconstruction by a dual-rotating compensator Mueller matrix polarimeter. *Thin Solid Films* **2014**, *571*, 653–659. [CrossRef]

31. Collins, R.W.; Koh, J. Dual rotating-compensator multichannel ellipsometer: Instrument design for real-time Mueller matrix spectroscopy of surfaces and films. *J. Opt. Soc. Am. A* **1999**, *16*, 1997–2006. [CrossRef]
32. Liu, S.; Chen, X.; Zhang, C. Development of a broadband Mueller matrix ellipsometer as a powerful tool for nanostructure metrology. *Thin Solid Films* **2015**, *584*, 176–185. [CrossRef]
33. Moharam, M.G.; Pommet, D.A.; Grann, E.B.; Gaylord, T.K. Stable implementation of the rigorous coupled-wave analysis for surface-relief gratings: Enhanced transmittance matrix approach. *J. Opt. Soc. Am. A* **1995**, *12*, 1077–1086. [CrossRef]
34. Li, L. New Formulation of the Fourier modal method for crossed surface-relief gratings. *J. Opt. Soc. Am. A* **1997**, *14*, 2758–2767. [CrossRef]
35. Liu, S.; Ma, Y.; Chen, X.; Zhang, C. Estimation of the convergence order of rigorous coupled-wave analysis for binary gratings in optical critical dimension metrology. *Opt. Eng.* **2012**, *51*, 081504. [CrossRef]
36. Press, W.H.; Teukolsky, S.A.; Vetterling, W.T.; Flannery, B.P. *Numerical Recipies: The Art of Scientific Computing*, 3rd ed.; Cambridge University Press: Cambridge, UK, 2007; Chapter 15.
37. Herzinger, C.M.; Johs, B.; Mcgahan, W.A.; Woollam, J.A.; Paulson, W. Ellipsometric determination of optical constants for silicon and thermally grown silicon dioxide via a multi-sample, multi-wavelength, multi-angle investigation. *J. Appl. Phys.* **1998**, *83*, 3323–3336. [CrossRef]
38. Forouhi, A.R.; Bloomer, I. Optical properties of crystalline semiconductors and dielectrics. *Phys. Rev. B* **1988**, *38*, 1865–1874. [CrossRef] [PubMed]
39. Jellison, G.E., Jr.; Modine, F.A. Parameterization of the optical functions of amorphous materials in the interband region. *Appl. Phys. Lett.* **1996**, *69*, 371–373. [CrossRef]

Article

A Mesh-Based Monte Carlo Study for Investigating Structural and Functional Imaging of Brain Tissue Using Optical Coherence Tomography

Luying Yi, Liqun Sun *, Mingli Zou and Bo Hou

State Key Laboratory of Precision Measurement Technology & Instruments, Department of Precision Instrument, Tsinghua University, Beijing 100084, China; yily15@mails.tsinghua.edu.cn (L.Y.); zml17@mails.tsinghua.edu.cn (M.Z.); houb15@mails.tsinghua.edu.cn (B.H.)

* Correspondence: sunlq@mail.tsinghua.edu.cn; Tel.: +86-010-6278-3033

Received: 12 August 2019; Accepted: 19 September 2019; Published: 25 September 2019

Featured Application: This work provides an effective analysis tool for OCT-based brain imaging and provides an approach to improve the quantitative accuracy of chromophores in tissue for the experimental study of brain functional sensing. What is more, these methods are also suitable for other complex 3D bio-tissues.

Abstract: Optical coherence tomography (OCT) can obtain high-resolution three-dimensional (3D) structural images of biological tissues, and spectroscopic OCT, which is one of the functional extensions of OCT, can also quantify chromophores of tissues. Due to its unique features, OCT has been increasingly used for brain imaging. To support the development of the simulation and analysis tools on which OCT-based brain imaging depends, a model of mesh-based Monte Carlo for OCT (MMC-OCT) is presented in this work to study OCT signals reflecting the structural and functional activities of brain tissue. In addition, an approach to improve the quantitative accuracy of chromophores in tissue is proposed and validated by MMC-OCT simulations. Specifically, the OCT-based brain structural imaging was first simulated to illustrate and validate the MMC-OCT strategy. We then focused on the influences of different wavelengths on the measurement of hemoglobin concentration C, oxygen saturation Y, and scattering coefficient S in brain tissue. Finally, it is proposed and verified here that the measurement accuracy of C, Y, and S can be improved by selecting appropriate wavelengths for calculation, which contributes to the experimental study of brain functional sensing.

Keywords: optical coherence tomography; Monte Carlo simulation; structural imaging; functional sensing

1. Introduction

Applications of optical coherence tomography (OCT) in brain tissue include structural imaging and functional sensing. Structural imaging includes macroscopic imaging [1,2] and angiography of the brain tissue [3]. Common detectable brain functional activities are shown in Figure 1 [4,5], and the associated optical processes are also indicated. Among them, scattering and absorption can be obtained by spectroscopic OCT [6,7], and Doppler shift can be measured by Doppler OCT [8]. This paper focuses on the applications of OCT in brain structural imaging as well as in the measurement of absorption and scattering coefficients of brain tissue.

Both numerical and in vivo testbeds are useful for brain tissue imaging. In vivo testbeds provide the ultimate confirmation of performance, while numerical testbeds allow a more controlled evaluation of imaging modalities and optimization of their parameters. Monte Carlo (MC) simulation is the golden means to simulate the interaction between light and tissue [9,10]. Many efficient MC models have been developed to study OCT signals of different tissues [11–14] or polarization-sensitive OCT [11,15]. However, there are few literatures to analyze OCT signals in brain tissue using MC technology. Therefore, as a slight supplement to this field, we detail how to use MC simulation to analyze the applications of OCT in brain tissue structural imaging and functional sensing. A model of mesh-based Monte Carlo [14,16,17] for OCT (MMC-OCT) is used in this paper, which is useful for any complex three-dimensional (3D) bio-tissues. This work provides an analysis method and related data for brain simulation as well as help for experimental study.

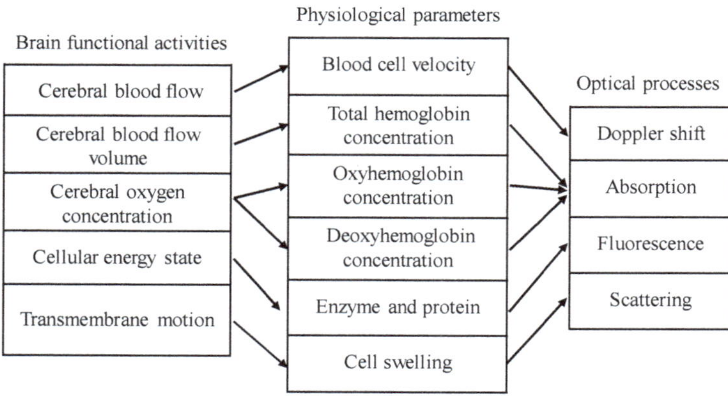

Figure 1. Relationship among common brain functional activities, physiological parameters, and optical processes. The total hemoglobin concentration C is equal to the sum of the oxyhemoglobin concentration and deoxyhemoglobin concentration. Oxygen saturation Y is the ratio of the oxyhemoglobin concentration to the total hemoglobin concentration.

Quantitative accuracy is undoubtedly an important indicator in the measurement of scattering and absorption of brain tissue. In the spectroscopic OCT, extinction coefficients of different wavelengths are obtained by using the short-time Fourier transform (STFT) and exponential fitting [7], and the extinction coefficient is the sum of the absorption coefficient and scattering coefficient S. It is a common choice to use a visible light waveband for detection of scattering and absorption [6,18], and the main absorption in this range comes from hemoglobin for brain tissue [19,20]. Thus, the total hemoglobin concentration C and oxygen saturation Y of the brain tissue can be calculated by solving linear equations using absorption coefficients of several different wavelengths [4], where C and Y are defined in Figure 1. The extinction coefficients of multiple wavelengths can be obtained in spectroscopic OCT, so the results obtained by fitting the oversampled data will undoubtedly be more accurate considering the noise and experimental errors in the actual applications. However, the spectral resolution and spatial resolution are contradictory [21], and the more positions the window function traverses, the longer the calculation time. In addition, the narrower the width of the window function in the spectral domain, the worse the signal-to-noise ratio (SNR). For measurements of blood vessels with high hemoglobin concentrations (~2 mmol) [22], two or three wavelengths can be used to resolve the correct functional parameters [23,24] while achieving high spatial resolution. In the process of solving C, Y, and S from the fitted extinction coefficients of three wavelengths, we found that the error of the extinction coefficient has a great influence on the calculated results, and the values of C, Y, and S calculated using different wavelengths are distinct for the certain error. Therefore, the selection of wavelengths that are used to calculate scattering and absorption is explored in this paper. It is proposed and validated here that the measurement accuracy of C, Y, and S can be improved by selecting the appropriate wavelengths for calculation.

The remainder of this paper is organized as follows. We present the basic principles of structural imaging and functional sensing of OCT as well as MMC-OCT models in Section 2. In Section 3, we first detail how to use MMC-OCT simulation to study structural imaging of tissue, and then the influences of different wavelengths on quantitative accuracy of scattering and absorption coefficients are discussed. Finally, in Section 4 we provide a brief summary and conclusion.

2. Methods

2.1. Structural and Functional Imaging Based on OCT

The spectroscopic OCT includes the spectroscopic time domain OCT (TD-OCT) and spectroscopic Fourier domain OCT (FD-OCT), both of which are implemented by windowed Fourier transform [21,25]. Here we mainly introduce the principle of structural imaging and functional sensing based on FD-OCT. In FD-OCT, it is only necessary to perform a Fourier transform on the interference signal collected by the spectrometer or detectors to obtain the structural image, i.e.,

$$I(k) = S(k)\left|aR + \int_0^\infty a(z)\exp\{i2k\mathbf{n}(k)z\}dz\right|^2 \quad (1)$$

and

$$H(z) = F\{I(k)\} = F\{S(k)\} \otimes \left\{a_R^2\delta(z) + a_R[a(z) + a^*(-z)] + AC(I_{au}(k)\right\}, \quad (2)$$

where $S(k)$ is the power spectrum of the light source; a_R is the amplitude of the reflected reference light; k is the wavenumber; $\mathbf{n}(k)$ is the refractive index of the sample at k; and $a(z)$ is the amplitude of the backscattered light at the z position of the sample. AC represents the self-coherent signal of the sample, which can usually be ignored because it is small.

During the calculation above, $\mathbf{n}(k)$ is treated as a real number. However, it is in fact a complex valued function $\mathbf{n}(k) = n(k) + i\kappa(k)$, defining both the group delay ($n(k)$) and absorption ($\kappa(k)$) within a medium [26]. Therefore, ignoring the conjugate and direct-current terms, Equation (2) should be expressed as

$$H(z) = F\{I(k)\} = F\{S(k)\} \otimes \{a_R a(z)\exp(-2nz\mu_t(z))\} \quad (3)$$

in which μ_t is the extinction coefficient of tissue; and n is the average refractive index of tissue. When a STFT or wavelet transform is applied to Equation (1), we can obtain the depth-resolved spectrum $H(z, k_n)$, shown as

$$H(z, k_n) = F\{I(k)g(k_n; \Delta k)\} = A\{\mu_b(z, k_n)\exp(-2nz\mu_t(z, k_n))\} \quad (4)$$

where A represents the system parameter and μ_b is the backscattered coefficient [7] and $g(k_n; \Delta k)$ is a window function having a width Δk centered at a mean scattering wave number k_n. Thus, the spectra of backscattering coefficient μ_b and extinction coefficient μ_t as a function of z and k_n are obtained.

From Equation (4) we can see that there is an inherent trade-off between spectral resolution and spatial resolution. For Gaussian windows, the resolution in z-space and k-space is related by Equation $\Delta k\Delta z = 1/4\pi$, i.e., $\Delta z = \lambda^2/(2\Delta\lambda)$ [21].

For brain tissue, the main absorption comes from water and hemoglobin. For simplicity, chromophores other than hemoglobin such as fat, melanin, and lipids are neglected [19]. Consequently, the extinction coefficient of the sample can be defined as

$$\mu_t(z,\lambda) = C(z)\cdot\ln(10)\cdot[Y\cdot\varepsilon_O(z,\lambda) + (1-Y)\cdot\varepsilon_D(z,\lambda)] + \mu_s(z,\lambda). \quad (5)$$

ε_O and ε_D are the molar extinction coefficients—the extinction coefficient divided by ln(10) per unit molar concentration—of oxyhemoglobin and deoxyhemoglobin, respectively [27]. μ_s is the scattering coefficient. The unit of C is mmol, the unit of ε_O and ε_D is $\text{mmol}^{-1}\ \text{mm}^{-1}$, and the unit of μ_s and μ_t is mm^{-1}. Functional sensing involves the measurement of C, Y, and μ_s based on the absorption and scattering spectra (Figure 2). Figure 2 shows the curves of μ_a and μ_s as a function of wavelength, where $\mu_a = C\ln(10)\varepsilon_o$ and C = 2 mmol [22].

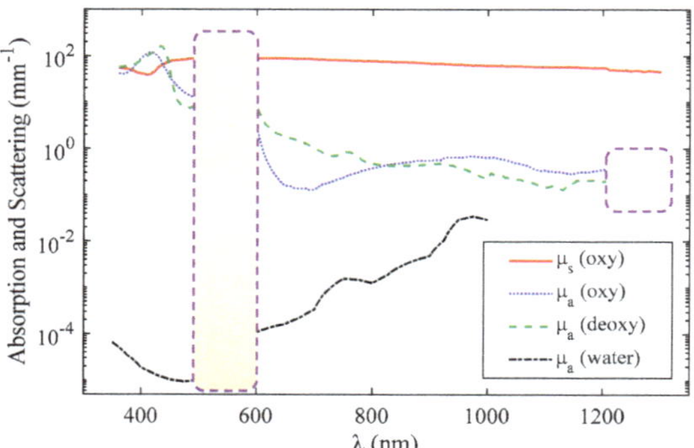

Figure 2. Wavelength dependence of oxyhemoglobin and deoxyhemoglobin absorption (blue and green, respectively) [22], whole blood scattering (red) [22], and water absorption (black) [20].

For the visible-light waveband (500–600 nm), the magnitude of the absorption and scattering coefficients are of the same order, and it can be approximately considered that the scattering coefficient is not affected by the wavelength when using the method introduced in [6]. Therefore, the extinction coefficient can be expressed as

$$\mu_t(z,\lambda) = C(z)\cdot\ln(10)\cdot[Y\cdot\varepsilon_O(z,\lambda) + (1-Y)\cdot\varepsilon_D(z,\lambda)] + \mu_s(z). \tag{6}$$

Consequently, it is only necessary to obtain the extinction coefficients of three different wavelengths at a certain depth z, and then, C, Y, and μ_s can be calculated by solving the ternary equation.

In the near-infrared light waveband (600–1100 nm), the scattering coefficient is affected more by the wavelength than in the visible range, and the scattering coefficient is approximately linear with the wavelength or satisfies a specific functional relationship $a \times \lambda^b$ [28,29], where both a and b are constants. Therefore, the extinction coefficient can be expressed as

$$\mu_t(z,\lambda) = C(z)\cdot\ln(10)\cdot[Y\cdot\varepsilon_O(z,\lambda) + (1-Y)\cdot\varepsilon_D(z,\lambda)] + (\mathrm{a} + \mathrm{b}\lambda) \tag{7}$$

when the linear relationship is considered. Accordingly, C, Y, and μ_s can be calculated by solving the quaternary equation.

In the range of >1200 nm, the absorption coefficients of oxyhemoglobin and deoxyhemoglobin are almost indistinguishable as shown by the purple dashed box in Figure 2, so the functional sensing is difficult to achieve.

This article focuses on functional sensing in the visible-light waveband. Within this range, the parameter C, Y, and S can be obtained by solving a ternary linear equation, expressed as Equation (6), or can be obtained by the fitting strategy using oversampled data. The latter has a higher probability of getting the correct parameters. However, if functional parameters with acceptable errors can be obtained using data at three wavelengths, it is advantageous where high spatial resolution is required, such as imaging small-sized microvessels in brain tissue.

2.2. Mesh-Based Monte Carlo for Optical Coherence Tomography (MMC-OCT)

Throughout the simulation of this paper, a fast mesh-based Monte Carlo (MMC) photon migration algorithm for imaging in 3D complex tissue was used [16,17], and a schematic that clearly illustrates the simulation process is shown in Figure 3. Before proceeding with the steps described in Figure 3, the "MMCLAB" package needs to be downloaded [30], which contains all the MATLAB m-files mentioned in Figure 3.

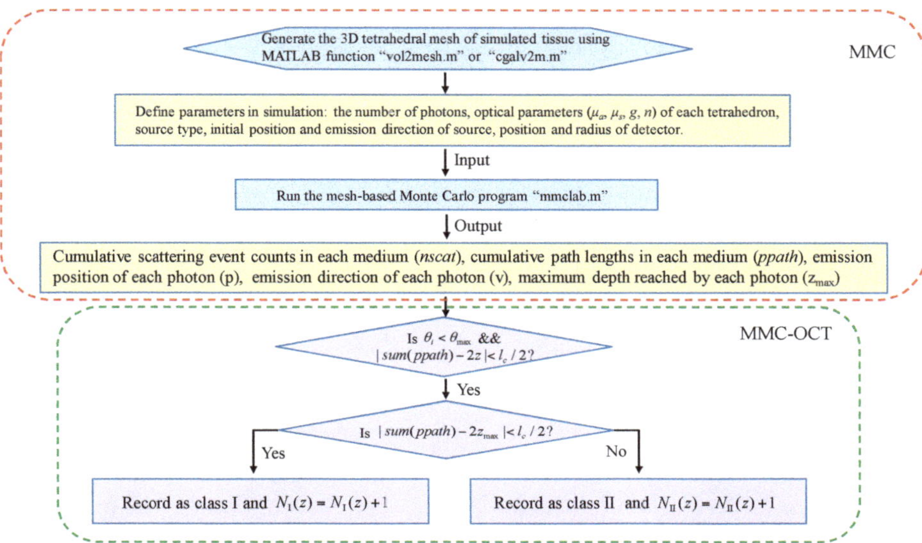

Figure 3. Flow chart describing the simulation process of mesh-based Monte Carlo optical coherence tomography (MMC-OCT). μ_a and g are the absorption coefficient and anisotropy factor, respectively; θ_i is the angle between the emission direction of the detected photon and z axis; θ_{max} is the collecting angle; l_c is axial resolution (i.e., the coherence length); z is the probing depth; *sum* (*nscat*) and *sum* (*ppath*) represent the sum of scattering events and the sum of path lengths in all media, respectively. N_I (z) and N_{II} (z) are the numbers of class I photon and class II photon that are reflected/backscattered at the z-depth, respectively.

In simulations, the light source was a pencil beam, and a circular detector with a radius of 10 µm was placed at the same position as the emission position of the photon packet. The initial set number of photons and θ_{max} were 5×10^{13} and 3°, respectively. After running Monte Carlo MATLAB program "mmclab", the parameters *nscat*, *ppath*, v, and z_{max} of each detected photon are obtained. It should be noted that the mesh of the tissue is subdivided in the depth direction in order to obtain the depth coordinates at which the scattering event occurs. Then, according to the condition $|sum(ppath) - 2z| < l_c/2$, the photons within the collecting angle θ_{max} are divided into different depths z having a thickness of l_c. To classify the photon further, on satisfaction of $|sum(ppath) - 2z_{max}| < l_c/2$, the photon is recorded as a class I photon, otherwise it is recorded as a class II photon [10,13].

We directly obtained the backscattered field of the tissue detected by the OCT system represented as Equation (3) through the above steps, without simulating the complete imaging process of the OCT. The resolution of z-space is equal to l_c, which is determined by the spectral bandwidth of the light source, and μ_t is the average extinction coefficients of the tissue within the spectral bandwidth.

Similarly, for the analysis of functional imaging of FD-OCT, the depth-resolved spectrum H (z, k_n) represented by Equation (4) was directly obtained. From Equation (4), we found that the effect of window function g in the STFT is equivalent to that of a light source with a narrower spectral width. Therefore, we controlled the z-space resolution by setting l_c, and the expression of l_c is $\Delta z = \lambda^2/(2\Delta\lambda)$ [21]. μ_t is the extinction coefficient of the tissue over the spectral bandwidth of the window function.

All simulations were run using a desktop computer with an Intel Core 3-GHz 64-bit processor (Intel Corporation, Santa Clara, CA, USA) and 16 GB of RAM running the Windows operating system. It, however, has to be pointed out that the simulation time of MMC-OCT is long. Simulation of a 300 × 300 × 500 μm gray matter (discretized using a mesh of 15,720 nodes and 87,996 elements) for 5×10^{11} photons took approximately 2.5 h per A-scan. However, the simulation time can be greatly reduced by improving the program. It has been proven that tetrahedral mesh-based MC and importance sampling can significantly reduce computation time, which can obtain more accurate OCT signals using fewer photons [31]. In addition, because of the inherently parallel nature of MC simulation, we can reduce simulation time by implementing the MMC-OCT program on a graphics processor unit (GPU).

3. Results and Discussions

We first generated a segmented human brain atlas as shown in Figure 4 [32], from which we know that the brain tissue has a hierarchical structure and some cylindrical structures (vessels) embedded in it.

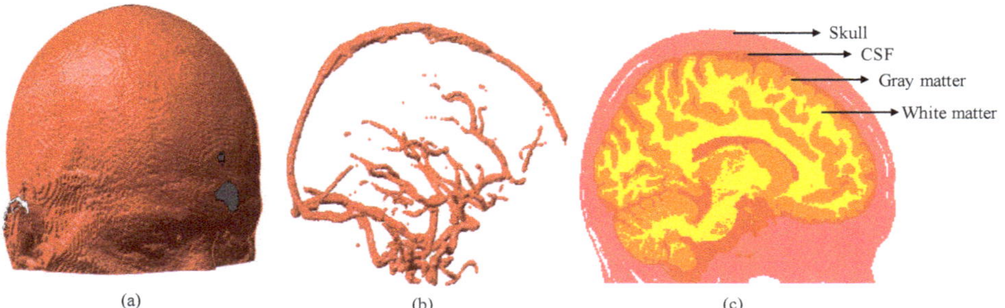

Figure 4. Three-dimensional (3D) brain atlas (181 × 217 × 181 mm): (**a**) 3D model of brain; (**b**) 3D model of blood vessels segmented from brain model shown in (**a**); and (**c**) 2D section of brain extracted at plane $x = 90$ mm. CSF represents the cerebrospinal fluid.

3.1. Simulation Results of Brain Structural Imaging

In this section, the structural imaging of hierarchies and vascular structures present in the brain tissue was simulated. The parameters of the hierarchical structure [16,33] and vessel structure [22] in the simulation are listed in Table 1. The n and g of tissue were 1.37 and 0.9, respectively. The simulated hierarchy structure consisted of the skull, cerebrospinal fluid (CSF), gray matter and white matter, and its 3D volumetric finite-element (FE) mesh is shown in Figure 5a. Figure 5b shows the 3D volumetric mesh of gray matter with a blood vessel embedded in it, and the position of the blood vessel is indicated on the figure. The numbers in Figure 5a,b represent the number of grids, and the grid resolution was 10 μm. In the simulation, the center wavelength was 630 nm and l_c was 10 μm, which is equivalent to using a light source having a spectral bandwidth of 13 nm.

First, in order to verify the correctness and effectiveness of the MMC-OCT model in OCT-based structural imaging, all parts of the hierarchy structure were compressed, and compressed sizes are shown inside parentheses in the "Thickness" column of Table 1. The backscattered signal of the compressed hierarchical structure (Figure 5c,d) shows the simulation result of the vessel structure obtained with actual tissue parameters. It is clearly seen that the backscattered signals reflect the structural distribution of the two tissues, and the dimensions of each part are identical to the preset structural dimensions. Moreover, the results show that the larger the scattering coefficient, the larger the amplitude of the backscattered signal and the faster the signal attenuates, which is completely consistent with the actual situation [7] and proves the validity of the MMC-OCT used in this paper.

Table 1. Geometric parameters and optical parameters of simulated tissues at 630 nm wavelength.

Tissue Type	Thickness (mm)	μ_a (mm^{-1})	μ_s (mm^{-1})
	Hierarchical Structure		
Skull	7 (0.15)	0.019	7.8
CSF	2 (0.05)	0.004	0.009
Gray matter	4 (0.1)	0.02	9
White matter	34 (0.2)	0.08	40.9
	Vessel Structure		
Medium (Gray matter)	0.5	0.02	9
Vessel	0.1 (Diameter)	0.28	88.55

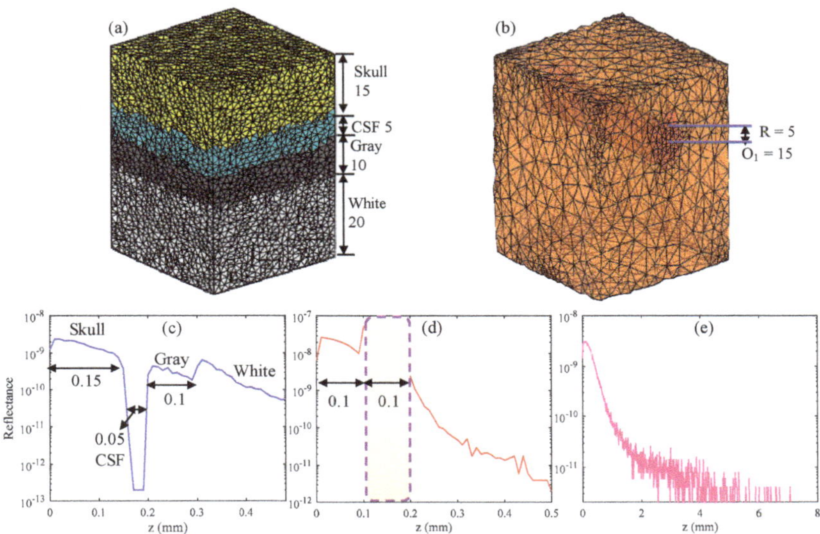

Figure 5. (**a**,**b**) 3D volumetric finite-element (FE) mesh of simulated tissue, in which R and O are the radius and center of the blood vessel, respectively. (**d**) Backscattered signal of the vessel tissue. (**c**,**e**) show the backscattered signal of the compressed and uncompressed hierarchical structures, respectively.

Figure 5e shows the simulation result of the brain hierarchy structure obtained using actual tissue parameters shown outside parentheses in the "Thickness" column of Table 1. It can be found that the backscattered signal attenuates rapidly with an increase in depth and the light cannot penetrate the skull, which is consistent with reality [4]. The imaging depth of SD-OCT is generally 1–2 mm, owing to the large attenuation coefficient of tissue, which is not enough for human brain tissue [4]. But for some in vitro studies, we may only focus on a part of the brain, such as the cerebral cortex or the vascular network. In addition, brain imaging of small animals also provides significant reference value for human brain research. In the aforementioned applications, the MMC-OCT can be used to simulate their imaging process, thus providing important theoretical support for brain imaging.

3.2. Simulation Results of Brain Functional Sensing

As we mentioned in the introduction, the application of spectroscopic OCT in brain functional sensing is usually to quantify three parameters: total hemoglobin concentration C, blood oxygen saturation Y, and scattering coefficient S. After obtaining the OCT signals of different wavelengths using MMC-OCT, C, Y, and S can be calculated using the quantitative method introduced in Section 2.1. In this section, we studied brain functional sensing through theoretical analysis and MMC-OCT simulation. Specifically, we explored factors that affect the quantitative accuracy of C, Y, and S using three wavelengths.

In the visible light waveband (500–600 nm), the absorption coefficient of hemoglobin is much larger than that in the near-infrared waveband, as shown in Figure 2, so functional sensing is very advantageous. Therefore, we chose a visible waveband to simulate functional sensing of the brain tissue. To obtain parameters C, Y, and S, we usually need to solve the following three-variable linear equations:

$$\begin{aligned} C[\varepsilon_{1D} + Y(\varepsilon_{1O} - \varepsilon_{1D})] + S = \mu_{t1} \\ C[\varepsilon_{2D} + Y(\varepsilon_{2O} - \varepsilon_{2D})] + S = \mu_{t2}, \\ C[\varepsilon_{3D} + Y(\varepsilon_{3O} - \varepsilon_{3D})] + S = \mu_{t3} \end{aligned} \tag{8}$$

in which subscripts 1, 2, and 3 represent three different wavelengths. ln(10) is multiplied to ε_O and ε_D. The extinction coefficients μ_{t1}, μ_{t2}, and μ_{t3} that are obtained by exponential fitting of the backscattered signals include the experimental error and data fitting error. The solutions to Equation (8) are

$$C = \frac{(\Delta\varepsilon_o' - \Delta\varepsilon_D')\Delta\mu_t - (\Delta\varepsilon_o - \Delta\varepsilon_D)\Delta\mu_t'}{\Delta\varepsilon_D\Delta\varepsilon_o' - \Delta\varepsilon_D'\Delta\varepsilon_o} \tag{9}$$

$$Y = \frac{\Delta\varepsilon_D\Delta\mu_t' - \Delta\varepsilon_D'\Delta\mu_t}{(\Delta\varepsilon_o' - \Delta\varepsilon_D')\Delta\mu_t - (\Delta\varepsilon_o - \Delta\varepsilon_D)\Delta\mu_t'} \tag{10}$$

and

$$S = \frac{(\varepsilon_{1o}\varepsilon_{2D} - \varepsilon_{2o}\varepsilon_{1D})\mu_{t3} + (\varepsilon_{3o}\varepsilon_{1D} - \varepsilon_{1o}\varepsilon_{3D})\mu_{t2} + (\varepsilon_{2o}\varepsilon_{3D} - \varepsilon_{3o}\varepsilon_{2D})\mu_{t1}}{\Delta\varepsilon_o\Delta\varepsilon_D' - \Delta\varepsilon_D\Delta\varepsilon_o'}, \tag{11}$$

where $\Delta\varepsilon_D = \varepsilon_{2D} - \varepsilon_{1D}$, $\Delta\varepsilon_o = \varepsilon_{2o} - \varepsilon_{1o}$, $\Delta\mu_t = \mu_{t2} - \mu_{t1}$, and the parameters with (′) as a superscript are the results of subtracting the corresponding parameters of the second wavelength from the parameters of the third wavelength. The variation of the three parameters C, Y, and S caused by the errors of μ_{ti} (i = 1, 2, and 3) can be expressed as

$$|dC| = |\frac{(\Delta\varepsilon_o' - \Delta\varepsilon_D')d\Delta\mu_t - (\Delta\varepsilon_o - \Delta\varepsilon_D)d\Delta\mu_t'}{\Delta\varepsilon_D\Delta\varepsilon_o' - \Delta\varepsilon_D'\Delta\varepsilon_o}|, \tag{12}$$

$$|dY| = |\frac{\Delta\varepsilon_D\Delta\varepsilon_o' - \Delta\varepsilon_D'\Delta\varepsilon_o}{[(\Delta\varepsilon_o' - \Delta\varepsilon_D')\frac{\Delta\mu_t}{\Delta\mu_t'} - (\Delta\varepsilon_o - \Delta\varepsilon_D)]^2}d(\frac{\Delta\mu_t}{\Delta\mu_t'})|, \tag{13}$$

and

$$|dS| = |\frac{(\varepsilon_{1o}\varepsilon_{2D} - \varepsilon_{2o}\varepsilon_{1D})d\mu_{t3} + (\varepsilon_{3o}\varepsilon_{1D} - \varepsilon_{1o}\varepsilon_{3D})d\mu_{t2} + (\varepsilon_{2o}\varepsilon_{3D} - \varepsilon_{3o}\varepsilon_{2D})d\mu_{t1}}{\Delta\varepsilon_o\Delta\varepsilon_D' - \Delta\varepsilon_D\Delta\varepsilon_o'}|, \tag{14}$$

in which $d\Delta\mu_t$, $d(\Delta\mu_t/\Delta\mu_t')$, and $d\mu_t$ all represent errors of μ_{ti}, and their signs and magnitudes are uncertain in practice.

It can be seen from Equations (12)–(14) that the calculation errors of C and S have nothing to do with Y and are only related to the errors of μ_{ti} and the absorption parameters of the selected wavelengths. However, the calculation error of Y varies with μ_{ti}, that is, it is related to C, Y, and S. In addition, we find that the change trends of $|dC|$ and $|dS|$ are the same, while that of $|dY|$ is opposite to the other two. In other words, we have no way to choose the optimal three wavelengths to minimize the errors of the C, Y, and S simultaneously, but can only make compromises according to the actual situation. For example, if the requirement for the accuracy of C is higher in practical application, the denominator of $|dC|$ can be larger and the numerator smaller by choosing appropriate wavelengths, while if the requirement for the accuracy of Y is higher, the opposite choice can be made.

Figure 6 shows the errors of C, Y and, S under two different wavelength choices, and the corresponding absorption parameters are shown in the pictures. It should be noted that these parameters are not actual values shown in Figure 2, but are obvious cases set to explain the above rules. In the simulation, C was 100 μm, Y was 70%, and S was 7 mm^{-1}. The C of the brain tissue is approximately 100 μm [19]. Here, for convenience, we assumed that μ_{t1} had error $d\mu_{t1}$, and μ_{t2} and μ_{t3} had no errors. In Figure 6a, when $d\mu_{t1}$ is 0.1 mm^{-1}, $dC/C = 0$ and $dS/S = 0.028$, while dY/Y is up to 0.9. In Figure 6b, when $d\mu_{t1}$ is 0.1 mm^{-1}, $dC/C = 3.3$ and $dS/S = 1.16$, while dY/Y is only 0.2. Therefore, we can choose the appropriate wavelengths to solve the scattering and absorption parameters according to the actual needs.

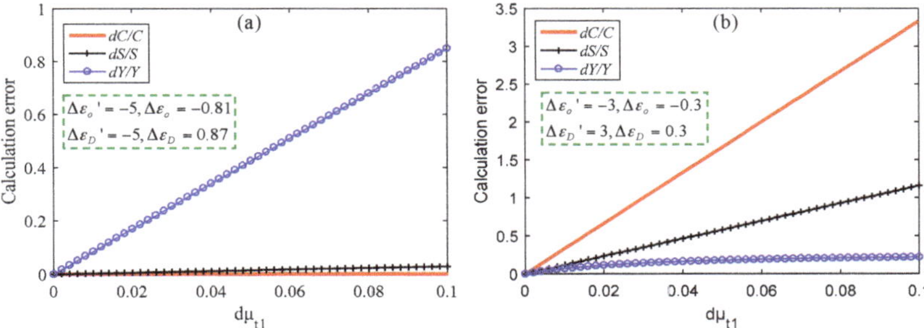

Figure 6. Calculation errors of three parameters when dC is small (**a**) and when dY is small (**b**).

Next, functional sensing of brain tissue was simulated using MMC-OCT. The most popular components of brain tissue for researchers are often the cerebral cortex (i.e., gray matter) and blood vessels, so we obtained backscattered signals at different wavelengths of vascular tissue. The blood vessel had a diameter of 50 μm and was embedded in the gray matter tissue with a depth of 150 μm. Firstly, the appropriate wavelengths were selected based on the spectral data shown in Figure 2. We set a light source with a center wavelength of 570 nm and a spectral width of 50 nm. The window function of the STFT has a width in the spectral domain of 15 nm, and there is no overlap between the window functions at the two adjacent wavelengths. Between 540 nm and 600 nm, we used Equations (12)–(14) to calculate the theoretical error of the three parameters obtained using different wavelength combinations, and each combination consists of three wavelengths with an interval of 15 nm. The error for each μ_t is a random number within 0–0.1 mm^{-1}; $Y = 0.7$, $C = 2$ mmol, and $S = 70$ mm^{-1}. The curve in Figure 7a shows the sum of the errors of the three parameters calculated using different wavelength combinations, and the abscissa is the minimum of the three wavelengths. The number of calculations is 200.

Within the spectral bandwidth of the source, the optimal and worst combinations are 560 nm, 575 nm, 590 nm and 553 nm, 568 nm, 583 nm, respectively. Figure 7a–c shows the calculation errors of C, Y, and S, respectively. Since the error of the extinction coefficient is random, we compare the maximum of the 200 calculation errors of the two combinations. The maximum values of the dC/C, dY/Y, and dS/S of the optimal combination are reduced by 17.6, 5.8, and 14.5 times, respectively, compared to those of the worst combination. The optical parameters of the two combinations are shown in Table 2, and Figure 8 shows the simulation results using MMC-OCT. When using the method described in [6], it can be approximated that the scattering coefficient is not affected by the wavelength, so we set the scattering coefficient for each wavelength to 70 mm^{-1}.

The simulation results for the 560 nm wavelength are summarized in Figure 8. Figure 8a shows the continuous wave (CW) fluence extracted at plane $y = 0$ μm, and the black dotted box in the figure indicates the position of the blood vessel. The blue curve in Figure 8b is the backscattered signal, and the signal in the depth range of 70 μm–100 μm is selected to calculate the extinction coefficient by exponential fitting, the results of which are shown in Figure 8c. The fitting coefficients and goodness are shown in the picture, and the R^2 value of the fitting results is 0.9969. The fitting results of two combinations and C, Y, and S calculated from the fitting results are listed in Table 3. The parameters marked "The" in Table 3 are the theoretical calculation values, and those labeled "Rea" are the simulation values. It can be seen that the two are basically the same, and the maximum error is 0.06 mm^{-1}. Of course, this does not mean that the error in the actual experiment is also at this level, but rather illustrates the feasibility of the MMC-OCT in simulating the process of obtaining the extinction coefficient.

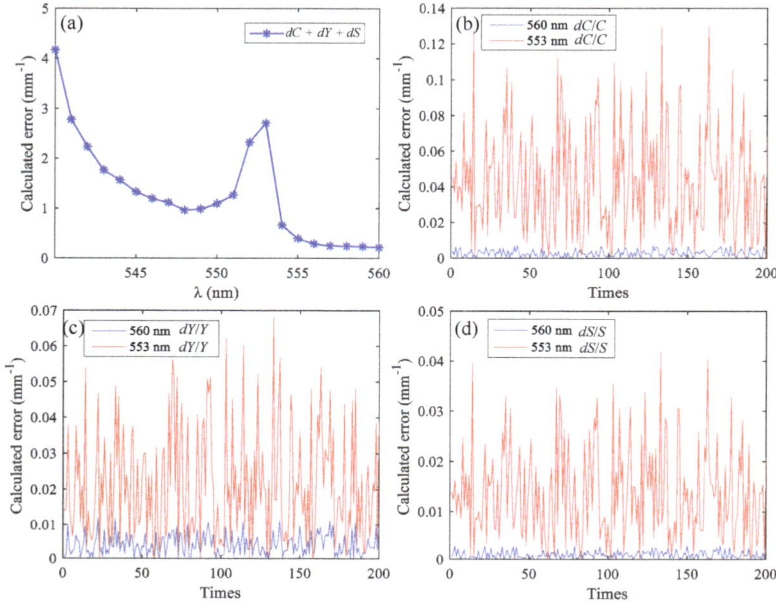

Figure 7. (**a**) The sum of the errors of the three parameters calculated using different wavelength combinations; calculation errors of C (**b**), Y (**c**), and S (**d**) calculated using the optimal wavelength combination and the worst wavelength combination.

Table 2. Optical parameters of optimal and worst wavelength combinations for functional sensing.

λ (nm)	ε_O ($mmol^{-1}$·mm^{-1})	ε_D ($mmol^{-1}$·mm^{-1})	S (mm^{-1})
	Optimal Combination		
560	4.1519	5.7674	70
575	5.8742	4.5500	70
590	2.1107	3.0857	70
	Worst Combination		
553	4.7611	5.7666	70
568	4.8988	5.2246	70
583	4.7178	3.8402	70

Table 3. Fitting results and calculated C, Y, and S for two combinations.

Combinations	The μ_{t1}	The μ_{t2}	The μ_{t3}	Rea μ_{t1}	Rea μ_{t2}	Rea μ_{t3}	C	Y	S_1
560, 575, 590	91.352	95.222	81.067	91.338	95.268	81.047	2.003	0.704	69.978
553, 568, 583	93.315	93.01	90.514	93.291	93.171	90.531	2.482	0.767	64.746

The errors of C, Y, and S obtained by the optimized combination of wavelengths are much smaller than those obtained by the worst combination, and the calculation errors of C, Y, and S are reduced by 160, 16.75, and 238.8 times, respectively.

The results in Table 3 show that we can obtain very accurate optical parameters of blood vessels by reasonably selecting three wavelengths. However, the error between the theorical extinction coefficient and the simulation result is very small, not exceeding 0.06 mm^{-1}, so the result is ideal. In actual experiments, when the method described in [34] is used, the accuracy of the exponentially fitted extinction coefficient can reach 0.8%, which is about 0.72 mm^{-1} for blood. Therefore, we set the maximum error of extinction coefficient to 1 mm^{-1}. At Y = 70%, dC/C, dY/Y, and dS/S calculated by the optimized combination are 0.0859, 0.0829, and 0.0296, respectively. Accordingly, we can conclude that when imaging vascular structure, the optimization scheme shown in Figure 7 can be used to reasonably select three wavelengths to calculate the optical parameters of the tissue while ensuring high spatial resolution.

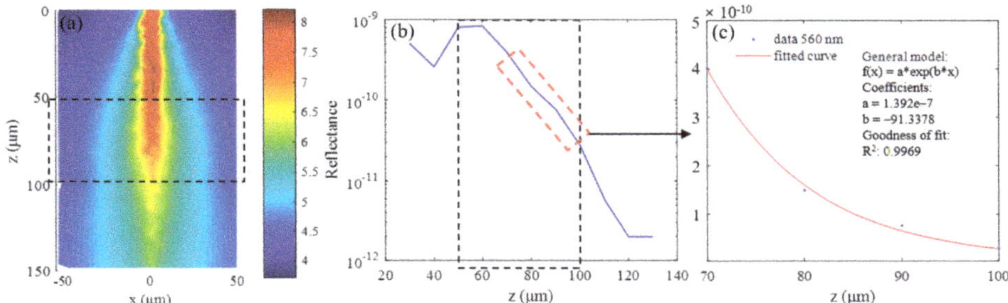

Figure 8. Simulation results for 560 nm wavelength: (**a**) continuous wave (CW) fluence extracted at plane y = 0 μm; (**b**) backscattered signals; and (**c**) fitting results when class II photons are removed.

It should be noted that the above optimization process was performed under the condition that Y = 0.7. The hemoglobin concentration of the blood vessel is high, and the absorption coefficient in the visible-light waveband is large, so the optimization results for different Y are approximately the same.

Next, the functional sensing of the cerebral cortex was discussed, where the spatial resolution requirements are reduced compared to imaging small-sized blood vessels. In the simulation, wavelengths of 540 nm, 546 nm, and 576 nm were selected, and the window function of the STFT has a width of 6 nm in the spectral domain. Optical parameters of gray matter at these wavelengths are listed in Table 4. There were two reasons for this choice. First, the scattering coefficients of these three wavelengths were approximately the same, as shown in Figure 2; second, we made a compromise by considering the calculation errors of C and Y. It should be noted that this selection of wavelengths was obtained under the condition that Y = 70% using the optimization method described in Figure 7. The n and g of tissue were 1.37 and 0.9, respectively. The scattering coefficients S in the Table 4 were obtained by dividing the scattering coefficients in Figure 2 by 10 so that they were approximately the same as the scattering coefficients in gray matter [35].

Table 4. Optical parameters of gray matter at selected wavelengths for functional sensing.

λ (nm)	ε_O ($mmol^{-1} \cdot mm^{-1}$)	ε_D ($mmol^{-1} \cdot mm^{-1}$)	S (mm^{-1})
540	5.89	4.81	7.014
546	5.66	5.38	7.016
576	5.91	4.45	7.010

The simulated gray matter was a 300 × 300 × 500 (μm) cube, and simulation results are shown in Figure 9. Figure 9a shows the CW fluence extracted at plane y = 0 μm. The black data points in Figure 9b represent backscattered signals, and the interval of z is 25 μm. The green curve in Figure 9b is the fitting result, and the fitting coefficients and goodness are shown in the picture. The R^2 value of the fitting results is 0.9908. The fitting result in Figure 9b was obtained after removing the class II photons. If class II photons are not removed, the fitting μ_t is small, as shown in Figure 9c because the optical path of the class II photon is larger. For comparison, the data in Figure 9c were normalized. The theoretical calculation values and simulation values of the three parameters are shown in Figure 9d. The points marked "The" in the legend are the theoretical calculation values, and the points labeled "Rea" are the simulation values.

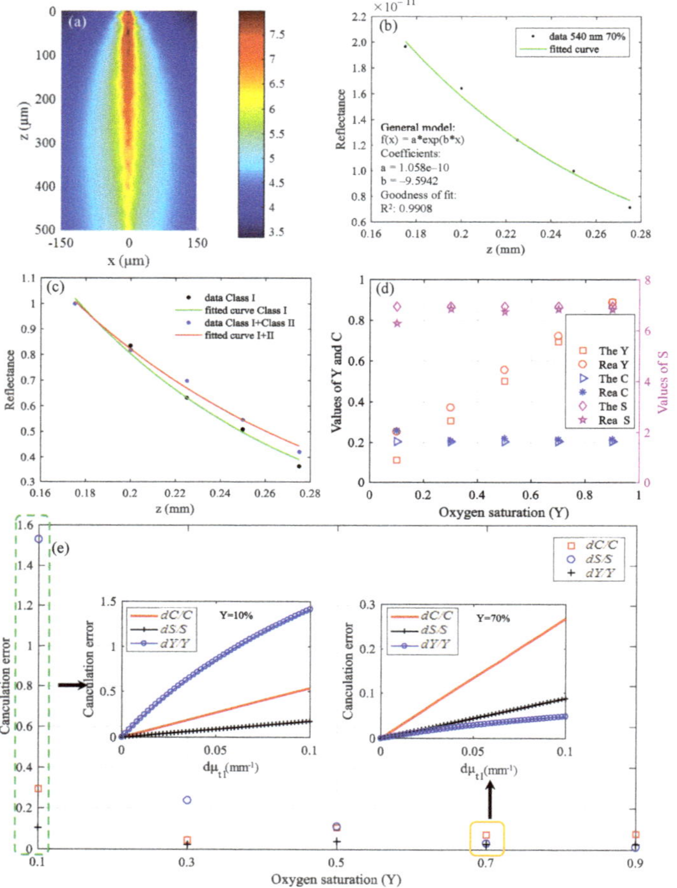

Figure 9. Simulation results in visible waveband: (**a**) CW fluence extracted at plane $y = 0$ μm; (**b**) backscattered signals and fitting results; (**c**) fitting results when class II photons are removed and when class II photons are not removed; (**d**) theoretical calculation values and simulation values of C, Y, and S; and (**e**) differences between theoretical calculation values and simulation values.

It can be seen from Figure 9d that the simulation values are basically in accordance with the theoretical calculation values except that the deviation of Y at $Y = 10\%$ is a little large, which is because the calculation error of Y varies with the value of μ_t, as described in Equation (13). The results of Figure 9e are the differences between the theoretical calculation values and simulation values. The pre-estimated calculation errors in the case that $Y = 10\%$ and $Y = 70\%$ are shown in Figure 9e, which can be seen to be consistent with the results of the actual simulation. If we predict that the Y of tissues is relatively high, then the choice of wavelengths is reasonable now. On the contrary, the current wavelengths selection is unreasonable if we predict that the Y of tissues is relatively low, and other wavelengths should be selected to make the errors of C, Y, and S smaller. When optimized for the case that Y = 10%, the wavelengths of 546 nm, 552 nm, and 570 nm were selected. When the error of the extinction coefficient is the same, dC/C, dY/Y, and dS/S were 0.235, 0.036, and 0.082, respectively, which were reduced by 1.25, 14.7 and 1.28 times, respectively, compared with the results of Figure 9e.

Therefore, in practical application, if we have a prior rough estimate of Y and C of the brain tissue to be measured, we can then reasonably choose the wavelengths to improve the measurement accuracy.

The simulation results show that the correct optical parameters of tissues with low concentrations of hemoglobin can be obtained by rational selection of three wavelengths. However, the error of the extinction coefficient obtained by the simulation is very small, and the maximum value does not exceed 0.05 mm^{-1}. The error of C reaches about 30% when the error of μ_t is 0.1 mm^{-1}. Therefore, when the sample has a low hemoglobin concentration and does not require high spatial resolution, there is no doubt that more accurate results can be obtained by fitting the data of multiple wavelengths.

In summary, we briefly describe the optimization strategy for functional sensing of brain tissue as follows: When imaging small-sized blood vessels, the optimization scheme shown in Figure 7 can be used to reasonably select three wavelengths to resolve C, Y, and S. When imaging the cerebral cortex and the spatial resolution requirement is not high, the method of fitting multi-wavelength data should be used; when the spatial resolution requirement is high, it is necessary to roughly estimate the range of Y and C in advance, and reasonably select three wavelengths to resolve C, Y, and S.

At the end of this section, as a little supplement to this work, functional sensing in a near-infrared waveband was simulated. The simulated tissue was gray matter with 100-μm depth in which a blood vessel with a 100-μm diameter is embedded. Similarly, considering the errors of C and Y simultaneously, we chose three wavelengths at 750 nm, 800 nm, and 900 nm for calculation. The window function of the STFT has a width of 50 nm in the spectral domain. Their optical parameters are shown in Table 5. The parameters of blood were the same as those in Figure 2, and the S values of gray matter in Table 5 were still obtained by dividing the scattering coefficients in Figure 2 by 10. In the near-infrared waveband, the scattering coefficient is approximately linear with the wavelength, as described in Equation (7). The C values of gray matter and vessel were 200 μm and 4 mM, respectively, and the Y values were both 70%.

Table 5. Optical parameters of gray matter and blood vessel at selected wavelengths.

λ (nm)	ε_O (mmol^{-1}·mm^{-1})	ε_D (mmol^{-1}·mm^{-1})	S-Gray (mm^{-1})	S-Vessel (mm^{-1})
750	0.052	0.176	7.625	76.25
800	0.083	0.102	7.3	73
900	0.122	0.1	6.65	66.5

The simulation results are shown in Figure 10. Figure 10a shows the CW fluence extracted at plane y = 0 μm. The structure parameters of the simulated tissue are also indicated. The blue curve in Figure 10b is the backscattered signal, and the fitting results of the gray matter and vessel are shown in the picture. The final fitting results and calculated C, Y, and S from the fitting results are listed in Table 6. We can see that the calculated C and Y are incorrect because the absorption of gray matter is too small in the near-infrared waveband. By contrast, the calculation results for blood vessels are much more accurate.

In the simulations of this paper, we did not simulate the OCT signals of all wavelengths within the spectral bandwidth of the light source but thought that the extinction coefficient is homogeneous within a certain spectral width, which undoubtedly leads to certain errors. In the visible-light waveband, the maximum wavelength bandwidth set in this work was 15 nm, which has proven to be reasonable [36]. In addition, the impacts of the STFT process on functional sensing were not studied. We will optimize the MMC-OCT program to simulate the complete imaging process of the spectroscopic OCT in future researches.

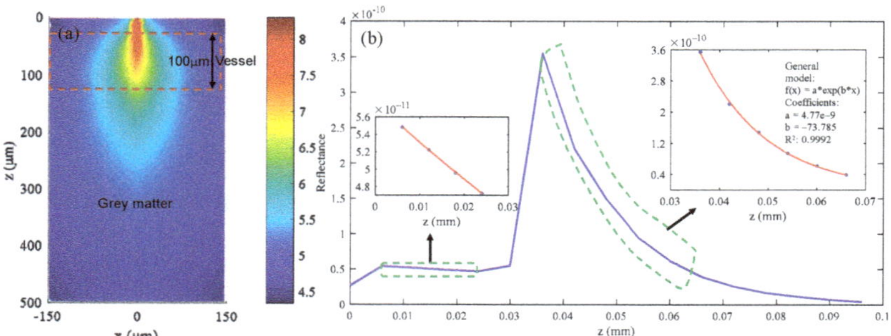

Figure 10. (**a**) CW fluence extracted at plane $y = 0$ μm; (**b**) the backscattered signal collected by the detector and the fitting results.

Table 6. Fitting results and calculated C, Y, and S in near-infrared waveband.

Tissue	The μ_{t1}	The μ_{t2}	The μ_{t3}	Rea μ_{t1}	Rea μ_{t2}	Rea μ_{t3}	C	Y	S_1
Vessel	77.072	73.814	67.56	77.084	73.785	67.524	4.14	0.65	76.18
Gray matter	7.6661	7.3407	6.703	7.712	7.385	6.725	−0.21	0.55	7.75

4. Conclusions

In this study, we analyzed the application of OCT to the fields of brain tissue structural and functional imaging using a model of mesh-based Monte Carlo for OCT (MMC-OCT). We first verified that the MMC-OCT model used in this paper is an effective analytical tool for brain researches through simulations of structural imaging. Then, through theoretical analysis and simulation of functional sensing, we found that the measurement accuracy of hemoglobin concentration C, oxygen saturation Y, and scattering coefficient S can be optimized by reasonable selection of wavelengths. Finally, we proposed the following optimization schemes for imaging vascular structure and cerebral cortex in brain tissue: When imaging small-sized blood vessels, we can reasonably select three wavelengths to resolve C, Y, and S. When imaging the cerebral cortex and the spatial resolution requirement is not high, the method of fitting with multi-wavelength data should be used; when the spatial resolution requirement is high, the measurement accuracy of C, Y, and S can be improved by roughly predicting their ranges in advance to select the appropriate wavelengths for calculation. What is more, these methods are also suitable for other complex 3D bio-tissues. In particular, the method of wavelength selection is also applicable to other optical imaging modalities such as photoacoustic imaging [37], near infrared multispectral imaging [19], etc. In the future, we will provide the whole set of codes and a small guidance for reproducing the results of simulations presented in this manuscript, which provides an analytical method for studying the brain tissue and other complex 3D bio-tissues.

Author Contributions: Conceptualization, L.Y.; formal analysis, L.Y.; methodology, L.Y. and L.S.; software, L.Y., M.Z. and B.H.; supervision, L.S.; validation, L.Y. and L.S.; writing—original draft, L.Y.; writing—review and editing, L.S., M.Z. and B.H.

Funding: This research received no external funding

Acknowledgments: L.Y., M.Z., and B.H. acknowledge the State Key Laboratory of Precision Measurement Technology and Instruments, Department of Precision Instruments, Tsinghua University for support through a PhD studentship.

Conflicts of Interest: The authors declare no conflict of interest.

References

1. Xie, Y.; Harsan, L.A.; Bienert, T.; Kirch, R.D.; Von, E.D.; Hofmann, U.G. Qualitative and quantitative evaluation of in vivo SD-OCT measurement of rat brain. *Biomed. Opt. Express* **2017**, *8*, 593–607. [PubMed]

2. Choi, W.J.; Wang, R.K. Optical coherence tomography imaging of cranial meninges post brain injury in vivo. *Chin. Opt. Lett.* **2017**, *15*, 16–19.
3. Tearney, G.J.; Fukumura, D.; Jain, R.K.; Vakoc, B.J.; Lanning, R.M.; Tyrrell, J.A.; Padera, T.P.; Bartlett, L.A.; Stylianopoulos, T.; Munn, L.L. Three-dimensional microscopy of the tumor microenvironment in vivo using optical frequency domain imaging. *Nat. Med.* **2009**, *15*, 1219.
4. Hillman, C.E.M. Optical brain imaging in vivo: Techniques and applications from animal to man. *J. Biomed. Opt.* **2007**, *12*, 051402. [CrossRef] [PubMed]
5. Villringer, A.; Chance, B. Non-invasive optical spectroscopy and imaging of human brain function. *Trends Neurosci.* **1997**, *20*, 435–442. [PubMed]
6. Chong, S.P.; Merkle, C.W.; Leahy, C.; Radhakrishnan, H.; Srinivasan, V.J. Quantitative microvascular hemoglobin mapping using visible light spectroscopic Optical Coherence Tomography. *Biomed. Opt. Express* **2015**, *6*, 1429–1450. [PubMed]
7. Ji, Y.; Vadim, B. Imaging a full set of optical scattering properties of biological tissue by inverse spectroscopic optical coherence tomography. *Opt. Lett.* **2012**, *37*, 4443–4445.
8. Ahn, Y.C.; Jung, W.; Chen, Z. Quantification of a three-dimensional velocity vector using spectral-domain Doppler optical coherence tomography. *Opt. Lett.* **2007**, *32*, 1587–1589.
9. Zhu, C.; Liu, Q. Review of Monte Carlo modeling of light transport in tissues. *J. Biomed. Opt.* **2013**, *18*, 50902.
10. Wang, L.; Jacques, S.L.; Zheng, L. MCML–Monte Carlo modeling of light transport in multi-layered tissues. *Comput. Methods Programs Biomed.* **1995**, *47*, 131–146.
11. Yao, G.; Wang, L.V. Monte Carlo simulation of an optical coherence tomography signal in homogeneous turbid media. *Phys. Med. Biol.* **1999**, *44*, 2307–2320. [PubMed]
12. Shi, B.; Zhuo, M.; Wang, L.; Liu, T. Monte Carlo modeling of human tooth optical coherence tomography imaging. *J. Opt.* **2013**, *15*, 075304.
13. Periyasamy, V.; Pramanik, M. Importance sampling-based Monte Carlo simulation of time-domain optical coherence tomography with embedded objects. *Appl. Opt.* **2016**, *55*, 2921–2929. [PubMed]
14. Hartinger, A.E.; Nam, A.S.; Isabel, C.C.; Vakoc, B.J. Monte Carlo modeling of angiographic optical coherence tomography. *Biomed. Opt. Express* **2014**, *5*, 4338–4349. [CrossRef] [PubMed]
15. Igor, M.; Mikhail, K.; Vladimir, K.; Risto, M. Simulation of polarization-sensitive optical coherence tomography images by a Monte Carlo method. *Opt. Lett.* **2008**, *33*, 1581–1583.
16. Fang, Q. Mesh-based Monte Carlo method using fast ray-tracing in Plücker coordinates. *Biomed. Opt. Express* **2010**, *1*, 165–175. [PubMed]
17. Yao, R.; Intes, X.; Fang, Q. Generalized mesh-based Monte Carlo for wide-field illumination and detection via mesh retessellation. *Biomed. Opt. Express* **2015**, *7*, 171–184.
18. Yi, J.; Chen, S.; Backman, V.; Zhang, H.F. In vivo functional microangiography by visible-light optical coherence tomography. *Biomed. Opt. Express* **2014**, *5*, 3603–3612. [PubMed]
19. Uludag, K.; Kohl, M.; Steinbrink, J.; Obrig, H.; Villringer, A. Cross talk in the Lambert-Beer calculation for near-infrared wavelengths estimated by Monte Carlo simulations. *J. Biomed. Opt.* **2002**, *7*, 51–59. [PubMed]
20. George, M. Optical constants of water in the 200-nm to 200-μm wavelength region. *Appl. Opt.* **1973**, *12*, 555–563.
21. Leitgeb, R.; Wojtkowski, M.; Kowalczyk, A.; Hitzenberger, C.K.; Sticker, M.; Fercher, A.F. Spectral measurement of absorption by spectroscopic frequency-domain optical coherence tomography. *Opt. Lett.* **2000**, *25*, 820–822. [CrossRef] [PubMed]
22. Bosschaart, N.; Edelman, G.J.; Aalders, M.C.G.; Leeuwen, T.G.V.; Faber, D.J. A literature review and novel theoretical approach on the optical properties of whole blood. *Laser Med. Sci.* **2013**, *29*, 453–479. [CrossRef]
23. Smith, M.H.; Denninghoff, K.R.; Hillman, L.W.; Chipman, R.A. Oxygen Saturation Measurements of Blood in Retinal Vessels during Blood Loss. *J. Biomed. Opt.* **1998**, *3*, 296–303. [PubMed]
24. Delori, F.C. Noninvasive technique for oximetry of blood in retinal vessels. *Appl. Opt.* **1988**, *27*, 1113–1125. [PubMed]
25. Hermann, B.; Bizheva, K.; Unterhuber, A.; Povazay, B.; Sattmann, H.; Schmetterer, L.; Fercher, A.; Drexler, W. Precision of extracting absorption profiles from weakly scattering media with spectroscopic time-domain optical coherence tomography. *Opt. Express* **2004**, *12*, 1677–1688.

26. Faber, D.J.; Aalders, M.C.G.; Mik, E.G.; Hooper, B.A.; van Gemert, M.J.C.; van Leeuwen, T.G. Oxygen saturation-dependent absorption and scattering of blood. *Phys. Rev. Lett.* **2004**, *93*, 028102. [CrossRef] [PubMed]
27. Wang, L.V.; Wu, H.I.; Masters, B.R. Biomedical Optics, Principles and Imaging. *J. Biomed. Opt.* **2008**, *13*, 049902.
28. Matcher, S.J.; Cope, M.; Delpy, D.T. In vivo measurements of the wavelength dependence of tissue-scattering coefficients between 760 and 900 nm measured with time-resolved spectroscopy. *Appl. Opt.* **1997**, *36*, 386–396.
29. Hong, G.; Diao, S.; Chang, J.; Antaris, A.L.; Chen, C.; Zhang, B.; Zhao, S.; Atochin, D.N.; Huang, P.L.; Andreasson, K.I. Through-skull fluorescence imaging of the brain in a new near-infrared window. *Nat. Photonics* **2014**, *8*, 723–730.
30. The Program was Downloaded from Open Source Website and Modified. Available online: http://mcx.space/mmc (accessed on 2 November 2018).
31. Siavash, M.; Lima, I.T.; Sherif, S.S. Monte Carlo simulation of optical coherence tomography for turbid media with arbitrary spatial distributions. *J.Biomed. Opt.* **2014**, *19*, 046001.
32. Collins, D.L.; Zijdenbos, A.P.; Kollokian, V.; Sled, J.G.; Kabani, N.J.; Holmes, C.J.; Evans, A.C. Design and construction of a realistic digital brain phantom. *IEEE Trans. Med. Imaging* **1998**, *17*, 463–468. [PubMed]
33. Umeyama, S.; Yamada, T. Monte Carlo study of global interference cancellation by multidistance measurement of near-infrared spectroscopy. *J.Biomed. Opt.* **2009**, *14*, 064025. [PubMed]
34. Kholodnykh, A.I.; Petrova, I.Y.; Larin, K.V.; Massoud, M.; Esenaliev, R.O. Precision of measurement of tissue optical properties with optical coherence tomography. *Appl. Opt.* **2003**, *42*, 3027–3037. [PubMed]
35. Yaroslavsky, A.N.; Schulze, P.C.; Yaroslavsky, I.V.; Schober, R.; Ulrich, F.; Schwarzmaier, H.J. Optical properties of selected native and coagulated human brain tissues in vitro in the visible and near infrared spectral range. *Phys. Med. Biol.* **2002**, *47*, 2059–2073. [CrossRef] [PubMed]
36. Ji, Y.; Qing, W.; Liu, W.; Backman, V.; Zhang, H.F. Visible-light optical coherence tomography for retinal oximetry. *Opt. Lett.* **2013**, *38*, 1796–1798.
37. Yin, G.; Yi, Y. Imaging of hemoglobin oxygen saturation using confocal photoacoustic system with three optical wavelengths. In Proceedings of the Seventh International Conference on Photonics and Imaging in Biology and Medicine, Wuhan, China, 24–27 November 2008; Volume 7280.

Article

Topological Charge Detection Using Generalized Contour-Sum Method from Distorted Donut-Shaped Optical Vortex Beams: Experimental Comparison of Closed Path Determination Methods

Daiyin Wang [1], Hongxin Huang [2,*], Haruyoshi Toyoda [2] and Huafeng Liu [1,*]

[1] College of Optical Science and Engineering, Zhejiang University, Hangzhou 310027, China; wangdaiyin@126.com

[2] Central Research Laboratory, Hamamatsu Photonics K. K., Hamamatsu 434-8601, Japan; toyoda@crl.hpk.co.jp

* Correspondence: huanghx@crl.hpk.co.jp (H.H.); liuhf@zju.edu.cn (H.L.)

Received: 30 August 2019; Accepted: 17 September 2019; Published: 20 September 2019

Featured Application: This study will be valuable to researchers working in optical metrology and in the diagnosis of optical communication links through long-distance free-space propagation.

Abstract: A generalized contour-sum method has been proposed to measure the topological charge (TC) of an optical vortex (OV) beam using a Shack–Hartmann wavefront sensor (SH-WFS). Moreover, a recent study extended it to be workable for measuring an aberrated OV beam. However, when the OV beam suffers from severe distortion, the closed path for circulation calculation becomes crucial. In this paper, we evaluate the performance of five closed path determination methods, including watershed transformation, maximum average-intensity circle extraction, a combination of watershed transformation and maximum average-intensity circle extraction, and perfectly round circles assignation. In the experiments, we used a phase-only spatial light modulator to generate OV beams and aberrations, while an SH-WFS was used to measure the intensity profile and phase slopes. The results show that when determining the TC values of distorted donut-shaped OV beams, the watershed-transformed maximum average-intensity circle method performed the best, and the maximum average-intensity circle method and the watershed transformation method came second and third, while the worst was the perfect circles assignation method. The discussions that explain our experimental results are also given.

Keywords: wavefront sensor; spatial light modulator; contour-sum method; topological charge; orbital angular momentum

1. Introduction

Recently, optical vortex (OV) beams, owing to their unique properties, have attracted more and more interest and have been utilized in a wide range of fields—from scientific research to advanced technology applications, such as optical communications [1–4], optical metrology [5–7], and optical trapping and manipulation [8–10]. Many specialties of OV beams are due to their phase singularity in the wavefront function, where the intensity drops to zero and the phase is undefined [11,12]. Moreover, the phase along a closed path enclosing the singularity point varies from 0 to $2n\pi$, where n is an integer known as the topological charge (TC) or the orbital angular momentum (OAM). OV beams with different TC values perform diverse characteristics and consequently are used as information carriers in state-of-art optical communication systems, which are used to generate sufficient force to manipulate the molecules and so on. To meet the requirements of these applications, the determination of how to precisely measure the TC value of an OV beam is an important issue, and therefore many methods, such as interferometry-based methods [13], diffraction-based methods [14–16], and model decomposition-based methods [17,18] have been proposed and comprehensively studied.

As one of the key devices used in adaptive optics systems [19], Shack–Hartmann wavefront sensors (SH-WFS) have also been utilized to determine the TC value of OV beams [12,20–23]. An SH-WFS consists of a lenslet array and an image sensor, and directly measures the phase slope of the incident wavefront at each lenslet position. TC determination with an SH-WFS is simple and direct, and the contour-sum method (CSM) has been proposed based on the principle that the net TC value in an area is proportional to the line integral of the phase slope along the closed path circumscribing the area [12].

The basic form of the contour-sum method uses the pre-assigned closed paths, e.g., the closed path associated with the central 2×2 or 3×3 lenslet area to calculate the discrete line integral, which is employed to determine the TC value of an OV beam [20,22]. This approach is successful in measuring OV beams with nearly uniform or quasi-uniform intensity distribution. However, it becomes deficient under some conditions, especially when the OV beams to be measured have donut-shaped intensity profiles, where phase singularities are embedded in the low-intensity region (dark region), and the phase slope measurement could be invalidated. This condition commonly exists in many OV applications such as OV-based optical communication and OV-based optical metrology. In view of this, in a previous study, we generalized the contour-sum method to be workable for a closed path with an arbitrary shape, and proposed a maximum average-intensity circle (MAIC) method to extract a closed path with only valid phase slope data from the annular intensity profile [24]. We experimentally investigated the MAIC method with both aberration-free OV beams as well as the OV beams distorted by simulated atmospheric turbulence, and we concluded that the proposed MAIC method has good robustness against aberrations. Moreover, we found that the closed path used for circulation calculation has a notable influence on the determined TC value when the OV beams are severely distorted by aberrations. Considering that the closed path is vital for the CSM, we also briefly demonstrated the superiority of the MAIC method against the use of perfectly round circles (PRC) as the closed path [24].

With the aim of further improving the measurement accuracy under the condition of severe distortion and enriching the discussion of the influence of closed paths in circulation calculations, in this paper, we compare the performance of several closed path determination methods: perfectly round circles assignation, watershed transformation, maximum average-intensity circle extraction, and the combination of the watershed transformation with maximum average-intensity circle extraction.

This paper is organized as follows. In Section 2, we present a brief of the generalized contour-sum method (GCSM), and in Section 3, we describe five closed path determination methods for the GCSM. In Section 4, we show the optical setup for the proof-of-principle experiments. In Section 5, we present some experimental results and discussions. The comparisons were conducted under scenarios of both aberration-free OV beams and OV beams distorted by simulated atmospheric aberrations. A summary and conclusion are given in Section 6.

2. Generalized Contour-Sum Method

The contour-sum method was firstly proposed by Fried and Vaughn in 1992 [12] to prove that there is a branch cut (phase singularity) in the phase function of a light field with strong intensity variation. Since then, it has been adopted to detect OV beams using the phase slopes measured by an SH-WFS [20]. The essence of this method is that the circulation value of the phase gradient along a closed path enclosing the singularity point has a relationship with the TC value n, which is an integer and can be expressed as [20]:

$$n = \frac{1}{2\pi} \oint_C \nabla\phi \cdot d\vec{r}, \tag{1}$$

where C is a closed path, $\nabla\phi$ is the phase gradient, and $d\vec{r}$ is an infinitesimal displacement along the closed path C.

In practice, in order to determine the TC, we must discretize the line integral according to the geometrical configuration of sampling points, and accordingly rewrite Equation (1) as:

$$Cir = \frac{A}{2\pi}\sum_{k=1}^{K}\vec{S}_k\cdot\vec{l}_k, \tag{2}$$

where K is the number of sampling points along the closed path, and S_k and l_k are the phase slope and discretized contour path of the k-th sampling point, respectively. In Equation (2), a constant A is introduced to compensate for the error caused by the discretization [24], and Cir is the TC value to be measured.

Specially, with the use of an SH-WFS, the discretization is realized by the lenslet array, and the phase slopes are simply obtained by the displacements of the focusing spots produced by the corresponding lenslets. Thus, the discretized contour paths connect the centers of the lenslet areas, forming the closed path. Figure 1 illustrates the generalized contour-sum method. In Figure 1, the square indicates the lenslet area, and the area marked with downward diagonal lines is the element forming the closed path. The discretized-contour path l_k is represented by the red vector that connects the centers of the two adjacent elements in the closed path, and the blue vector S_k is the phase slope, which is the average of the phase slopes at the two adjacent elements.

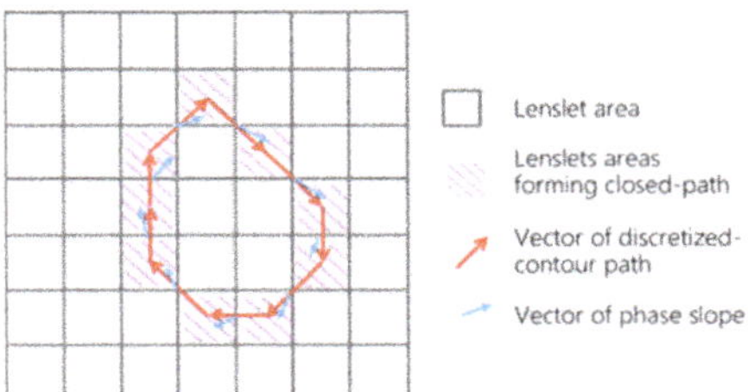

Figure 1. Graph illustrating the generalized contour-sum method.

The raw data from the SH-WFS measurement consist of a Hartmanngram of multiple focused spots. The five main steps of the GCSM are (1) preprocessing the Hartmanngram, including thresholding and segmentation; (2) calculating the phase slope of the individual lenslet area locations according to the SH-WFS working principle; (3) summing the intensity values of all pixels in each lenslet area to obtain the intensity sum value; (4) extracting a closed path from the intensity sum map; and (5) calculating the circulation, which is TC value of the OV beam.

In this algorithm, the closed path should be extracted from the intensity sum map, instead of from the Hartmanngram itself. This is because the intensity distribution behind the individual lenslet areas reflects the average distortion, but not the intensity distribution of the sub-wavefront incident into the lenslets. Since closed path determination is a crucial step, we focus on exploring the most proper closed path determination method in the rest of this paper.

3. Methods for Closed Path Determination

Basically, there are two strategies to determine a closed path from a given intensity sum map. One is to adaptively extract the closed path along the ridge of the intensity sum map. The reason why we search for the ridge elements is to ensure that all the measured phase slope data along the closed path are valid. Another strategy is to generate a closed path with a perfectly round shape, whose diameter and center could be adaptively varied within a given range. Hence, in this section, we explain closed path determination methods, following each strategy in detail. We introduce three methods that conform to the ridge-extracting strategy, which are watershed transformation (WT), MAIC extraction, and the combination of watershed transformation and MAIC extraction (WT-MAIC), as well as two methods that conform to the perfectly round circle generation strategy, which are the fixed-center perfectly round circle method (FC-PRCM) and the shifting-center perfectly round circle method (SC-PRCM).

3.1. Watershed Transformation Method

Watershed transformation (WT) is a technique to extract ridges from an elevation map. This idea originated from the field of topography, and was firstly introduced as an image processing method by Beucher and Lantuéjoul [25]. Since then, many modifications and improvements to the method have been proposed [26–29]. The kernel of watershed transformation is to treat the two-dimensional grayscale input image as a topographic map, as shown in Figure 2, with the grayscale value of each individual point representing the elevation. After this transformation, the image is intuitively divided into several catchment basins, each of which corresponds to a regional minimum in the elevation dimension. The boundaries between diverse catchment basins are considered as the ridges, which are commonly termed watershed lines.

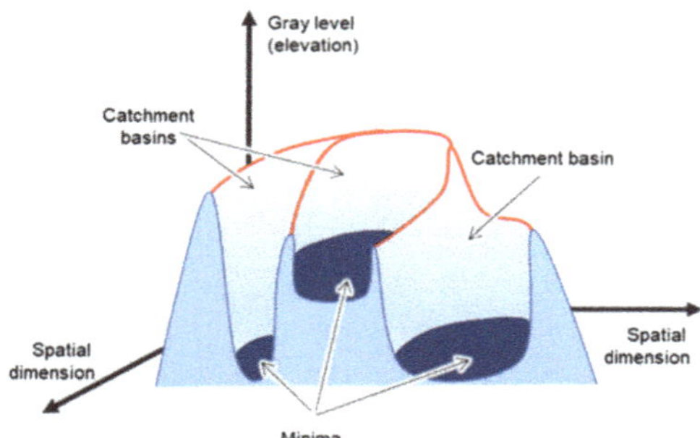

Figure 2. A diagram to illustrate the topographic map of an image after performing watershed transformation. The red curves are the watershed lines.

To perform watershed transformation on an image, flooding is the most commonly used strategy, and Meyer's flooding algorithm is the corresponding extensively used algorithm [27]. Its core idea is to flood the entire topographic map, which is equivalent to raising the zero-elevation plane with time. With the flooding, the catchment basin corresponding to the global minimum is first filled with water, following which the other catchment basins start to fill with water one by one. After a certain amount of time, water from different catchment basins meet, and we build barriers to prohibit this phenomenon. The resulting barriers comprise the watershed lines, which correspondingly segment the input image into different regions.

Specifically, to apply watershed transformation on an intensity sum map to determine a closed path, we first perform preprocessing on the intensity sum map, such as thresholding and filtering [26], and then we perform the watershed transformation using the built-in function 'watershed' of MATLAB [30]. Moreover, considering that we aim to determine the net TC value in an OV beam, we should obtain only one closed path out of the multiple watershed lines. This can be realized by merging small regions into their neighboring large regions until there is only one region.

3.2. Maximum Average-Intensity Circle Extraction

The maximum average-intensity circle (MAIC) method has been proposed to extract a closed path from an intensity sum map for TC determination [24], as is briefly explained here. The basic process is to iteratively search the local peaks from the 2D intensity sum map and to connect these peak elements to form a closed path. The searching begins with the global maximum in the intensity sum map. The local maximum element from among several candidates of the eight neighbors of the current element is selected as the next element.

3.3. Watershed Transformed Maximum Average-Intensity Circle Extraction

Both WT and MAIC methods are simple and intuitive ways to obtain a closed path for circulation calculation, and they have equally excellent performance when the OV beam has little to no aberration. However, when the OV beam is severely distorted by aberration, the intensity sum map of the OV beam will have a shape far from a perfect circle, and may even split into several parts, resulting in both methods losing their efficiency. To further improve the performance under this severe aberration condition, we here propose a new method that combines WT and MAIC. In this method, we first perform watershed transformation on the intensity sum map, and then search for the ridge from among the elements on the watershed lines based on the MAIC algorithm. With the output of watershed transformation previously obtained, the influence of the element along the ridge and the noise can be significantly reduced, and thereby, a more proper closed path can be extracted. For simplicity, we named this method the watershed-transformed MAIC method (WT-MAICM).

3.4. Perfectly Round Circle Assignation

As mentioned above, besides the ridge-extracting strategy, adaptively generating an appropriate perfectly round circle according to the intensity sum profile of an OV beam is also a closed path determination strategy that is worthy of discussion. In general, the center position and radius are two integral parameters to generate a perfectly round circle. However, given an intensity sum profile, it is difficult to directly determine the most appropriate perfectly round circle. Consequently, in our previous research, we proposed to generate a group of concentric perfect circles, and chose the one for which the measured TC value had the maximum absolute value as the determined perfect round circle. The center position of the concentric perfect circles was determined as the nearest lenslet position from the centroid of the intensity sum map, while the lower and upper bounds of the radius variation range was determined by the radii of the inscribed and circumscribed circles of the intensity sum profile [24].

The performance of this generated perfectly round circle method (PRCM) as the closed path determination method to measure the TC value of a distorted OV beam is generally worse than that of the MAIC method [21,24]. The reasons are listed as follows:

- The method determining the center position of the generated perfectly round circles by the centroid of intensity sum map is not optimal, because the nonuniformity along the azimuthal direction in the intensity sum distribution significantly affects the centroid calculation.
- The center position as well as the radius are all forced to be integers.
- The generated circles restricted to perfectly round shapes will unavoidably go through the low-intensity region, which means that invalid phase slope data will be obtained.

Notwithstanding, the generated circle should be fixed to a perfectly round shape to agree with the annular intensity sum profile and the position. Moreover, the radius should be forced to be an integer. This is because if we adaptively vary the shape or precisely calculate the position and radius and determine the phase slopes of the non-integer position by interpolation, the incurred complexity will go against our original goal of maintaining simplicity in the calculations. However, regarding the center position determination, the performance may be improved by shifting the center position within a proper region, referring to the radius variation. Although this dramatically increases the closed path determination time, we can accept a tradeoff between TC determination speed and accuracy if such a tradeoff manifests an improvement in the determination performance.

Consequently, in this paper, we propose a new, modified method for the generation of perfectly round circles based on the previously proposed method, where the modification is merely to shift the center position of the concentric perfectly round circles within an advisable region, as is further discussed in Section 5. For the sake of differentiation, we named this method the shifting-center perfectly round circle method (SC-PRCM) and the previously proposed perfectly round circle method the fixed-center perfectly round circle method (FC-PRCM).

4. Experimental Setup

To verify the performance of these methods of closed path determination for the GCSM, we built an experimental setup, as shown in Figure 3 [24]. As shown in Figure 3, a collimated laser beam of wavelength 632.8 nm passed an aperture with a diameter of 4 mm, and was incident on a liquid crystal on silicon-spatial light modulator (LCOS-SLM). The LCOS-SLM was used to transform the incident beams into the optical vortex beams as well as bring aberrations into the beam. The beam reflected back from the LCOS-SLM was converged by a lens with a focal length of 2 m, and was split by a beam splitter into two beams. Finally, we used an SH-WFS to record the Hartmanngram and a complementary metal-oxide semiconductor (CMOS) camera to check and record the intensity profile of the OV beam. The LCOS-SLM was set at the front focal plane of the lens, while the SH-WFS as well as the CMOS camera were both located at the back focal plane of the lens.

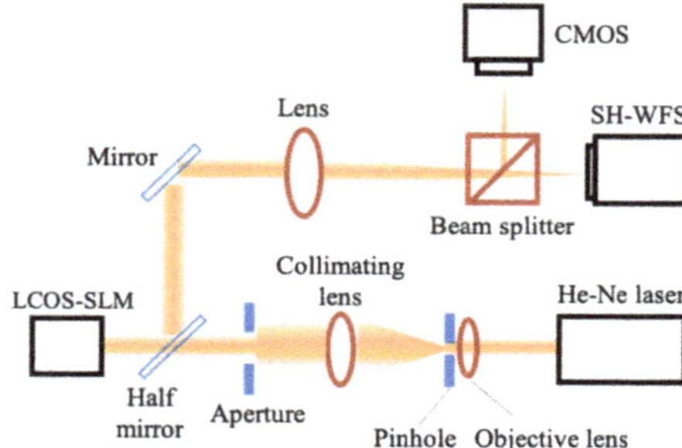

Figure 3. Schematic diagram of the experimental setup.

The LCOS-SLM (Hamamatsu, X10468-01) was a pure phase modulator consisting of 792 × 600 pixels with a pixel size of 20 × 20 μm [31]. The SH-WFS consisted of two elements: a square grid lenslet array with a pitch size of 200 μm and focal length of 11 mm, and a high-speed intelligent vision sensor with 512 × 512 pixels and a pixel size of 20 × 20 μm [32]. The SH-WFS was mounted on a mechanical platform that could be moved along the horizontal direction. The movement was precisely controlled by a stepping motor system. The CMOS camera was 2592 × 2048 pixels and had a pixel size of 4.8 × 4.8 μm.

5. Results and Discussion

5.1. Performance Comparison Based on Aberration-Free OV Beams

To compare the above five closed path determination methods, we first evaluated the performance under the aberration-free OV beam condition. In the experiments, we displayed various computer-generated holograms (CGHs) with a spiral phase structure on the LCOS-SLM to generate an OV beam with TC values ranging from ± 1 to ± 20. In order to enrich the data amount as well as eliminate the randomness, for each TC value, we repeatedly recorded the Hartmanngrams at 15 different SH-WFS positions by laterally moving the mechanical platform. Hence, we obtained a total of 600 Hartmanngrams (40 TC values and 15 SH-WFS positions). Figure 4 shows an example of the spiral phase pattern displayed on LCOS-SLM, and the corresponding Hartmanngram as well as the intensity profile image respectively recorded by SH-WFS and CMOS camera.

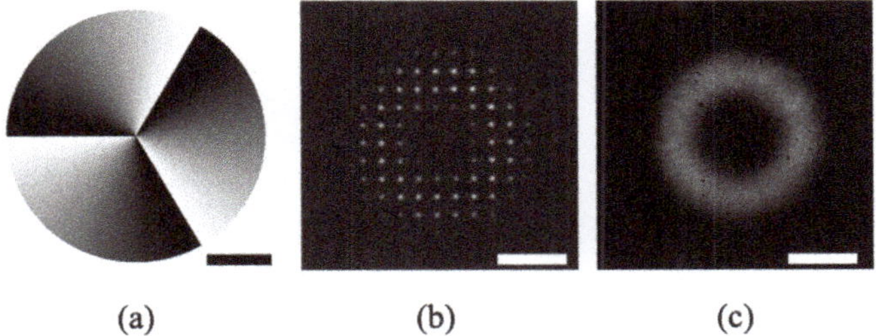

Figure 4. Example of (**a**) spiral phase pattern, (**b**) Hartmanngram, and (**c**) intensity profile recorded by the complementary metal-oxide semiconductor (CMOS) camera. The bar indicates 1 mm.

As described in Section 2, for each recorded Hartmanngram, we first performed the necessary preprocessing, and then summed all the pixel values in each lenslet region to obtain the intensity sum map and measured the phase slopes at the lenslet positions according to the SH-WFS working principle. After that, we used each of the studied five methods to determine the closed path, and then utilized the generalized counter-sum method to calculate the TC value.

Considering that the theoretical TC value is an integer, for TC measurement, if the absolute difference between the measured TC value and the theoretical TC value is lower than 0.5, the measured TC value will be identical to the theoretical TC value after rounding—we define this as a well-determined TC measurement. Based on this definition, we introduced a parameter, the well-determined TC measurement ratio (WTCMR), which is the percentage of well-determined TC measurements within a given set of TC measurements, to quantitatively compare the performance of the diverse closed path determination methods.

The performance evaluation results of the individual closed path determination methods are given in Figures 5 and 6. In Figure 5, each value is the WTCMR of a group of 30 measurements (15 positions for both positive and negative TC values), while in Figure 6, the WTCMR is the statistical average of all 600 measurements. Figure 6 also shows the corresponding processing speed of the individual methods in terms of frames per second, which was measured on an Intel Core i7 computer with a CPU frequency of 3.7 GHz and 16 GB of RAM.

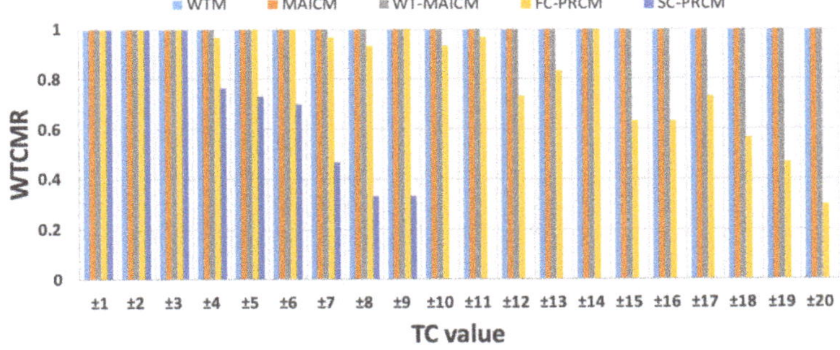

Figure 5. Well-determined topological charge (TC) measurement ratio (WTCMR) distributions of the five closed path determination methods when the TC value changes from ±1 to ± 20.

As shown in Figures 5 and 6, the WT, MAIC, and WT-MAIC methods performed perfectly, since the WTCMR was shown to be 100% for all tested TC values. However, the WTCMR of the FC-PRCM and the SC-PRCM was not always 100%, even when measuring aberration-free OV beams. Moreover, the performance of these methods deteriorated with the increase of the TC value. The SC-PRCM was the worst method in terms of both the WTCMR and the computing speed.

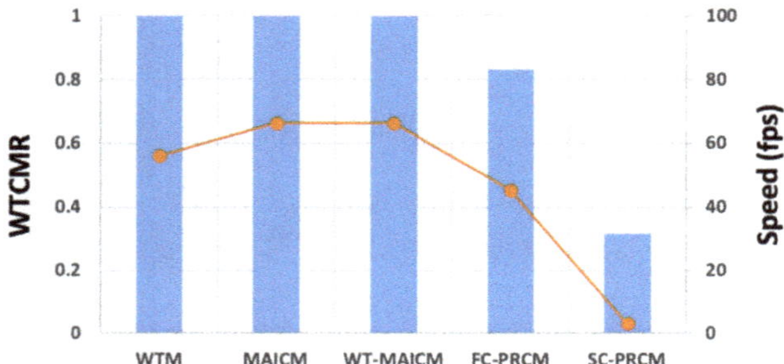

Figure 6. TC determination performances of five closed path determination methods in terms of the WTCRM (bars) and speed in frames per second (fps) (points). Each bar or point is the statistical average of all 600 measurements under the corresponding closed path determination method.

Considering that the TC determination progress is the same apart from the closed path being determined by diverse closed path determination methods, the reason for the differences in performance was exclusively due to the differences in the determined closed paths. To concretely demonstrate these differences, as well as the origin of the differences and why this factor significantly changes the TC determination performance, we present some examples in Figure 7. In this figure, the columns from left to right indicate the results using the WTM, MAICM, WT-MAICM, FC-PRCM, and SC-PRCM, and the rows from top to bottom are the inputted OV beams of TC values of 10 (top row), 15 (middle row), and 20 (bottom row). In Figure 7, each row has the same intensity sum map, and the red circle superimposed on the intensity sum map is the closed path extracted by the individual method.

From the figure, it can be seen that the WTM, MAICM, and WT-MAICM can properly extract closed paths passing through the ridge elements, thus obtaining precise TC values. On the other hand, for the FC-PRCM and SC-PRCM, the closed path might go through the low-intensity sum regions where the measured phase slopes are invalid. When the TC value of the OV beam varied from 10 to 15 to 20, the effects of the nonuniformity of the intensity sum map along the azimuthal direction increased. As a result, the closed path deviated more from the ridge elements and went through the low-intensity regions, generating a large error in the measured TC values. As for the SC-PRCM, its performance deteriorated more compared with that of the FC-PRCM. This reveals that simply shifting the center position of the concentric perfectly round circles does not improve, but rather reduces the performance. We believe that this is because shifting the center position of perfectly round circles may deviate the centers even further from the real position of the vortex, and thus they become liable to be affected by the invalid slope data in the low-intensity regions. Considering that we chose the generated perfectly round circle with the maximum absolute calculated TC value as the determined closed path, the SC-PRCM mostly output a measured TC value larger than the theoretical value. Moreover, in combination with the poor performance of the FC-PRCM, we found that the unified closed path determination criterion in the PRCM strategy, which is to select the generated perfectly round circle with the maximum absolute calculated TC value as the determined closed path, is not entirely rational, although it is simple.

The above finding was also supported by analyzing the errors of the measured TC values. Figure 8 shows the histogram distributions of the absolute error (AE) of TC measurements, wherein the x-axis is the interval of the AE, which is the absolute difference between the measured TC value and the theoretical TC value, and the y-axis is the frequency of the 600 measurements. From the distribution, we can view the striking difference between the performances of the WTM, MAICM, and WT-MAICM, and those of the FC-PRCM and SC-PRCM. Almost all of the AEs of the former three methods are concentrated within the value interval of 0 to 0.2; on the contrary, the AEs are distributed in a partly uniform pattern with one-sixth within the value interval of over 0.5 when using the FC-PRCM, while the majority are in the value interval of over 0.5 when using the SC-PRCM.

This phenomenon profoundly influences the correct rate of TC determination when we shrink the confidence interval of the well-determined TC measurements. For example, when using AE < 0.3 as a criterion for the well-determined TC measurement definition, the WTCMR values of the WTM, MACIM, and WT-MAICM are still over 99%; however, those of the FC-PRCM and SC-PRCM drop to 63.3% and 19.7%, respectively.

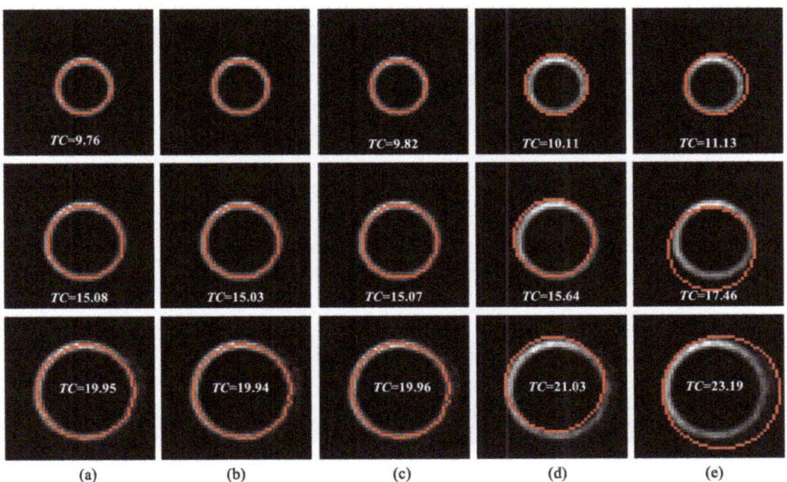

Figure 7. Examples showing the differences of the closed paths (the red circles) and the TC values determined by (**a**) the watershed transformation method (WTM), (**b**) the maximum average-intensity circle method (MAICM), (**c**) the WT-MAICM, (**d**) the fixed-center perfectly round circle method (FC-PRCM), and (**e**) the shifting-center (SC-PRCM). The theoretical TC values are 10 (top row), 15 (middle row), and 20 (bottom row). In each image, the background is the intensity sum map, and the red circle is the closed path determined by the individual methods. The corresponding measured TC value is labeled in each image.

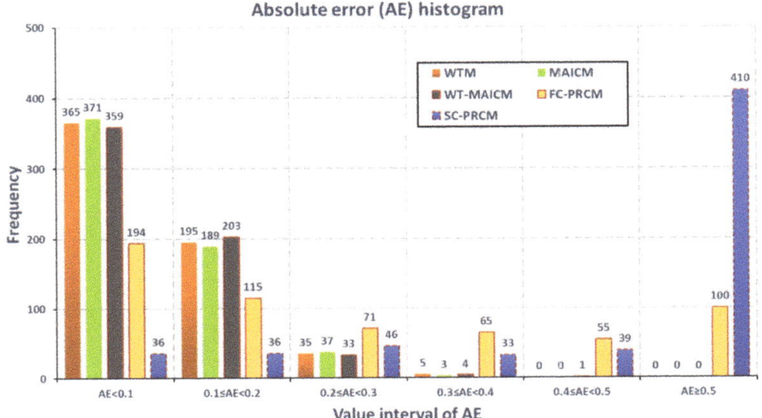

Figure 8. Absolute error (AE) histograms of 600 TC measurements separately using WTM, MAICM, WT-MAICM, FC-PRCM, and SC-PRCM. The number over each bar indicates the number of the measurements whose AE values are within the corresponding interval.

According to the previous discussions, we concluded that the FC-PRCM and SC-PRCM are not appropriate as closed path determination methods, and that generating a perfectly round circle is not a plausible nor practical closed path determination strategy. Therefore, we started to just involve the other three closed path determination methods in the following comparison.

5.2. Performance Comparison Based on Distorted OV Beams

Many practical applications of OV beams, e.g., free-space optical communication and optical remote metrology, require the determination of the TC value of the OV beam after its propagation over a certain distance in the atmosphere. Consequently, it is vital to evaluate and compare these methods for OV beams distorted by atmospheric turbulence. Generally, a turbulent atmosphere can be treated as an inhomogeneous refractive index media, featured by the structure parameter C_n^2 [33], and its impact on the light beam propagating through it can be accumulated as phase screens [34]. Supposing the turbulence satisfies the Kolmogorov model, the phase screen can be simulated through the Zernike polynomial [35]. The coefficients of the Zernike polynomial are related to the normalized correlation length r_0/D, where r_0 is the Fried parameter and D is the beam size [36].

In the experiments, considering that the beam size used in our optical system was 4 mm, we chose r_0 to be 1, 2, 3, 4, and 6 mm, and the normalized correlation length r_0/D was chosen to be 0.25, 0.5, 0.75, 1, and 1.5. Under each r_0/D value, we performed 50 random realizations of phase screens, which were then separately superposed with the spiral phase pattern (SPPs) whose TC values were 1, 5, and 10. Therefore, we generated 750 phase patterns altogether (three TC values, five r_0/D values, and 50 phase screens). Afterwards, we displayed the phase patterns one by one on the LCOS-SLM and recorded the corresponding Hartmanngrams. The SH-WFS did not move herein, because we had already generated 50 diverse phase patterns under each condition for repeated testing. For each Hartmanngram, we respectively used WTM, MAICM, and WT-MAICM as the closed path determination method, and determined the TC value by the generalized contour-sum method.

The experimental results evaluated by the WTCMR are given in Figure 9, where Figure 9a–c corresponds to a set of 50 measurements for the individual TC, and Figure 9d is the mean of these measurements.

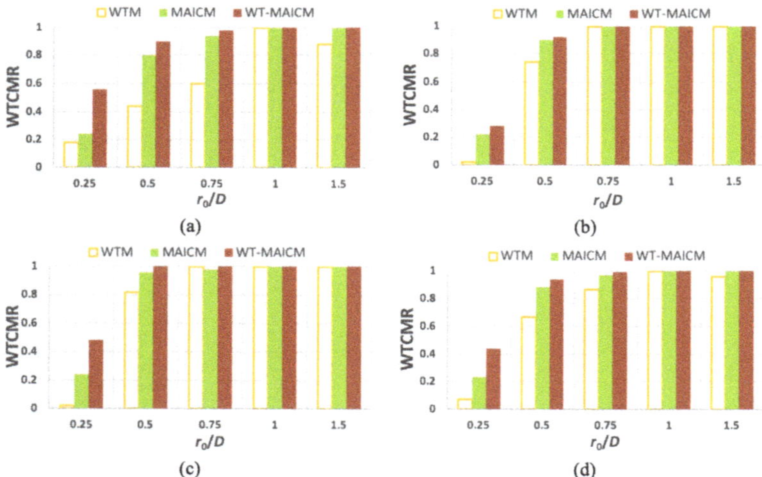

Figure 9. TC determination performance of three methods under different atmospheric turbulence conditions, where (**a**) is under $TC = 1$, (**b**) is under $TC = 5$, (**c**) is under $TC = 10$, and (**d**) is the average of (**a**), (**b**), and (**c**).

Based on the experimental results, we found that the general tendency of TC measurement accuracy increased along with the increase of r_0/D, which means the atmospheric turbulence becomes gentler. Moreover, when changing the closed path determination method from WTM to MAICM to WT-MAICM, the TC determination accuracy improved, especially when the atmospheric turbulence was severe. Supposing the accredited WTCMR to be over 0.9, we found that the limitations of r_0/D for WTM, MAICM, and WT-MAICM were around approximately 0.75–1, 0.5–0.75, and 0.25–0.5, respectively.

Figure 10 is an example specifically illustrating why the WT-MAICM has a better performance than the WTM and MAICM. Here, the OV beam to be measured has a TC value of 10 and a turbulence strength parameter r_0/D of 0.25. The top, middle, and bottom rows show the closed path determination processes and corresponding results when utilizing WTM, MAICM, and WT-MAICM separately. From the significantly different determined closed paths, we can see that when the turbulence is severe, the distorted OV beam can have extremely nonuniform regions. These nonuniform regions can trap the MAICM searching process into a local loop (middle row), or lead the WTM-determined closed path to be more than one part (top row), which conflicts with the aim of extracting only one closed path; these results eventually deteriorate the performance of the WTM and MAICM.

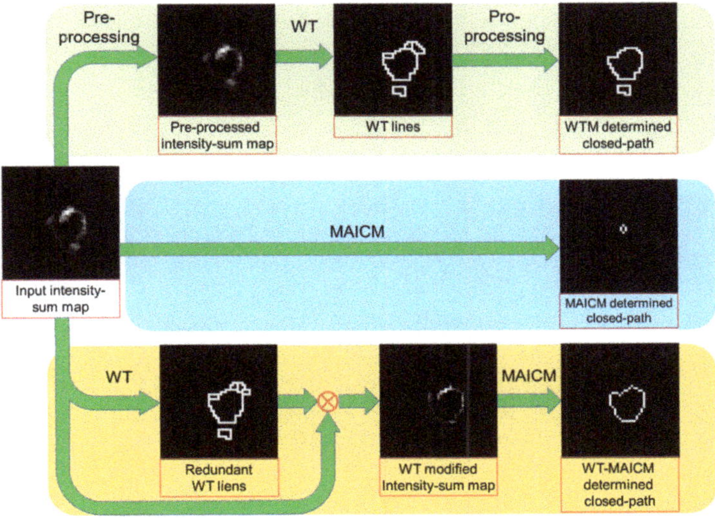

Figure 10. Closed path determination processes and results using the WTM (top row), MAICM (middle row), and WT-MAICM (bottom row).

6. Conclusions

In this paper, we presented an experimental comparison of five closed path determination methods for the use of GCSM to detect the TC of an OV beam. These five methods are the previously proposed MAICM and FC-PRCM, and three newly proposed methods: the WTM, WT-MAICM, and SC-PRCM. These methods come from two strategies: WTM, MAICM, and WT-MAICM belong to the strategy of ridge extraction, and FC-PRCM and SC-PRCM are derived from the strategy of PRC assignation. The codes for the algorithms are available from the authors by email. The methods were tested with an optical setup that used a LCOS-SLM as an OV beam, and an aberration generator and an SH-WFS located at the far-field plane to measure the OV beams. Two types of evaluation experiments were performed. One was under the condition that the OV beam had hardly any aberration, and the other was under the condition that the OV beam was distorted by simulated atmospheric turbulence. The experimental results indicate that the methods with ridge extraction outperform those based on PRC assignation in terms of both the well-determined detection rate and the processing speed.

Under the condition of an aberration-free OV beam, the WTM, MAICM, and WT-MAICM performed excellently and achieved WTCMR values of 100%. On the other hand, the WTCMR values of the FC-PRCM and SC-PRCM were not always 100% and the performance was reduced with the increase of the TC value. What is more, the SC-PRCM was found to be the worst method in terms of both the WTCMR and the processing speed. There are a number of reasons that can explain the poor performance of the PRC assignation strategy, but the major factors are that the generated circle is prone to deviate from the OV center and go through regions with low intensity as well as invalid phase slope data, which are caused by nonuniformity in the intensity sum map along the azimuthal direction. The performance of the SC-PRCM was worse than that of the FC-PRCM, indicating that simply shifting the center of the PRC is not a suitable solution, and the maximum absolute TC value hunting criterion in the PRC assignation strategy lacks rationality.

In the case of measuring distorted OV beams, the WTM, MAICM, and WT-MAICM show certain differences in term of the WTCMR, especially when the turbulent strength is high. Among these three methods, the WT-MAICM shows the strongest robustness against distortion, and the WTM was found to be the weakest. The limits of r_0/D to achieve WTCMR > 90% were around 0.75–1, 0.5–0.75, and 0.25–0.5 for the WTM, MAICM, and WT-MAICM, respectively, in terms of normalized correlation length. Overall, these results reveal that adaptively determining the closed path is necessary for GCSM in detecting the TC from a distorted OV beam.

Author Contributions: Conceptualization, D.W. and H.H.; funding acquisition, H.L.; methodology, D.W. and H.H.; project administration, H.H.; resources, H.T.; supervision, H.L. and H.T.; validation, D.W. and H.H.; writing—original draft, D.W.; writing—review and editing, D.W. and H.H.

Funding: Shenzhen Innovation Funding (No: JCYJ20170818164343304, JCYJ20170816172431715); National Natural Science Foundation of China (No: U1809204, 61525106, 61427807, 61701436); National Key Technology Research and Development Program of China (No: 2017YFE0104000, 2016YFC1300302).

Acknowledgments: We gratefully acknowledge A. Hiruma and T. Hara for their support and encouragement throughout this study. Wang thanks the exchange program between Hamamatsu Photonics K.K. and Zhejiang University. The support from the above funding organizations is also gratefully acknowledged.

Conflicts of Interest: The authors declare no conflict of interest.

References

1. Bozinovic, N.; Yue, Y.; Ren, Y.; Tur, M.; Kristensen, P.; Huang, H.; Willner, A.E.; Ramachandran, S. Terabit-scale orbital angular momentum mode division multiplexing in fibers. *Science* **2013**, *340*, 1545–1548. [CrossRef] [PubMed]
2. Zhang, D.; Feng, X.; Huang, Y. Encoding and decoding of orbital angular momentum for wireless optical interconnects on chip. *Opt. Express* **2012**, *20*, 26986–26995. [CrossRef] [PubMed]
3. Wang, J.; Yang, J.Y.; Fazal, I.M.; Ahmed, N.; Yan, Y.; Huang, H.; Ren, Y.; Yue, Y.; Dolinar, S.; Tur, M.; et al. Terabit free-space data transmission employing orbital angular momentum multiplexing. *Nat. Photonics* **2012**, *6*, 488. [CrossRef]
4. Li, S.; Wang, J. Experimental demonstration of optical interconnects exploiting orbital angular momentum array. *Opt. Express* **2017**, *25*, 21537–21547. [CrossRef] [PubMed]
5. Chu, S. Laser Manipulation of Atoms and Particles. *Science* **1991**, *253*, 861–866. [CrossRef] [PubMed]
6. Otsu, T.; Ando, T.; Takiguchi, Y.; Ohtake, Y.; Toyoda, H.; Itoh, H. Direct evidence for three-dimensional off-axis trapping with single Laguerre-Gaussian beam. *Sci. Rep.* **2014**, *4*, 4579. [CrossRef] [PubMed]
7. Liphardt, J.; Dumont, S.; Smith, S.B.; Tinoco, I.; Bustamante, C. Equilibrium information from nonequilibrium measurements in an experimental test of Jarzynski's equality. *Science* **2002**, *296*, 1832–1835. [CrossRef] [PubMed]
8. Sato, S.; Fujimoto, I.; Kurihara, T.; Ando, S. Remote six-axis deformation sensing with optical vortex beam. In Proceedings of the Free-Space Laser Communication Technologies XX, San Jose, CA, USA, 19–24 January 2008.

9. Wang, W.; Yokozeki, T.; Ishijima, R.; Takeda, M.; Hanson, S.G. Optical vortex metrology based on the core structures of phase singularities in Laguerre-Gauss transform of a speckle pattern. *Opt. Express* **2006**, *14*, 10195–10206. [CrossRef] [PubMed]
10. Fujimoto, I.; Sato, S.; Kim, M.Y.; Ando, S. Optical vortex beams for optical displacement measurements in a surveying field. *Meas. Sci. Technol.* **2011**, *22*, 105301. [CrossRef]
11. Nye, J.F.; Berry, M.V. Dislocations in wave trains. *Proc. R. Soc. Lond. A Math. Phys. Sci.* **1974**, *336*, 165–190. [CrossRef]
12. Fried, D.L.; Vaughn, J.L. Branch cuts in the phase function. *Appl. Opt.* **1992**, *31*, 2865–2882. [CrossRef] [PubMed]
13. Rockstuhl, C.; Ivanovskyy, A.A.; Soskin, M.S.; Salt, M.G.; Herzig, H.P.; Dändliker, R. High-resolution measurement of phase singularities produced by compute-generated holograms. *Opt. Commun.* **2004**, *242*, 163–169. [CrossRef]
14. Chen, R.; Zhang, X.; Zhou, Y.; Ming, H.; Wang, A.; Zhan, Q. Detecting the topological charge of optical vortex beams using a sectorial screen. *Appl. Opt.* **2017**, *56*, 4868–4872. [CrossRef] [PubMed]
15. Hickmann, J.M.; Fonseca, E.J.; Soares, W.C.; Chávez-Cerda, S. Unveiling a truncated optical lattice associated with a triangular aperture using light's orbital angular momentum. *Phys. Rev. Lett.* **2010**, *105*, 053904. [CrossRef] [PubMed]
16. Prabhakar, S.; Kumar, A.; Banerji, J.; Singh, R.P. Revealing the order of a vortex through its intensity record. *Opt. Lett.* **2011**, *36*, 4398–4400. [CrossRef]
17. Schulze, C.; Naidoo, D.; Flamm, D.; Schmidt, O.A.; Forbes, A.; Duparré, M. Wavefront reconstruction by modal decomposition. *Opt. Express* **2012**, *20*, 19714–19725. [CrossRef]
18. Li, S.; Wang, J. Simultaneous demultiplexing and steering of multiple orbital angular momentum modes. *Sci. Rep.* **2015**, *5*, 15406. [CrossRef]
19. Platt, B.C.; Shack, R. History and principles of Shack-Hartmann wavefront sensing. *J. Refract. Surg.* **2001**, *17*, S573–S577. [CrossRef]
20. Chen, M.; Roux, F.S.; Olivier, J.C. Detection of phase singularities with a Shack-Hartmann wavefront sensor. *J. Opt. Soc. Am. A* **2007**, *24*, 1994–2002. [CrossRef]
21. Le Bigot, E.O.; Wild, W.J. Theory of branch-point detection and its implementation. *J. Opt. Soc. Am. A* **1999**, *16*, 1724–1729. [CrossRef]
22. Huang, H.; Luo, J.; Matsui, Y.; Toyoda, H.; Inoue, T. Eight-connected contour method for accurate position detection of optical vortices using Shack–Hartmann wavefront sensor. *Opt. Eng.* **2015**, *54*, 111302. [CrossRef]
23. Luo, J.; Huang, H.; Matsui, Y.; Toyoda, H.; Inoue, T.; Bai, J. High-order optical vortex position detection using a Shack-Hartmann wavefront sensor. *Opt. Express* **2015**, *23*, 8706–8719. [CrossRef] [PubMed]
24. Wang, D.; Huang, H.; Matsui, Y.; Tanaka, H.; Toyoda, H.; Inoue, T.; Liu, H. Aberration-resistible topological charge determination of annular-shaped optical vortex beams using Shack–Hartmann wavefront sensor. *Opt. Express* **2019**, *27*, 7803–7821. [CrossRef] [PubMed]
25. Beucher, S.; Lantuejoul, C. Use of watersheds in contour detection. In Proceedings of the International Workshop Image Processing, Real-Time Edge and Motion Detection/ Estimation, CCETT/IRISA, Rennes, France, 17–21 September 1979.
26. Meyer, F.; Beucher, S. Morphological segmentation. *JVCIP* **1990**, *1*, 21–46. [CrossRef]
27. Meyer, F. Topographic distance and watershed lines. *Signal. Process.* **1994**, *38*, 113–125. [CrossRef]
28. Vincent, L.; Soille, P. Watersheds in digital spaces: An efficient algorithm based on immersion simulations. *IEEE Trans. Pattern Anal. Mach. Intell.* **1991**, *13*, 583–598. [CrossRef]
29. Osma-Ruiz, V.; Godino-Llorente, J.I.; Sa´enz-Lecho´n, N.; Go´mez-Vilda, P. An improved watershed algorithm based on efficient computation of shortest paths. *Pattern Recognit.* **2007**, *40*, 1078–1090. [CrossRef]
30. Watershed Function in MathWorks. Available online: https://www.mathworks.com/help/images/ref/watershed.html?s_tid=srchtitle (accessed on 11 August 2019).
31. Inoue, T.; Tanaka, H.; Fukuchi, N.; Takumi, M.; Matsumoto, N.; Hara, T.; Yoshida, N.; Igasaki, Y.; Kobayashi, Y. LCOS spatial light modulator controlled by 12-bit signals for optical phase-only modulator. *Proc. SPIE* **2007**, *6487*, 64870Y.
32. Sugiyama, Y.; Takumi, M.; Toyoda, H.; Mukozaka, N.; Ihori, A.; Kurashina, T.; Nakamura, Y.; Tonbe, T.; Mizuno, S. A high-speed CMOS image sensor with profile data acquiring function. *IEEE J. Solid State Circuits* **2005**, *40*, 2816–2823. [CrossRef]

33. Tyson, R.K. *Principles of Adaptive Optics*; Academic: Salt Lake City, UT, USA, 1991.
34. Kolmogorov, A.N. A refinement of previous hypotheses concerning the local structure of turbulence in a viscous incompressible fluid at high Reynolds number. *J. Fluid Mech.* **1962**, *13*, 82–85. [CrossRef]
35. Roddier, N.A. Atmospheric wavefront simulation using Zernike polynomials. *Opt. Eng.* **1990**, *29*, 1174–1180. [CrossRef]
36. Fried, D.L. Statistics of a geometric representation of a wavefront distortion. *J. Opt. Soc. Am.* **1965**, *55*, 1427–1435. [CrossRef]

Article

Non-Contact Dermatoscope with Ultra-Bright Light Source and Liquid Lens-Based Autofocus Function

Dierk Fricke [1,*], Evgeniia Denker [2], Annice Heratizadeh [2], Thomas Werfel [2], Merve Wollweber [1] and Bernhard Roth [1,3]

[1] Hannover Centre for Optical Technologies (HOT), Leibniz University, 30167 Hannover, Germany; merve.wollweber@hot.uni-hannover.de (M.W.); bernhard.roth@hot.uni-hannover.de (B.R.)

[2] Hannover Medical School, MHH, 30625 Hannover, Germany; denker201@gmail.com (E.D.); Heratizadeh.Annice@mh-hannover.de (A.H.); Werfel.thomas@mh-hannover.de (T.W.)

[3] Cluster of Excellence PhoenixD, Leibniz University, 30167 Hannover, Germany

* Correspondence: dierk.fricke@hot.uni-hannover.de

Received: 29 March 2019; Accepted: 23 May 2019; Published: 28 May 2019

Featured Application: This work reports on a new non-contact dermatoscope targeting at improved examination and documentation of skin diseases through enhanced functionality compared to conventional contact-based systems.

Abstract: Dermatoscopes are routinely used in skin cancer screening but are rarely employed for the diagnosis of other skin conditions. Broader application is promising from a diagnostic point of view as biopsies for differential diagnosis may be avoided but it requires non-contact devices allowing a comparably large field of view that are not commercially available today. Autofocus and color reproducibility are specific challenges for the development of dermatoscopy for application beyond cancer screening. We present a prototype for such a system including solutions for autofocus and color reproducibility independent of ambient lighting. System performance includes sufficiently high feature resolution of up to 30 µm and feature size scaling fulfilling the requirements to apply the device in regular skin cancer screening.

Keywords: dermatoscopy; skin screening; biomedical imaging

1. Introduction

A dermatoscope is the standard instrument for a first examination of skin conditions. The whole field of dermatoscopy started with the use of epiluminescence microscopes for this purpose [1–3]. State-of-the-art dermatoscopy devices today are mostly contact based and often include a camera to capture digital images for documentation. Compared to a non-contact setup, a contact-based dermatoscope exhibits several disadvantages. State of the art contact-based dermatoscopes have built-in cameras to make the pictures digitally available. The skin contact may cause distortion of the skin geometry, which makes it more difficult to compare pictures made at different examinations. Also, the contact could suppress the perfusion hampering detection of small vessel structures. When thinking of infected lesions, the contact can also be painful and represents a larger invasion to the privacy of patients. A non-contact device circumvents all these problems. In addition, one can realise a larger field of view and the concept is much more compatible with automation approaches. Challenges of a non-contact design are the implementation of a variable focus and the realization of a lighting situation, which is independent from the surrounding light.

In order to exploit the advantages of the non-contact design with regard to automation, some challenges need to be addressed. For example, to obtain sharp images, focus control is required. The preferred solution for easy handling or automation is an autofocus system. For documentation and post-processing, the images should also be digitally available. Whereas digital image acquisition is already state-of-the-art in contact-based devices [4,5], the documentation is required to compare pictures from different examinations in order to identify changes in size or color of relevant skin areas. The size of a nevus is, for example, one criterion of the ABCDE (**A**symmetry, **B**order, **C**olor, **D**iameter, **E**volving) rule of dermatoscopy to visually identify malignant melanoma [6,7]. Therefore, it is necessary to record comparable images in non-contact mode without distortion of the skin geometry, as skin contact with standard devices can significantly distort the geometry and suppress blood perfusion correct assessment of skin lesions if often difficult [8]. As the hemoglobin of the blood, besides melanin, is one of the two dominant dye molecules in the skin, the suppression results in a color change of the skin area under study. The comparison of skin colors is not only important for skin cancer screening but also interesting, for example, for inflammatory skin diseases (e.g., erythema in the Psoriasis Area and Severity Index (PASI)). To enable to calibrate the camera sensor in a way that colors can be identified and compared between images made at different examinations and possibly under different ambient light situations, a bright light source is needed.

A dermatoscope provides 2D information about light intensity and color of the imaged location, i.e., skin area. As this is the same information generated by the human eye, the resulting data is comparably intuitive for the dermatologists. In addition, to obtain information from deeper layers of the skin, cross polarization can be used (see Section 2.3) [9]. Other techniques like optical coherence tomography (OCT) [10–12] or high frequency ultrasound [11,13] normally generating a 2D depth images are based on the optical thickness (for OCT) or the time of the propagation of the acoustic waves in the medium (for ultrasound), respectively. For these modalities, by scanning a 2D area, 3D information is generated. The resulting data, however, do not provide information about the color of the tissue and special training is required for their correct interpretation. Also, compared to dermatoscopes, OCT and high frequency ultrasound devices are much more expensive. Thus, there is a need for simple, easy-to-operate and yet accurate dermatoscope systems capable of providing the functionality required for proper skin examination without distortion of the geometry, i.e., in non-contact mode.

In this work, we present a novel non-contact dermatoscope featuring properties advantageous for more reliable and comparable skin examinations in the future. The light source of the developed prototype is an ultra-bright white light-emitting diode (LED), which allows for full control of the lighting situation to be independent from the surrounding light. Also, by realizing a calibrated display, it is possible to display the natural colors of the skin area. Furthermore, the non-contact approach enables a much larger field of view. While pigmented nevi that are examined in melanoma screening are often 0.5–2-cm wide, inflammatory lesions are generally several centimeters wide. Thus, their examination requires a much larger field of view of the imaging system compared to a device designed for skin cancer screening only. The prototype described here has a field of view of about 17×13 mm^2 and can, in the future, be combined with a second camera to record images of even larger skin areas.

1.1. State-of-the-Art Commercially Available Non-Contact Systems

A few non-contact devices for skin examination have already been reported. They can be categorized in (i) non-contact dermatoscopes, which have a high magnification (see for example [4,5]), and (ii) non-contact skin screening devices for overview images of the skin (see for example [14]).

So far, different designs for non-contact dermatoscopes are utilized. One type is realized as a handheld magnifying glass while another is configured as an adaptor for mobile phones. The latter use the camera of the mobile phone to take the images. Other devices rely on built-in camera and illumination sources. However, all these systems are applied at short working distances of a few centimeters. An important aspect in this context is (auto)focusing, which has to be considered for all non-contact devices. For example, non-contact devices with more than 2 cm working distance use spacers to avoid the need of autofocusing. Furthermore, the illumination conditions are more difficult to control for non-contact devices as environmental influences are more relevant. Thus, sufficiently bright light sources are needed to ensure reproducible illumination conditions [4,5].

Further concepts rely on imaging systems at large working distances, which provide only overview images of the skin with lower resolution compared to dermatoscopes. They usually make use of standard cameras with passive autofocusing techniques that employ phase or contrast detection. The cameras are mounted on a movable holder so that images of the whole body can be taken. As room light is used for illumination it is difficult to compare images taken under varying lighting conditions [14].

1.2. Prototype of the New Non-Contact Device

The main advantages of the non-contact prototype system presented in this work are the large working distance while providing the high-resolution level of contact-type dermatoscopes. At present, the prototype is optimized for a working distance of 45 cm ± 3 cm. This is realized with a liquid lens-based autofocus, which has not yet been used for this purpose so far. In principle, other working distances are also possible requiring only marginal changes to the current design as further discussed in Section 2.

As an illumination source, an ultra-bright LED light source with full polarization control is used. This choice makes the device independent from the surrounding light situation and allows reproducible color measurements. Also, the color of different skin sites and lesions can more easily be compared between different images. Such analysis is, for example, important for skin cancer screening as multiple colors and color changes can be indicative of melanoma [6]. Due to the design on a swivel arm the ultra-bright light source can easily be powered. By controlling the polarization state of the measured light, the physician can increase or decrease the contrast of the surface morphology as desired. The presented autofocus function is fast compared to iterative autofocus solutions, which are based on contrast analysis of the captured image. Also, the built-in infrared distance sensor can provide information about the scale in the taken image. The new non-contact system allows for automated imaging functionality so the physician is able to map the skin without manual activity.

First preclinical studies with an earlier version of our non-contact dermatoscope indicated the potential benefit of the system [8]. For example, in the investigation of inflammatory skin diseases such as lichen ruber planus and psoriasis subtle details of the lesions could be visualized and natural color appearance ensured. With the early stage of the system structural changes such as hypergranulosis [15] were visible for lichen ruber planus. In psoriasis, variations of the capillary vessels could be seen. For this inflammatory skin disease, characteristic blood vessels are opening in the upper skin. These vessels are normally seen in the histopathology of the diseased skin. With the non-contact prototype, it was possible to observe these vessels as round structures in the image. This success could lead to a reduction of necessary biopsies for differential diagnosis. In the future, the newly designed setup described here could be part of an automated skin scanner where the patient is screened at a first examination stage without the need for a dermatologist to be present. However, the system described in [8] did not feature an autofocus function. Also it was mounted on a stable substrate board so that the patients had to move in front of it. Thus, not all regions of interest on a patient's body could be reached and handling by the dermatologist was still limited. Also, image acquisition took longer so that fewer patients could be examined. Using the new system presented in this work, even lesions on hard-to-reach areas of the body could be examined with high resolution.

2. Materials and Methods

Figure 1 shows the non-contact dermatoscope developed in this work. The setup consists of a camera with liquid lens-based autofocus and an illumination unit. An infrared laser distance sensor is attached besides the illumination unit. The camera is connected to a computer for control of the setup and image post processing. A movable polarizer is mounted in front of the liquid lens. The illumination unit consists of an ultra-bright white LED-chip and a reflector that redirects the light emitted by the LED and illuminates the region of interest homogeneously. A fixed polarizer is mounted in front of the reflector. In the subsections below, the different parts of the prototype are described in detail.

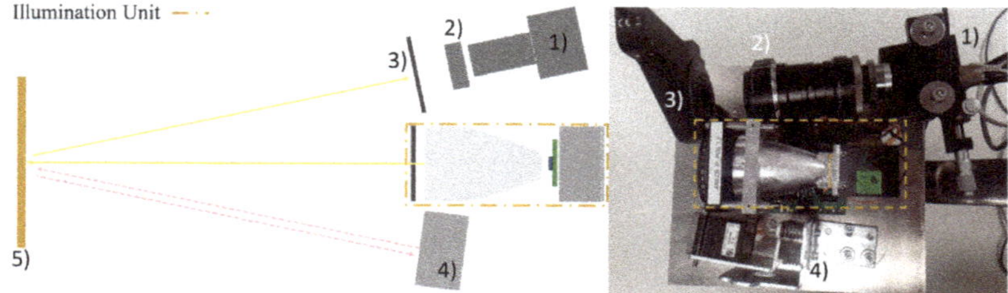

Figure 1. Schematic drawing (**left**) and photograph of the prototype of the non-contact dermatoscope. The imaging unit consists of the camera itself (**1**) a liquid lens (**2**) in front of the camera lenses and a rotatable polarizer (**3**). An infrared laser-based distance measurement device (**4**) besides the illumination unit is used to control the shape of the liquid lens and, thus, the focus length. The illumination unit is set up of an ultra-bright white LED source mounted on a radiator and a reflector. The light is polarized by a polarizer mounted in front the reflector. This setup is able to illuminate a target (**5**), measure the distance to this target and take images for different polarization settings.

The setup is mounted on an aluminum plate that is connected to the swivel arm mounted on a table that makes it possible to adjust its position in the room and with respect to the patients. The camera, the polarizers and the liquid lens are connected to a computer, which controls the system and performs digital image processing. Component costs for the presented non-contact dermatoscope are approximately 4000 €.

2.1. Camera with Liquid Lens-Based Autofocus

We integrated a liquid lens-based autofocus as a novel solution for the variable focus problem in a freely movable non-contact dermatoscope. A liquid lens has the advantage of fast tunability of the focal length. The liquid lens consists of a reservoir filled with an opto-fluid delimited by a transparent membrane consisting of silicon or thin glass. The reservoir is enclosed by a Piezo ring, which can increase the pressure in the reservoir if an electrical voltage is applied to it. The pressure leads to different curvatures of the membrane, which results in different focal lengths of the lens. The schematic setup of the camera is shown in Figure 1. The radius of curvature of the lens is controlled by the distance sensor. The system is adjusted so the skin area of interest is always in the focus. This is ensured by actively measuring the distance between the target and the camera to follow the target even if the latter is not illuminated or does not show visible sharp edges.

The camera (FL3-GE-28S4C-C, FLIR Integrated Imaging Solutions, Inc., Richmond, BC, Canada) has a resolution of 1928 × 1448 and a maximum image rate of 15 frames per second (FPS). It is equipped with an ICX687 CCD color sensor by Sony (Minato, Tokio, Japan) which has a sensor size of 1/1.8′′ and a pixel size of 3.69 µm. To record a dermatoscopic view of the skin, a zoom lens (NMV-75M1, Navitar, Rochester, NY, USA) with a focal length of 75 mm is mounted in front of the camera. In addition, the liquid lens (EL-16-40-TC, Optotune AG, Dietikon, Switzerland) is mounted in front of the zoom objective with a self-made adapter. It is an electrically tunable large aperture lens (with a clear aperture of 16 mm). The lens is controlled by the Optotune Lens Driver 4i via the "Lens Driver Controller" software. In order to realize an autofocus system, the tunable lens was combined with a distance sensor (DT35-B15851, Sick AG, Waldkirch, Germany). The lens driver can process an analog input signal from the distance sensor to set the correct focus length for a sharp image of the target. The distance sensor is based on an infrared laser distance sensor operating at 827 nm. It has an accuracy of 0.5 mm in the relevant range below 1 m, depending on the reflectivity of the target and the measurement speed, which is set. This setup realizes a fast autofocus as it does not contain moving parts. It is calibrated with the lens driver controller software. Here, a look-up table can be configured which relates every measured distance to a specific voltage for the liquid lens. The lens needs around 5 ms to respond and around 25 ms to settle to the new focal length. The exact settle time depends on the voltage difference applied to the lens. The distance sensor has a time delay of 6.5 ms between the measured event and the output signal in the "fast" mode. This results in a total delay from the event to the settling of the lens of approximately 31.5 ms and allows to adjust the focus approximately 30 times a second, which is twice the frame rate of the camera. As patients usually only move slowly and slightly during dermoscopy, the autofocus is able to focus on the target all the time.

As the camera and the distance sensor are not coaxially aligned, the functional distance of the setup is limited. The available focus distance is 45 ± 3 cm. If the distance sensor was arranged coaxially to the camera, for example, by aid of a dichroic mirror, it would be possible to focus over a much larger distance. A wider focal range could be realized in combination with an illumination unit that is able to homogeneously illuminate a larger area than the actual version with the same light intensity Though the used liquid lens is one of the biggest commercially available lenses by now, its aperture is much smaller than the aperture of the camera lens. This makes it necessary to choose the aperture of the camera to be smaller to avoid strong aberrations from the edge of the liquid lens. A small aperture also leads to a relatively low light exposure of the camera sensor, which needs to be compensated by larger exposure times. In dermatoscopy, exposure times are limited by the natural movement of the patient can be avoided, such as blurring or other movement artifacts. Due to the ultra-bright LED illumination, acceptable exposure times of 75 ms can be achieved.

2.2. Ultra-Bright LED White Light Source

For illumination, a phosphor based ultra-bright LED white light source (CBT-90 White LED, Luminus Inc., Sunnyvale, CA, USA) is employed. The source is phosphor based and converts blue light from an LED-chip into the yellow spectral range. The emitted light mixture exhibits a white spectrum, which is generally more continuous compared to a mixture of red, green, and blue (RGB) LED chips.

The source has a color rendering index (CRI) of 76 (for comparison: lighting in surgical rooms requires a CRI > 85 [16]). In general, the closer the CRI of a light source is to the value of 100 the better the source can render the colors of real-world objects, i.e., skin lesion in our case. This then corresponds to the color perception during daylight.

The LED is mounted on a self-made reflector that was simulated and designed with OpticStudio (Zemax LLT, Kirkland, WA, USA) aiming at homogeneous illumination of a target at the highest possible intensity in the distance of interest (ca. 45 cm from the liquid lens). The bright illumination has the advantage, that the illumination of the target is well-defined and less influenced by the surrounding light. This is important for the color adjustment of the camera, which is of interest as the color of a lesion is a relevant parameter in dermatoscopy. For example, it is also taken into account in the ABCDE rule of dermatoscopy [6,7]. Also, it is of interest to evaluate the temporal evolution of a lesion. Furthermore, in the field of inflammatory skin diseases, different color variations need to be identified, e.g., shades of red. Another useful application area could be the color identification of hematoma allowing to predict the age of injuries of patients [17,18]. For detection of changes of the colors of lesions, the comparability of images taken at different examination sessions and different times must be ensured. Also, different disease stages could be assigned by evaluating the color (e.g., redness in the PASI as well as the Eczema Area and Severity Index (EASI), respectively). To fulfill these requirements, in particular, to allow for a more natural color representation comparable to the perception of a physician, color calibration is important. This ultimately also increases the acceptance of the system by the dermatologists.

2.3. Full Polarization Control

The concept of using polarized light for skin examination is well known in the literature [19–21]. There are two polarizers in the setup presented in this work: one in front of the illumination unit and one in front of the liquid lens.

The polarizer in front of the camera is mounted on a rotation stage such that the operator can switch between cross polarization and parallel polarization by means of custom software. In cross-polarization mode, the orientation of the two polarizers is 90 degrees. Because the directly reflected light does not change its polarization for the most part, this light will be filtered out. Only light that is scattered in the skin is detected in this case. If the orientation of the polarizers is parallel to each other, the effect is the opposite: the scattered light is filtered out and the light that is directly reflected will be detected. An image taken with parallel setting of the polarizers can provide information about the surface topography and structure while a cross polarization image will provide information from deeper skin layers. Due to the rotation stage, also any orientation between cross and parallel can be set. In Figure 2, images of a scar taken with parallel and cross-polarization are shown. On the picture taken with a parallel polarization setup, facets of the skin on the scar appear to be larger than those of the surrounding skin areas. The image with the crossed polarization shows a darker reddened area around the scar, which is not clearly visible in the other image. The scar itself appears brighter and whiter.

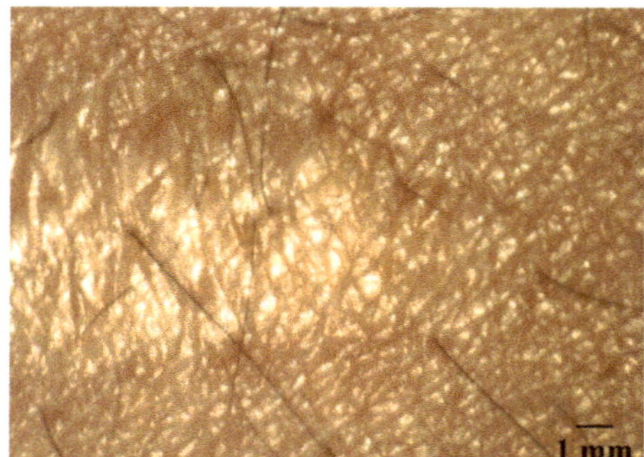

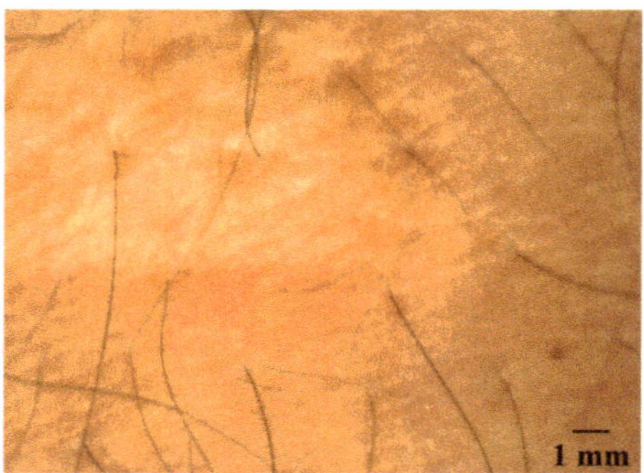

Figure 2. Image of a scar with parallel polarization (**left**) and cross-polarization (**right**). Due to the movement of the patient, the images do not show exactly the same part of the skin. The imaged area contains healthy skin and scar tissue. The scar tissue can be recognized by the larger facets in the left image and by the lighter color in the right image.

2.4. Computer Control and Digital Post Processing

The focal length of the tunable lens is controlled by the Optotune lens driver controller software. The autofocus is calibrated using a look-up table where an input voltage from the distance sensor is related to an output voltage for the liquid lens. The software calculates the resulting linear fit function automatically. The capturing of the image and the polarization control as well as image post-processing is done with a self-developed software based on LabVIEW (National Instruments AG, Austin, TX, USA). This software includes a basic database structure for the pictures.

3. Results

3.1. Camera Parameters

Relevant camera parameters were measured with and without liquid lens included in the setup to compare the system performance, i.e., focal length, resolution, depth of field (DOF), and image scale (detailed below).

3.1.1. Focus Distance

Without the liquid lens, the focal length of the optical system is limited to a range of 40 cm starting from a distance of approximately 43 cm from the end of the camera lenses to approximately 83 cm along the optical axis.

As mentioned, the current alignment of the components allows to ensure, that the laser spot of the distance measurement sensor is in the area imaged by the camera in a distance range of 45 cm ± 3 cm.

Ideally, the sensor aims for the center of the region of interest on the skin, which is also supposed to be in the center of the illuminated spot. As the camera and the image sensor are not on axis, this is only true for a certain distance, which was set to 45 cm. In the range of ±3 cm from this point, the laser spot from the distance sensor is still in the field of view of the camera, so that the signal correlates with the image signal. Leaving this area between the system and the patient, the autofocus works only for small sample curvatures in range of the depth of field. In the future, the laser could be co-aligned with the camera. In such an aligned setup, the range of the autofocus would be in the range of the possible focal length of the combination of the camera lenses and the liquid lens. With our setup, it was possible to focus objects at distances of approximately 17 cm from the liquid lens up to at least 8 m (limited by laboratory size). In this setup, only the alignment of the camera system and the illumination unit would limit the range of operation, however, the resolution decreases and the image size increases with distance. This effect has to be taken in to account to be able to specify a reasonable work range.

3.1.2. Resolution

In consultation with dermatologists from the Hannover Medical School (MHH), the goal was to be able to resolve 30-µm large structures on or within the skin. This ensures that all blood vessels that could be of diagnostic relevance can be resolved. A 1951 USAF resolution test chart was used to determine the maximum resolution of the prototype (see in Figure 3). In this case, the test chart was placed at a distance of 45 cm from the liquid lens or the camera lens, respectively. The first picture Figure 4 left shows an image taken with the described prototype. A magnified view of the groups four and five are shown next to the image. The second picture, Figure 4 right shows an image taken under the same conditions but without the liquid lens. Due to the measurement process, the images are mirrored. In the setup with the liquid lens, the element 2 in group 4 can still be resolved. This corresponds to 17.96 lines per millimeter leading to a resolution of 27.87 µm. In the image taken without the liquid lens, the element 4 in group 4 can still be resolved. This element contains 22.63 lines per millimeter and corresponds to a resolution of 22.1 µm. The results of this measurement with the system without liquid lens being slightly more accurate cannot easily be transferred to an in vivo measurement. In the latter case, the autofocus could compensate for blurring caused by transversal movements of the patient during the measurement. However, this result shows that the resolution of the setup with the liquid lens still enables reasonable resolution for the measurements on skin samples together with improved handling.

3.1.3. Depth of Field (DOF)

DOF, i.e., the region where objects are sharply imaged, is an important parameter for dermatoscopic systems because the human skin is not flat and nevi may even be raised up to several millimeters from the skin. The DOF here is given for a fixed focal length of the liquid lens. It is important to compensate for calibration errors of the autofocus system and possible measurement uncertainties of the distance sensor as it adds some tolerance if the focus is not exactly on the target. To measure the DOF, a special target was used which consists of a 45° inclined plane with a scale and line pairs (DOF 5-15 Depth of Field Target, Edmund Optics GmbH, Karlsruhe, Germany). The system performance can be seen in Figure 4.

The scale in the image is the depth of the target in mm. The DOF can be read off the scale and is at least 22 mm. This is sufficient for most dermatoscopic investigations and skin structures.

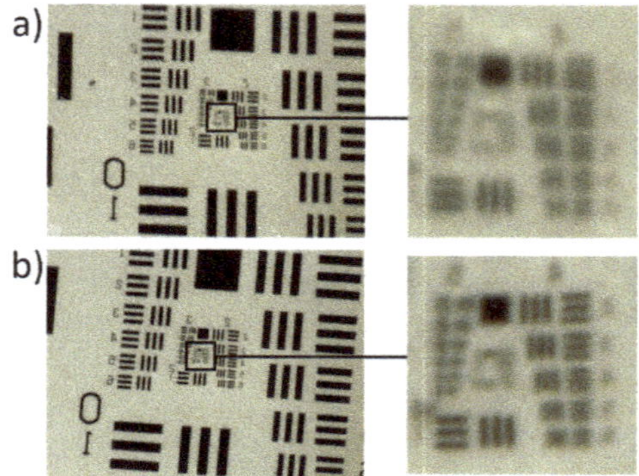

Figure 3. Images of a 1951 USAF resolution test chart at a distance of 45 cm with (**a**) and without (**b**) the liquid lens in front of the camera lens.

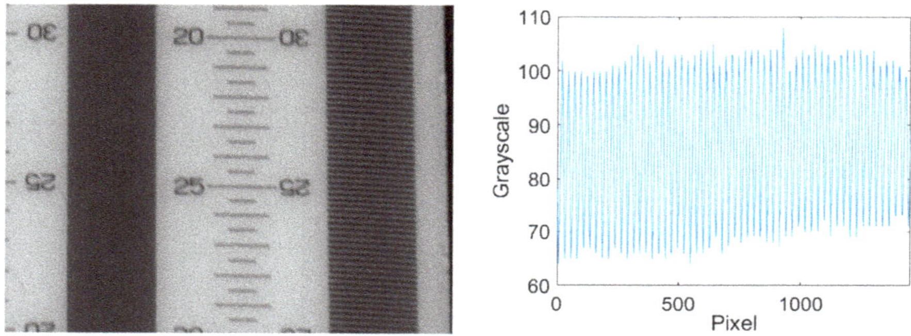

Figure 4. Image of the depth of field target (**left**) and the intensity profile over the left stripe pattern of the image (**right**).

3.1.4. Image Scale

In order to be able to determine the size of interesting skin features for images taken at different distances to the imaging system, a method for reliable scaling of the images was developed. The image scale was calibrated by taking pictures of a scale bar from various distances. These distances where correlated to the output voltage of the distance sensor. In this way the already mentioned look-up table was realized. The image size at different distances to the patient can be seen in Table 1. The scale allows comparing images taken at different examinations and measuring changes in size of the lesions with time. This is for example important for the widely known ABCDE-rule in melanoma screening or for follow-up examinations in the therapy of inflammatory skin diseases.

Table 1. Image sizes for different distances to the target.

Distance to the Patient in cm	Imaged Area of the Skin (Horizontal × Vertical) in mm	
41	15.4	11.6
43	16.2	12.2
45	17	12.8
47	17.8	13.4
49	18.2	13.7
51	19.1	14.4

3.2. Characteristics of the Illumination Unit

3.2.1. Illumination Intensity and Spectral Data

The reflector for shaping of the illuminated area, designed with OpticStudio as mentioned before, was milled and polished from aluminum. The light distribution was measured with a goniometer (PM-1200N-1, Radiant Vision Systems, Redmond, WA, USA) and the data exported again to OpticStudio. Here the light intensity at different distances was calculated. As the output of the illumination unit was measured without polarizer, the intensity had to be multiplied by a factor of 0.5 to obtain a good estimate. Furthermore, the illumination unit was measured with an integrating sphere (50 cm diameter, Mountain Photonics GmbH, Landsberg am Lech, Germany; with connected spectrometer: AvaSpec-2048L, Avantes BV, Apeldoorn, The Niederlande) to determine light intensity and spectral data for color correlated temperature (CCT) and CRI. The CCT of the illumination unit is 6350 K with a CRI of 76. In Figure 5, the light distribution at a distance of 45 cm from the illumination unit is shown for an imaged area of 10 cm × 10 cm. The overall light flux through this area is 775 lm without or 378.5 lm with polarizer. The light flux on the area of interest (the area which is captured by the camera) at this distance, i.e., 17 mm × 13 mm, is 71.6 lm or 35.8 lm after the polarizer. This results in a polarized illuminance of 162,000 lx. To compare this value, the minimal illuminance for office work spaces in Europe according to work safety regulations is 500 lx [22]. The measured illuminance from room lighting on the table in our laboratory was 525 lx. If we consider the illuminance in a doctor's office to be 1000 lx, the illuminance of the area of interest by the presented prototype is 162 times higher. This makes the light situation comparable for every image recorded, as the influence of ambient light on the total illuminance on the area of interest is negligible.

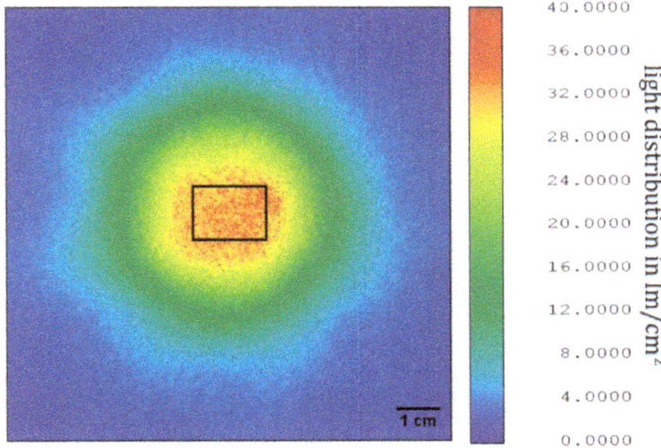

Figure 5. Light distribution at a distance of 45 cm from the illumination unit. The square shows the area of interest at this distance.

3.2.2. Homogeneous Illumination

A homogeneous illumination is important for both online assessment of an image by the practitioner and digital post processing of the images. In such post processing, often gray scale values are compared to each other and thresholds are used to segment images. This requires comparable gray scale values of the structures to be detected in the whole image. To measure the homogeneity, an image of a homogenous gray scale target (ColorChecker, X-Rite Incorporated, Grand Rapids, MI, USA) was taken.

The average gray scale value of both, the vertical and the horizontal intensity profile is 140.9 with a standard deviation of 2.64 (vertical) corresponding to a relative variance of 4.48% and 2.56 (horizontal) corresponding to a relative variance of 4.76%, respectively. With an average relative variance of 4.62% the illumination is sufficiently homogeneous for the use in a dermatoscopic device.

3.3. Image Colors

3.3.1. Comparable Color Representation

Usually, there is a difference between image colors obtained during examinations and true colors. Due to the relatively small aperture of the camera lens in combination with the ultra-bright LED light source, the system only detects the LED light reflected from the sample under study. This makes the system independent from the ambient light situation and ensures equal exposure times and measurement conditions for every measurement session. In the first row of Figure 6, two images of the same target but for different lighting conditions are shown. The first image was taken in the dark laboratory. The second was taken while the neon lights of the laboratory were turned on. Comparing gray scale values for equal exposure times shows that there is no significant difference in the pictures. The target used was a signal white (RAL 9003, RAL gGmbH, Bonn, Germany) card from the RAL color standard. In the second row, a picture of human skin is shown under the same lighting conditions. To quantify the difference, the gray scale values from the images in Figure 6a,b were subtracted from each other. Because small movements/displacements of the target between the images would lead to errors, an image registration was performed first to compensate for that. The average gray scale value of the resulting difference image was determined and is in the range of the noise of the individual images in Figure 6a,b which was estimated by the standard deviations of the respective gray scale values. The latter were calculated for a 100 pixels x 100 pixels square area in the middle of each image, which shows a homogeneous target. The registration and subtraction was also performed with the images from Figure 6c,d. The observed differences for this case are slightly larger compared to the first example and can be explained by the residual motion of the skin area between images. That means that the illumination situation is very well controlled and reproducible. If the illumination conditions are known, one calibration is enough to generate a camera color profile and calibrate all pictures taken by that camera. Colors then are comparable to the colors seen by eye and comparable to each other in different images taken with the calibrated system.

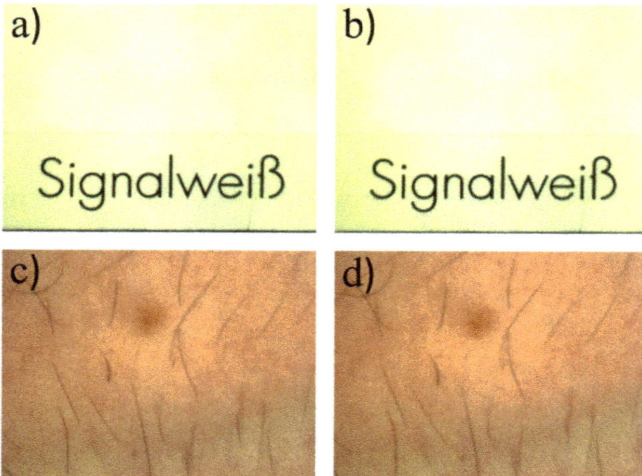

Figure 6. (**a**): Picture of the RAL color "Signalweiß" (signal white) taken in a dark laboratory and (**b**) while neon light is turned on in the same laboratory. (**c**) Picture of human skin with a nevus and a vein in the bottom part of the image taken in a dark laboratory. (**d**) Picture of the same area of human skin taken while neon light is turned on in the laboratory (at least 525 lx). Both pictures were taken with activated cross polarization and the hand was fixed, so that both images show the same area. The images look very similar. This shows that the system is mostly independent of ambient lighting conditions.

Because the light situation is controlled and the measurement parameters are the same for each measurement, the colors of the pictures taken in different examinations are comparable. They can be calculated from the relation between the different color channels.

3.3.2. True Color Rendering

If the system is calibrated once, the colors in the images appear closer to natural colors as can be seen in Figure 7. It shows images of a standard color reference target (ColorChecker Classic, X-Rite, Grand Rapids, MI, USA). Because the target is too large to fit on one picture, a composite image was assembled from single pictures of each tile of the ColorChecker. The calibration procedure uses the CIE (R,G,B) and LAB color spaces [23]. It aims at minimizing the color difference between the calibrated image and the optimal colors of the ColorChecker. Color values of the ColorChecker target are given for reference by the distributor.

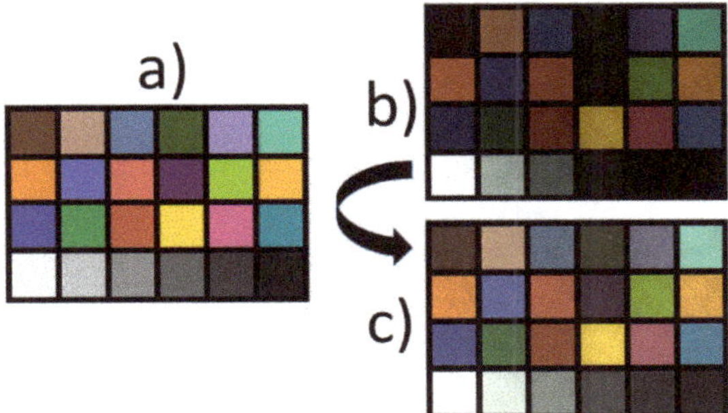

Figure 7. Color calibration: (**a**) shows a picture of the ideal RGB (red, green, blue) values of the ColorChecker Classic. (**b**) Shows an image the color checker and (**c**) after the image after color correction. The RGB values are now closer to the ideal RGB values in (**a**).

For a more realistic color perception a calibrated monitor is recommended. The calibration also facilitates the comparison of images obtained from different systems.

The result of the calibration can be seen in Figure 8.

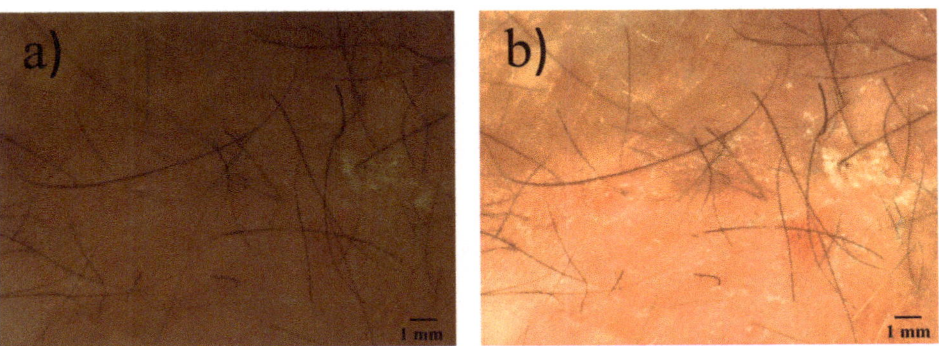

Figure 8. Two images of a patient suffering from Lichen simplex chronicus. (**a**) Uncalibrated image data. (**b**) Same image after color calibration. The image looks brighter and the skin color is more natural and closer to what the doctor is used to see with his bare eyes.

4. Discussion

The non-contact dermatoscope presented in this work features both an integrated autofocus function calibrated by using an infrared laser-based distance sensor as well as image post-processing. The large working distance is beneficial for routine examinations. If in a future version of the system the distance sensor will be aligned along the optical axis of the camera via a dichroitic mirror, for example, it will be able to focus over an even larger range. Also, a distance-variable image scale would be advantageous and can be included in our system. Furthermore, a color read out tool could be added to the software. With this tool, images taken during different examinations could more easily be compared which could provide, for example, additional information about the blood perfusion. Also, use in therapy monitoring is possible.

The color calibration of our system is not yet optimal as the color values in the calibrated image still deviate from the reference color values. The uncalibrated image is comparably dark. This is because the gain of the camera was set to zero, so that the images can more easily be compared after image noise reduction. Also, the exposure time was set to 75 ms as a compromise between image brightness and exposure time to avoid motion blur. An even brighter light source would be advantageous in this regard. Another option would be to track the gain for the images taken and include this value in the calibration algorithm.

With regard to CRI requirements, halogen light sources and high pressure xenon lamps, for example, can have CRI values of 100 and 93, respectively [24]. As we use an LED with a more discrete emission spectrum, the CRI cannot directly be compared with CRI of the more continuously emitting light sources mentioned above. The reason for this is, that small changes in a discrete spectrum can result in large changes in the CRI which does not correlate with the changes in color representation as perceived by an human observer or a camera [25]. Also, the spectral acceptance of the sensor employed, i.e., the camera system, has to be taken into account. As even the light situation in examination rooms can alter because of, for example, the daytime, no exact color matching is needed in this case. It is more important to ensure comparable colors in every picture taken by the system. The light source and image algorithms can further be improved by spectral tuning to enhance visibility of different features relevant for diagnostics in dermatology [26].

Finally, the hardware components of the system can be made more compact, so that the it can serve as a module in an automated skin-screening device. In such a scheme the patient lies on a medical lounger and an automated arm equipped with suitable imaging systems takes overview pictures of the skin and identifies nevi and other areas relevant to dermatology. Subsequently, the system described in this work would take high-resolution images of the selected skin areas. In case of nevi, an algorithm that evaluates the ABCDE criteria could then make a risk analysis so that the physician only needs to examine the suspicious ones in more detail. In further versions, external image analysis from outside the examination room could be implemented as well as deep learning approaches for fast and essentially real-time analysis as pointed out in [27].

5. Conclusions and Outlook

In this work, we present a prototype of a non-contact dermatoscope with a built-in autofocus system. The auto-focus is based on a liquid lens in combination with an infrared laser distance sensor. This setup is sufficiently fast for dermatoscopy enabling a focus refresh rate of 30 Hz for image acquisition, which is double the maximal frame rate of the camera. The resolution of the dermatoscopic system is slightly lower than the resolution of the system without the liquid lens, but it is still within the required minimal resolution of 30 μm. In the current alignment of the components, the system is optimized for a working distance of 45 cm ± 3 cm. If the surface to be examined is not strongly curved (over the range of the depth of field) the system delivers sharp images at even larger distances as the infrared laser spot is reflected on the image plane. The prototype is largely independent of the ambient light due to its ultra-bright LED-light source illuminating the area of interest on the target at a distance of 45 cm with a luminous flux of 35.8 lm, which is equivalent to 162,000 lx. This makes images taken under different ambient light conditions comparable. The system also allows for color calibration to obtain more realistic color representation

In the next step, the prototype will be tested in a clinical environment. This could reveal further diagnostic advantages. A study will be designed to collect a larger set of images from skin lesions correlated to specific diagnostic data. The focus will be placed on a number of common inflammatory diseases in order to identify (new) diagnostic features that are revealed by the dermatoscopic images and to better evaluate the diagnostic potential of the non-contact remote approach.

Author Contributions: Conceptualization, D.F. and M.W.; Funding acquisition, B.R.; Investigation, E.D. and D.F.; Project administration, B.R., Resources, A.H., T.W. and B.R.; Software, D.F.; Supervision, M.W. and B.R.; Writing—original draft, D.F.; Writing—review & editing, D.F., M.W. and B.R.

Funding: This project is funded by the Lower Saxony Ministry for Culture and Science (MWK) through the program Tailored Light and the Deutsche Forschungsgemeinschaft (DFG, German Research Foundation) under Germany's Excellence Strategy within the Cluster of Excellence PhoenixD (EXC 2122, Project ID 390833453).

Conflicts of Interest: The authors declare no conflicts of interest.

References

1. MacKIE, R.M. An aid to the preoperative assesment of lesions of the skin. *Br. J. Dermatol.* **1971**, *85*, 232–238. [CrossRef]
2. Pehamberger, H.; Steiner, A.; Wolff, K. In vivo epiluminescence microscopy of pigmented skin lesions. I. Pattern analysis of pigmented skin lesions. *J. Am. Acad. Dermatol.* **1987**, *17*, 571–583. [CrossRef]
3. Steiner, A.; Pehamberger, H.; Wolff, K. In vivo epiluminescence microscopy of pigmented skin lesions. II. Diagnosis of small pigmented skin lesions and early detection of malignant melanoma. *J. Am. Acad. Dermatol.* **1987**, *17*, 584–591. [CrossRef]
4. Optotechnik, H. HEINE Optotechnik|Home. Available online: https://www.heine.com/en_GB/home/ (accessed on 27 May 2019).
5. 3Gen. The Worldwide Leader in Handheld Dermoscopy for Skin Cancer Detection. Available online: https://dermlite.com/ (accessed on 27 May 2019).
6. Nachbar, F.; Stolz, W.; Merkle, T.; Cognetta, A.B.; Vogt, T.; Landthaler, M.; Bilek, P.; Braun-Falco, O.; Plewig, G. The ABCD rule of dermatoscopy. *J. Am. Acad. Dermatol.* **1994**, *30*, 551–559. [CrossRef]
7. Abbasi, N.R.; Shaw, H.M.; Rigel, D.S.; Friedman, R.J.; McCarthy, W.H.; Osman, I.; Kopf, A.W.; Polsky, D. Early diagnosis of cutaneous melanoma: revisiting the ABCD criteria. *JAMA* **2004**, *292*, 2771–2776. [CrossRef] [PubMed]
8. Meinhardt-Wollweber, M.; Heratizadeh, A.; Basu, C.; Günther, A.; Schlangen, S.; Werfel, T.; Schacht, V.; Emmert, S.; Haenssle, H.A.; Roth, B. A non-contact remote digital dermoscope to support cancer screening and diagnosis of inflammatory skin disease. *Biomed. Phys. Eng. Express* **2017**, *3*, 55005. [CrossRef]
9. Bassoli, S.; Seidenari, S. Benefits of polarized versus nonpolarized dermoscopy. *Expert Rev. Dermatol.* **2010**, *5*, 17–21. [CrossRef]
10. Drexler, W.; Fujimoto, J.G. (Eds.) *Optical Coherence Tomography: Technology and Applications*, 2nd ed.; Springer: Berlin, Germany, 2015.
11. Varkentin, A.; Mazurenka, M.; Blumenroether, E.; Meinhardt-Wollweber, M.; Rahlves, M.; Broekaert, S.M.C.; Schäd-Trcka, S.; Emmert, S.; Morgner, U.; Roth, B. Comparative study of presurgical skin infiltration depth measurements of melanocytic lesions with OCT and high frequency ultrasound. *J. Biophotonics* **2017**, *10*, 854–861. [CrossRef] [PubMed]
12. Varkentin, A.; Mazurenka, M.; Blumenröther, E.; Behrendt, L.; Rahlves, M.; Meinhardt-Wollweber, M.; Morgner, U.; Roth, B. Trimodal system for in vivo skin cancer screening with combined optical coherence tomography-Raman and co-localized optoacoustic measurements. *J. Biophotonics* **2018**, *11*. [CrossRef]
13. De Oliveira Barcaui, E.; Carvalho, A.C.P.; Valiante, P.M.N.; Barcaui, C.B. High-frequency ultrasound associated with dermoscopy in pre-operative evaluation of basal cell carcinoma. *An. Bras. Dermatol.* **2014**, *89*, 828–831. [CrossRef]
14. FotoFinder Systems GmbH, "FotoFinder: Bodystudio ATBM. Available online: https://www.fotofinder.de/en/products/bodystudio/atbm/ (accessed on 27 May 2019).

15. Ragaz, A.; Ackerman, A.B. Evolution, maturation, and regression of lesions of lichen planus. New observations and correlations of clinical and histologic findings. *Am. J. Dermatopathol.* **1981**, *3*, 5–25. [CrossRef] [PubMed]
16. DIN—German Institute for Standardization, DIN 5035-3. Artificial Lighting—Part 3: Lighting of Health Care Premises. Available online: https://www.beuth.de/de/norm/din-5035-3/88503242 (accessed on 27 May 2019).
17. Langlois, N.E.; Gresham, G.A. The ageing of bruises: A review and study of the colour changes with time. *Forensic Sci. Int.* **1991**, *50*, 227–238. [CrossRef]
18. Stephenson, T.; Bialas, Y. Estimation of the age of bruising. *Arch. Dis. Child.* **1996**, *74*, 53–55. [CrossRef] [PubMed]
19. Groner, W.; Winkelman, J.W.; Harris, A.G.; Ince, C.; Bouma, G.J.; Messmer, K.; Nadeau, R.G. Orthogonal polarization spectral imaging. A new method for study of the microcirculation. *Nat. Med.* **1999**, *5*, 1209–1212. [CrossRef] [PubMed]
20. Jacques, S.L.; Roman, J.R.; Lee, K. Imaging superficial tissues with polarized light. *Lasers Surg. Med.* **2000**, *26*, 119–129. [CrossRef]
21. Jacques, S.L.; Ramella-Roman, J.C.; Lee, K. Imaging skin pathology with polarized light. *J. Biomed. Opt.* **2002**, *7*, 329–340. [CrossRef] [PubMed]
22. DIN—German Institute for Standardization, DIN 12464-1 (2011-08): Light and Lighting—Lighting of Work Places—Part 1: Indoor Work Places. Available online: https://www.beuth.de/de/norm/din-en-12464-1/136885861 (accessed on 27 May 2019).
23. János, S. (Ed.) *Colorimetry: Understanding the CIE System*; John Wiley & Sons, Ltd.: Hoboken, NJ, USA, 2007; ISBN 978-0-470-04904-4.
24. Hunt, R.W.G. *The Reproduction of Colour*; John Wiley & Sons, Ltd.: Hoboken, NJ, USA, 2004.
25. Narendran, N.; Deng, L. Color rendering properties of LED light sources. *Proc. SPIE* **2002**, *4776*, 61–67.
26. Basu, C.; Schlangen, S.; Meinhardt-Wollweber, M.; Roth, B. Light source design for spectral tuning in biomedical imaging. *J. Med. Imaging* **2015**, *2*, 44501. [CrossRef] [PubMed]
27. Haenssle, H.A.; Fink, C.; Schneiderbauer, R.; Toberer, F.; Buhl, T.; Blum, A.; Kalloo, A.; Hassen, A.B.H.; Thomas, L.; Enk, A.; et al. Man against machine: Diagnostic performance of a deep learning convolutional neural network for dermoscopic melanoma recognition in comparison to 58 dermatologists. *Ann. Oncol. Off. J. Eur. Soc. Med Oncol.* **2018**, *29*, 1836–1842. [CrossRef] [PubMed]

Article

Single-Mode Polymer Ridge Waveguide Integration of Organic Thin-Film Laser

Marko Čehovski [1,†], Jing Becker [2], Ouacef Charfi [1,†], Hans-Hermann Johannes [1,*,†], Claas Müller [2] and Wolfgang Kowalsky [1,†]

[1] Institut für Hochfrequenztechnik, Technische Universität Braunschweig, Schleinitzstraße 22, 38106 Braunschweig, Germany [2] Freiburger Zentrum für interaktive Werkstoffe und bioinspirierte Technologien, Universität Freiburg, Georges-Köhler-Allee 105, 79110 Freiburg, Germany

* Correspondence: h2.johannes@ihf.tu-bs.de; Tel.: +49-531-391-2006 (ext. 8002)

† Current address: Cluster of Excellence PhoenixD (Photonics, Optics, and Engineering – Innovation Across Disciplines), 30167 Hannover, Germany.

Received: 6 March 2020; Accepted: 16 April 2020; Published: 18 April 2020

Abstract: Organic thin-film lasers (OLAS) are promising optical sources when it comes to flexibility and small-scale manufacturing. These properties are required especially for integrating organic thin-film lasers into single-mode waveguides. Optical sensors based on single-mode ridge waveguide systems, especially for Lab-on-a-chip (LoC) applications, usually need external laser sources, free-space optics, and coupling structures, which suffer from coupling losses and mechanical stabilization problems. In this paper, we report on the first successful integration of organic thin-film lasers directly into polymeric single-mode ridge waveguides forming a monolithic laser device for LoC applications. The integrated waveguide laser is achieved by three production steps: nanoimprint of Bragg gratings onto the waveguide cladding material EpoClad, UV-Lithography of the waveguide core material EpoCore, and thermal evaporation of the OLAS material Alq_3:DCM2 on top of the single-mode waveguides and the Bragg grating area. Here, the laser light is analyzed out of the waveguide facet with optical spectroscopy presenting single-mode characteristics even with high pump energy densities. This kind of integrated waveguide laser is very suitable for photonic LoC applications based on intensity and interferometric sensors where single-mode operation is required.

Keywords: integrated optics and photonics; integrated polymer optics; organic laser; integration; polymeric waveguide; Lab-on-a-Chip

---

1. Introduction

The light from only a few light sources is able to couple effectively into single-mode waveguides. Lasers are light sources with very high coupling efficiencies. For polymeric waveguides, however, not all laser types are suitable. Especially for LoC devices, single-mode operation is mostly required. The coupling of the laser light into the single-mode waveguide is mainly achieved by prism-coupling, grating couplers with a free-space set-up or butt-coupling with a lensed fiber [1]. An important step towards waveguide integrated lasers can be realized by combining the laser resonator, such as distributed feedback resonator (DFB) with the waveguide structure. The DFB wavelength selectivity and high efficiency as well as the ease of fabrication makes them very suitable for this kind of integration. OLAS are well suited for this application due to their low layer thickness, their flexibility and simple processing [2]. In the last decade, the integration of organic DFB lasers into waveguides and LoC systems has been demonstrated. Balslev et al. [3] realized a LoC system with an integrated organic laser source and a photodiode. The laser light source is achieved by the laser dye Rhodamine 6G dissolved in ethanol, which is purged through a microfluidic channel to a DFB grating. The integration of the laser source and the multi-mode SU-8 waveguides is realized by butt coupling.

Christiansen et al. [4] reported on active and passive SU-8 waveguides either doped or undoped with the laser dye Rhodamine 6G. First, the active waveguide is constructed by a lithography and imprint process. The following passive waveguide is subsequently built up and also butt coupled to the active waveguide. Mappes et al. [5] and Vannahme et al. [6] report on an integration of an OLAS into PMMA. The resonant DFB structure is formed by the aid of hot-embossing and the waveguides are manufactured by deep UV irradiation of the PMMA bulk material. Up to now, all the integrations are only realized with multi-mode waveguides or slab waveguides. However, it is a great approach for intensity based LoC applications where, for example, fluorescence excitation of an analyte in a microfluidic channel is used. Despite that, pure single-mode operations is required for interferometric based sensors such as optical based LoC-systems. Since higher modes propagate with different speed of light and have different evanescent wave characteristics, this leads to an inefficient interferometric interaction in the sensor as well as signal losses due to modal dispersion [1]. Recently Becker et al. [7] integrated DFB gratings onto a 2.0 m in width and 2.5 m in height few-mode ridge waveguide. This is realized by a straightforward combined nanoimprint and photolithography process (CNP process) using OrmoCore, a silicon-containing hybrid waveguide core material. On it, the OLAS is finally evaporated. In our study, we used a three-step fabrication process to integrate the OLAS into the 1.0 m in width and 1.0 m in height polymeric single-mode waveguide as schematically depicted in Figure 1a. Therefore, we used the photopatternable epoxies EpoClad (refractive index = 1.579 @650 nm) and EpoCore (refractive index = 1.593 @650 nm) for the cladding layer including the grating structure and the waveguide core, respectively. The waveguide materials are commercially available and distributed by Micro Resist Technology GmbH in Germany. The active organic material (Alq_3:DCM2) was then evaporated on top of the waveguide structure with the DFB gratings underneath. The final device contains five DFB gratings in parallel and five single-mode waveguides on top of each grating area, forming 25 different waveguide integrated lasers with five different emission wavelengths.

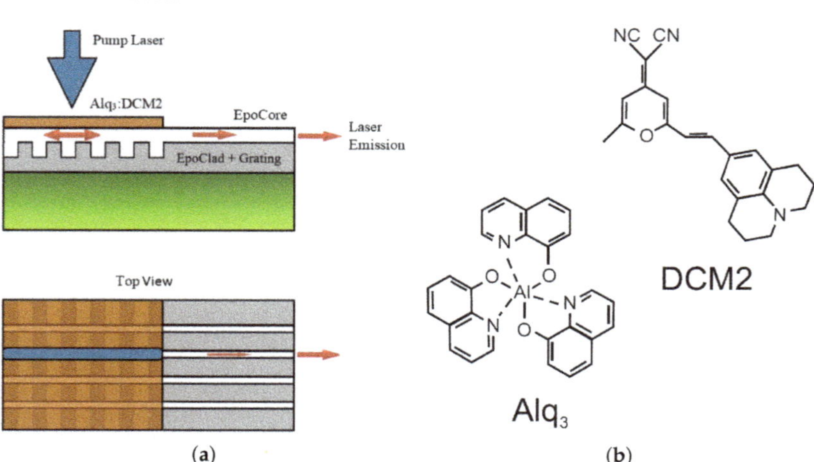

Figure 1. (**a**) Schematic sketch of the OLAS integrated into the polymeric single-mode waveguide; and (**b**) the structural formula of Alq_3 and DCM2.

2. The Polymeric Single-Mode Waveguide

Optical simulations using the finite difference eigenmode solver support the single-mode characteristic. Figure 2 shows the cross section of the single-mode ridge waveguide structure as described in the introduction as well as the TE0 and TE1 mode distribution with air (refractive index $\approx$1.0002 @650 nm) and EpoClad as the surrounding cladding medium.

For the ridge-type waveguide with air as upper cladding, the TE0 mode is asymmetric and extends into the lower cladding (see Figure 2a). The effective refractive index was calculated to be 1.561 @650 nm. Figure 2b shows the TE0 mode distribution for the symmetrical case where the waveguide core is surrounded by the cladding material making it a buried waveguide. The mode field maximum is located in the waveguide core and the minority of the field intensity is located to be outside and is called evanescent field. In this structure, the decay length of the evanescent field is about 1 m. For the buried waveguide, the effective refractive index of the TE0 mode is 1.58 @650 nm. In both cases, higher modes such as TE1 (see Figure 2c,d) are not capable of propagation in the waveguide core.

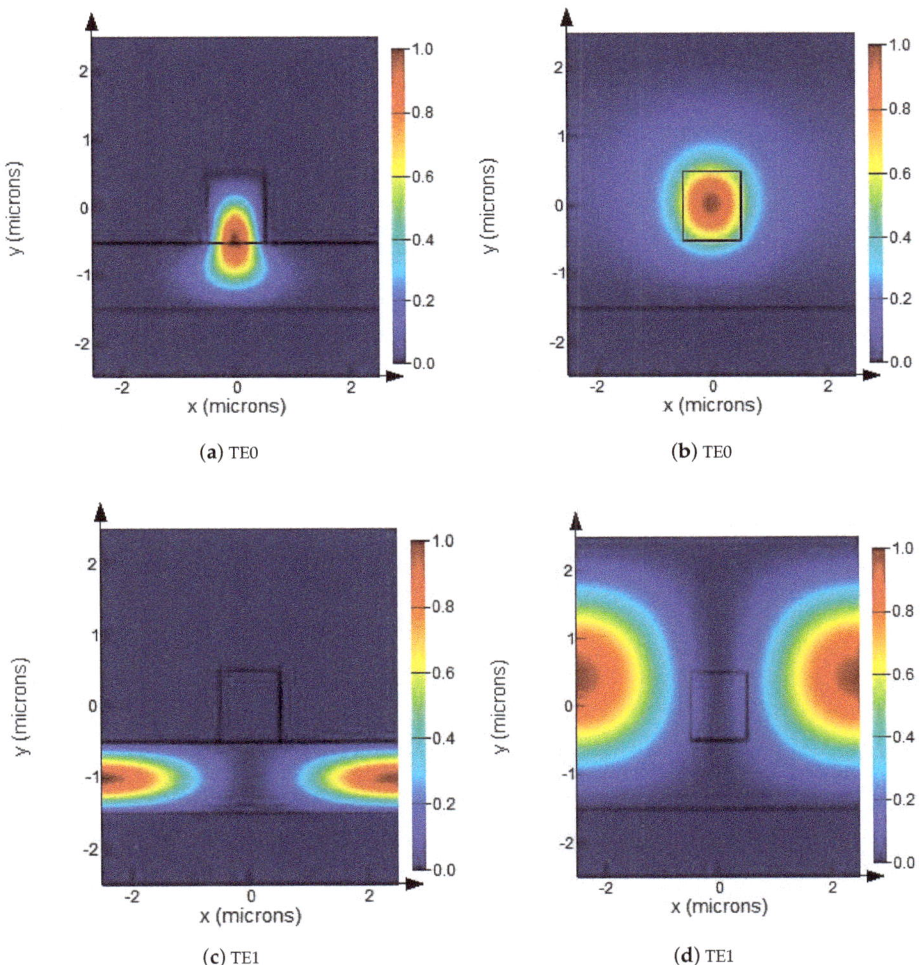

Figure 2. TE0 and TE1 mode distribution with: (**a**,**c**) air as upper cladding medium; and (**b**,**d**) EpoClad as upper cladding medium.

3. The Organic Thin-Film Laser

We used the well-known guest–host active organic laser material containing Tris-(8-hydroxyquinoline)aluminum (Alq_3) as the host material and the laser dye 4-(Dicyanomethylene)-2-methyl-6-julolidyl-9-enyl-4H-pyran (DCM2) as the guest material (see Figure 1b). Figure 3 shows the photoluminescence spectra (PL spectra) of Alq_3 and Alq_3:DCM2 with different DCM2 doping concentrations. The samples were evaporated and characterized on glass substrates.

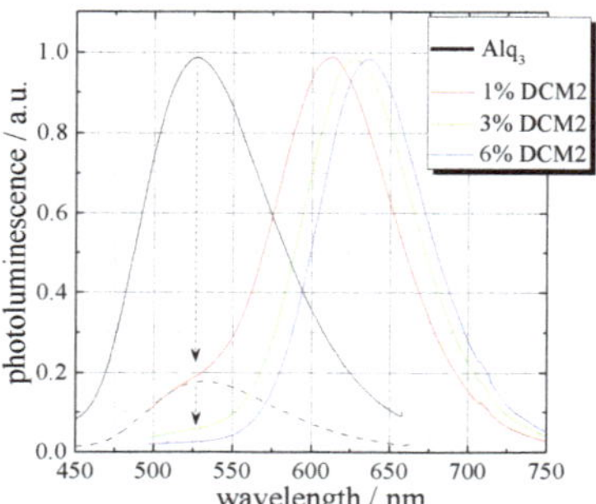

Figure 3. PL spectra of Alq_3 and Alq_3:DCM2 at different doping concentrations.

Figure 3 shows a redshift in the emission spectra of the Alq_3:DCM2 samples with increasing DCM2 doping concentration. The reason for this is the Solid State Solvation Effect (SSSE). It describes the polarity influence of the host matrix on the emitter molecule [8]. Furthermore, Figure 3 also shows a decreased Alq_3 emission in the PL spectra with increasing doping concentration of DCM2. The fraction of Alq_3 emission is about 18% (see dashed line in Figure 3) with the DCM2 doping concentration of 1%. When the doping concentration increases to 3%, the fraction of Alq_3 emission drops to 5%. With further increasing of the doping concentration, as shown in Figure 3, the Alq_3 emission fraction in the PL spectra drops even more. This is due to the Förster resonant energy transfer [9,10] where with decreasing donor/acceptor mean distance the probability of a resonant energy transfer increases. The great advantages of this material system are the broad optical laser tuning range of up to 115 nm, low lasing threshold down to approximately 3 J/cm^2, and high optical gain, which saturates at about 300 cm^{-1} [11–13]. Moreover, the Alq_3:DCM2 thin-film is characterized by its absolutely smooth surface after the vapor deposition process. The surface roughness was measured to be $R_a = 950$ pm (arithmetic mean value) and $R_q = 1.2$ nm (root mean square value). However, surface roughness, morphological variations as well as the pump process itself can cause slightly stochastic fluctuations and noise in the signal. Therefore, in the spectral analysis, the Karhunen–Loève Transformation can be applied [14]. Different kinds of resonating structures could be demonstrated to achieve lasing in organic thin-films, such as DFB resonators, micro disks, spheres, and distributed Bragg reflectors (DBR) [15–19]. The DFB resonator, however, stands out because of its ease of fabrication and its optical properties such as the extraordinary wavelength selectivity and high efficiency. The emission wavelength (λ_{Bragg}) of the distributed feedback laser can be varied by choosing the suitable grating periodicity Λ. It follows the Bragg condition:

$$\lambda_{\text{Bragg}} = \frac{2 \cdot n_{\text{eff}} \cdot \Lambda}{m}, \tag{1}$$

where n_{eff} is the effective refractive index of the allover waveguide structure, Λ is the grating period, and $m = 1, 2, 3, ...$, is the diffraction order.

4. Device Fabrication

First, a Polydimethylsiloxan (PDMS) stamp was made for replication of the DFB gratings. The grating area contains five DFB gratings arranged in parallel, with the dimension of 2 mm in width and 10 mm in length. The periodicity of the gratings was 195 nm to 215 nm with a period spacing of 5 nm. After the fabrication of the PDMS stamp, the bottom cladding with the grating structures was fabricated by a UV nanoimprint process. In this step, a mixture of EpoClad and γ-Butyrolactone

(GBL) with a mixing ratio of 2:1 was used. The GBL diluted EpoClad was spin-coated onto a silicon wafer with 4000 rpm for 30 s and then soft baked at 120 °C for 2 min on a hot plate. After the soft bake, the PDMS stamp was brought into contact with the still sticky EpoClad. A subsequent flood-exposure starts the UV-induced cross-linking of the polymer. After that, the substrate was baked again on a hot plate at 120 °C for 3 min. After the EpoClad was completely polymerized, the silicon-PDMS stamp set was taken from the hot plate and the PDMS stamp was removed without any residue. The layer thickness of EpoClad was about 1 m. Figure 4a shows the profile of one of the gratings with Λ = 195 nm and Figure 4b shows a good replication quality. This image was taken with an atomic force microscope. The measured average periodicity is 194.97 nm, which is extremely close to the desired value.

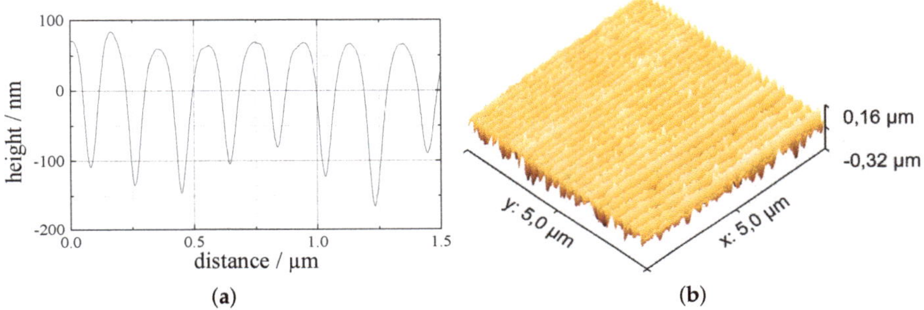

Figure 4. (**a**) Grating profile; and (**b**) 3D AFM image of the imprinted DFB grating with $\Lambda = 195$ nm into EpoClad.

The next fabrication step is structuring the waveguides onto the DFB gratings. In this case, a mixture of EpoCore and γ-Butyrolactone (GBL) in a mixing ratio of 2:1 was used. The mixture was spin-coated with 4600 rpm for 30 s onto the substrates with the pre-structured cladding layer. The substrates were first baked at 50 °C and then at 90 °C for 2 min each followed by a UV-lithography step using a chromium mask with the waveguide structures. After that, the substrates were put on the hot plate again for 2 min at 50 °C and for 3 min at 85 °C for a post-exposure bake. The non-cross-linked EpoCore was removed by dipping it for 10 s into the mr-Dev 600 developer. Finally, the substrates were rinsed in isopropyl alcohol to stop the development process. Figure 5 shows schematic sketch of the final device.

Before starting the measurements, the final device was carved at the break points with a diamond scriber so that the silicon wafer split apart along its crystal lattice. Figure 6 presents the SEM images of the EpoCore waveguide on top of the DFB gratings.

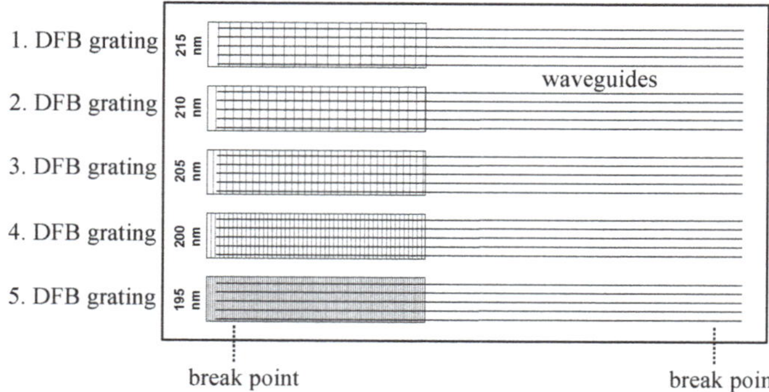

Figure 5. Schematic sketch of the final device after the lithography process containing DFB grating and waveguide structures.

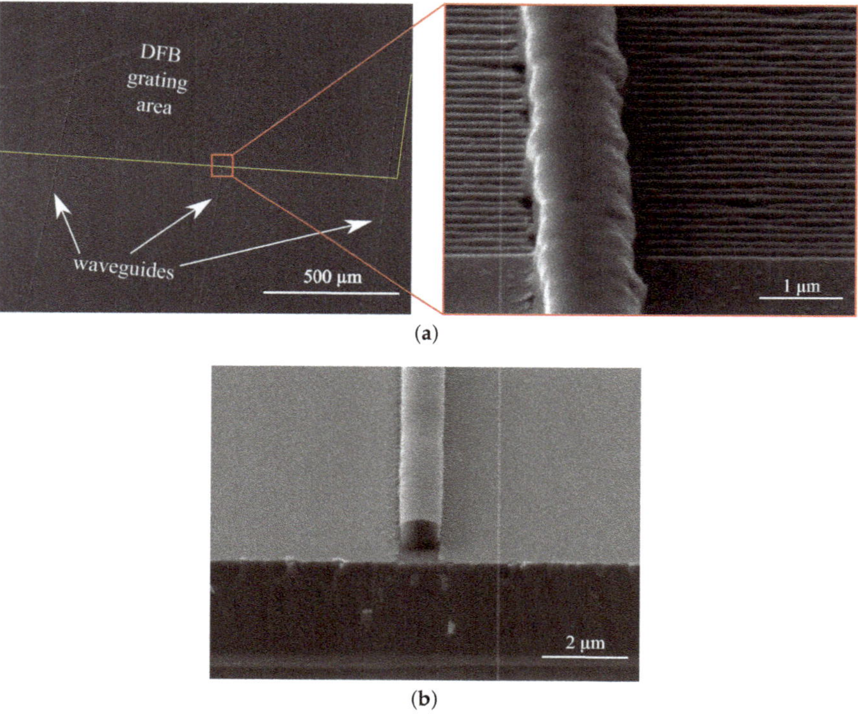

Figure 6. (**a**) SEM overview image of the whole waveguide and grating structure and a detailed view on the waveguides on top of the DFB grating; and (**b**) the waveguide end facet.

The figures show a good waveguide integration on top of the DFB gratings as well as a satisfying waveguide end facet for the optical analysis of the integrated waveguide laser. Figure 6a also gives a detailed look at the waveguide and shows some irregularities and an uneven surface. This can probably be addressed to the grating structure below. The waveguides outside this grating area show good optical quality and a smooth surface (cf. Figure 6b). After the fabrication of the DFB gratings and the waveguides, 200 nm of the guest–host laser active material system Alq_3:DCM2 with a DCM2 doping concentration of 6% was co-evaporated in an ultrahigh vacuum chamber with $p < 10^{-8}$ mbar by organic molecular beam deposition method.

5. Results and Discussion

The samples were measured and characterized by the optical set-up schematically shown in Figure 7.

The waveguide integrated OLAS was excited with the third harmonic ($\lambda_3 = 355$ nm) of a passively Q-switched Nd:YAG laser. The excitation energy was controlled by a neutral density filter. For an efficient excitation, the Gaussian beam out of the Nd:YAG laser was formed into a laser stripe by the cylindrical lenses L1, L2, and L3. Additional apertures form a flat-top profile out of the Gaussian beam. Thus, an excitation stripe of 0.15 mm in width and 3.0 mm in length was created. The dimensions of the laser stripe were measured with a Thorlabs BP209-VIS/M scanning-slit optical beam profiler. The energy out of the pump laser was measured with the Coherent LabMax-Top system. The waveguide integrated laser probes were stored and measured in a N_2-flushed chamber. Instead of this method, to avoid unnecessary degradation and oxidation under atmosphere conditions, a thin-film encapsulation can be applied on top of the waveguide integrated lasers [20]. The excited laser emission out of the end facet of the waveguide was measured by using the lens pair L4 and L5. An additional aperture at the N_2-flushed chamber in front of the Lens L4 was included to suppress the scattered light. The waveguide integrated lasers were analyzed by a precise monochromator (Triax 320, HORIBA Scientific) and a liquid nitrogen cooled CCD detection system (Symphony, HORIBA Scientific).

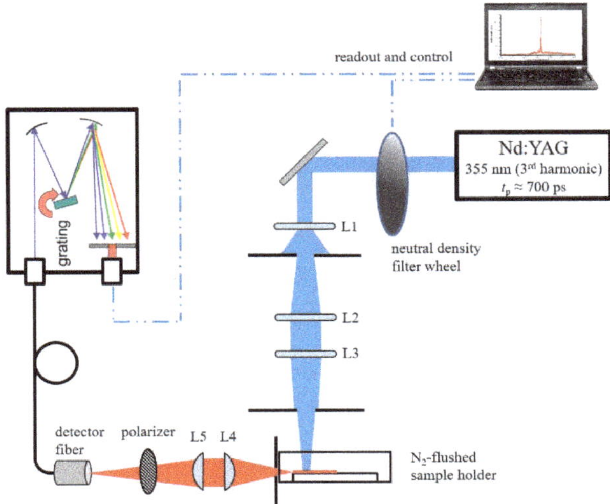

Figure 7. Optical set-up for laser performance measurements of the integrated OLAS. The laser emission was measured orthogonal to the pump beam.

The measurement of the OLAS out of the waveguide facet is depicted in Figure 8a, where lasing occurred with all grating periods. For the sake of simplicity, we took a closer look at just one laser line. Figure 8b shows the laser line at $\lambda_{\text{Bragg}} = 640.24$ nm. Using the Bragg condition in Equation (1) for the first order laser emission ($m = 1$) at $\Lambda = 205$ nm, an effective refractive index n_{eff} of about 1.562 can be determined. This result is comparable with the result of the mode propagation simulation. The effective refractive index n_{eff} of the fundamental mode for this waveguide was calculated to be approximately 1.561 @640 nm. The spectral width of the emission peak at the full width at half maximum (FWHM) was measured to be 200 pm. Furthermore, by increasing the excitation energy density up to 274 J/cm^2 (cf. Figure 8c), no further modes except the fundamental mode were found. In previous measurements on few-mode waveguides (waveguide facet dimension 2 × 2 m^2), up to five modes could be found with increased pump energy.

Deviating from the DFB theory [21,22], where two laser modes near the Bragg wavelength are generated due to the symmetry of the gratings in index-coupled DFB-lasers [23], we could only measure one laser line out of our waveguide integrated lasers. We suppose that this effect can be addressed to the replication process of the Bragg grating structure into the polymeric cladding layer. Possible slight thickness variations of the polymer layers due to the spin-coating technique as well as potential variations in the doping concentration of the active laser material can have an effect on the effective refractive index change in the structure leading to an asymmetrical behavior and therefore to single-mode operation of the waveguide integrated laser.

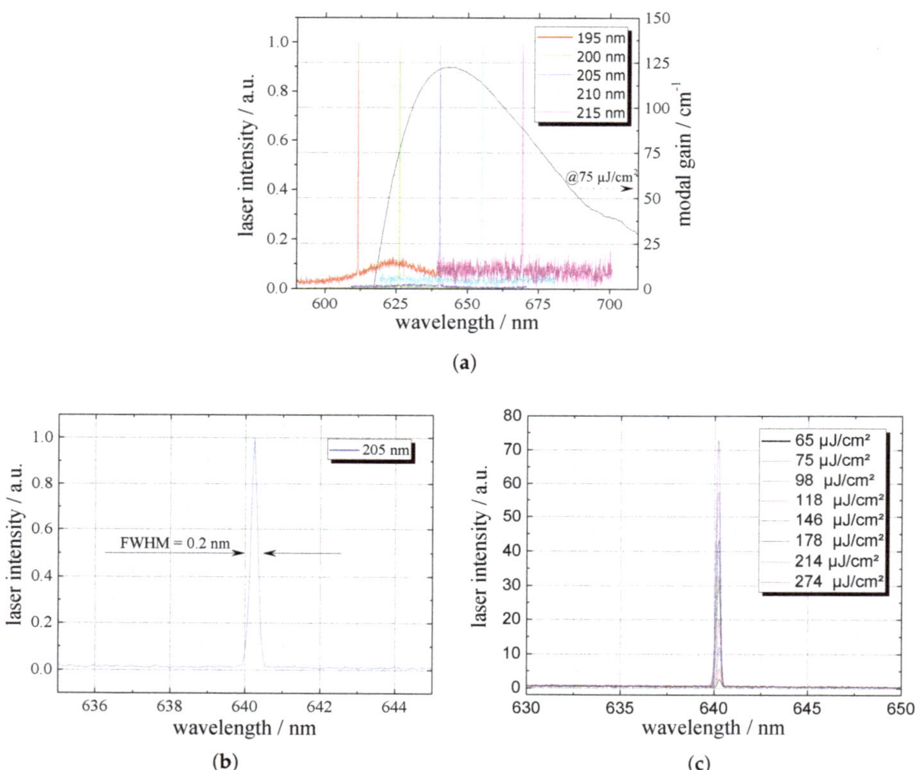

Figure 8. (**a**) Lasing lines out of the integrated OLAS with different DFB grating periods with the optical modal gain in the background; and (**b**) the spectral response of the integrated OLAS with (**c**) increasing excitation energy density.

Figure 9a shows the emitted intensity of the waveguide integrated lasers in dependence of the excitation energy density. An increased laser emission could be found by exceeding the lasers above their laser thresholds. Laser thresholds could be found to be in the range from approximately 25 J/cm^2 to approximately 170 J/cm^2. Moreover, with increasing the Bragg grating period, and therefore increasing the laser wavelength, the laser threshold decreases and the slope of the lasers increases. At a certain turning point, the laser threshold increases and the laser slope decreases. This can be attributed to the spectral response of the optical modal gain of this material (cf. Figure 8a). The turning point corresponds with the modal gain maximum where the stimulated cross section is the highest [12,13]. The spectral width of the modal gain increases with increasing excitation energy density. Thus, for shorter laser wavelengths, higher excitation energy is needed until the optical modal gain overcomes absorption and optical losses. For higher wavelengths, the optical modal gain decreases, which can be seen in the laser threshold of the laser with 666.82 nm laser wavelength. The lowest lasing threshold is one magnitude higher than classic OLAS without waveguides (cf. [2]) but still up to 210-times lower than the thresholds reported in [3,4]. By inserting a polarization filter in front of the detection fiber into the optical set-up (cf. Figure 7), a polarization measurement could be done to get the TE- and TM-proportions. The radiated emission was clearly linear polarized with a polarization extinction ratio (PER) of 15.5 : 1. The intensity difference between the values at 0° and 360° can be addressed to a slight degradation of the sample during this measurement.

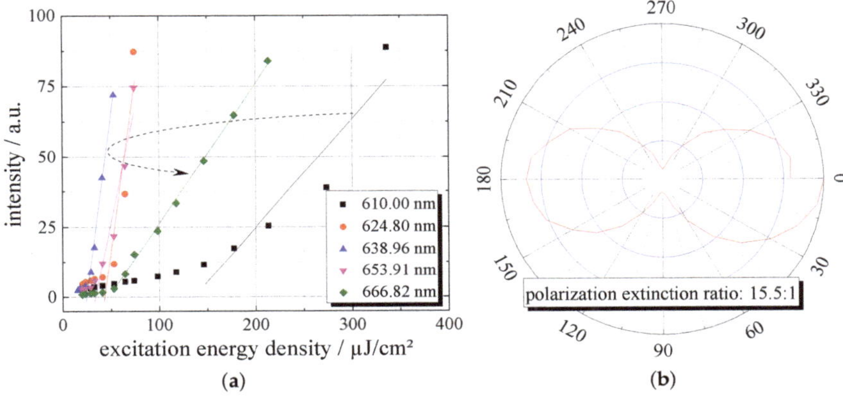

Figure 9. (**a**) Laser threshold of the waveguide integrated OLAS; and (**b**) the polar plot of the polarization properties. The lasing threshold corresponds to the optical modal gain for the active organic material.

6. Conclusions

In this study, EpoCore and -Clad based single-mode waveguides with integrated Bragg gratings were successfully fabricated. By depositing the organic laser material Alq_3:DCM2 on top of the waveguide structure, a monolithically single-mode waveguide laser device is achieved. The laser emission could be measured at all waveguide integrated lasers with a FWHM of 200 pm. No other higher modes could be observed even with excitation energy densities up to 274 J/cm^2. The laser threshold could be measured to be in a range of approximately 25 J/cm^2 to 170 J/cm^2. The LoC application ability for photonic or interferometric based sensors could be supported by the polarization extinction ratio of 15.5:1, which indicts a linear polarization. By using this type of laser with the described integration, light can be easily coupled into the single-mode waveguide. This type of laser integration allows an uncomplicated light coupling into a single-mode waveguide. With this work, a milestone towards the monolithically integration of organic lasers is achieved. Especially, optical sensor systems based on single-mode waveguides and, in particular, LoC systems could benefit from this work.

Author Contributions: conceptualization: M.Č.; software: M.Č.; validation: M.Č., J.B. and O.C.; investigation: M.Č. and J.B.; writing—original draft preparation: M.Č.; writing—review and editing: J.B. and H-H.J.; supervision: H-H.J., C.M. and W.K.; All authors have read and agreed to the published version of the manuscript.

Funding: The authors would like to thank the German Research Foundation (Deutsche Forschungsgemeinschaft, DFG) for funding this work within the PolySens Project (Project ID 410203759) as well as the Germany's Excellence Strategy within the Cluster of Excellence PhoenixD (EXC 2122, Project ID 390833453).

Acknowledgments: The authors would like to acknowledge the contribution of the SEM measurements used in this publication by M.Sc. Jan Gülink from the Institut für Halbleitertechnik, TU Braunschweig.

Conflicts of Interest: The authors declare no conflict of interest.

References

1. Estevez, M.; Alvarez, M.; Lechuga, L. Integrated optical devices for lab-on-a-chip biosensing applications. *Laser Photonics Rev.* **2012**, *6*, 463–487. [CrossRef]
2. Kuehne, A.J.C.; Gather, M.C. Organic Lasers: Recent Developments on Materials, Device Geometries, and Fabrication Techniques. *Chem. Rev.* **2016**, *116*, 12823–12864. [CrossRef] [PubMed]
3. Balslev, S.; Jorgensen, A.M.; Bilenberg, B.; Mogensen, K.B.; Snakenborg, D.; Geschke, O.; Kutter, J.P.; Kristensen, A. Lab-on-a-chip with integrated optical transducers. *Lab Chip* **2006**, *6*, 213–217. [CrossRef] [PubMed]
4. Christiansen, M.B.; Schøler, M.; Kristensen, A. Integration of active and passive polymer optics. *Optics Express* **2007**, *15*, 3931–3939. [CrossRef] [PubMed]
5. Mappes, T.; Vannahme, C.; Klinkhammer, S.; Woggon, T.; Schelb, M.; Lenhert, S.; Mohr, J.; Lemmer, U. Polymer biophotonic lab-on-chip devices with integrated organic semiconductor laser. In *Organic Semiconductors in Sensors and Bioelectronics II*; Shinar, R., Malliaras, G.G., Eds.; International Society for Optics and Photonics, SPIE: New York, NY, USA, 2009; Volume 7418, pp. 48–55. [CrossRef]
6. Vannahme, C.; Klinkhammer, S.; Kolew, A.; Jakobs, P.J.; Guttmann, M.; Dehm, S.; Lemmer, U.; Mappes, T. Integration of organic semiconductor lasers and single-mode passive waveguides into a PMMA substrate. *Microelectron. Eng.* **2010**, *87*, 693–695. [CrossRef]
7. Becker, J.; Čehovski, M.; Caspary, R.; Kowalsky, W.; Müller, C. Distributed feedback ridge waveguide lasers fabricated by CNP process. *Microelectron. Eng.* **2017**, *181*, 29–33. [CrossRef]
8. Madigan, C.F.; Bulović, V. Solid State Solvation in Amorphous Organic Thin Films. *Phys. Rev. Lett.* **2003**, *91*. [CrossRef] [PubMed]
9. Kozlov, V.G.; Bulovic, V.; Burrows, P.E.; Baldo, M.; Khalfin, V.B.; Parthasarathy, G.; Forrest, S.R.; You, Y.; Thompson, M.E. Study of lasing action based on Foerster energy transfer in optically pumped organic semiconductor thin films. *J. Appl. Phys.* **1998**, *84*, 4096–4108. [CrossRef]
10. Berggren, M.; Dodabalapur, A.; Slusher, R.E. Stimulated emission and lasing in dye-doped organic thin films with Foerster transfer. *Appl. Phys. Lett.* **1997**, *71*, 2230–2232. [CrossRef]
11. Schneider, D.; Rabe, T.; Riedl, T.; Dobbertin, T.; Kröger, M.; Becker, E.; Johannes, H.H.; Kowalsky, W.; Weimann, T.; Wang, J.; et al. Ultrawide tuning range in doped organic solid-state lasers. *Appl. Phys. Lett.* **2004**, *85*, 1886–1888. [CrossRef]
12. Čehovski, M.; Döring, S.; Rabe, T.; Caspary, R.; Kowalsky, W. Combined optical gain and degradation measurements in DCM2 doped Tris-(8-hydroxyquinoline)aluminum thin-films. In *Organic Photonics VII*; Cheyns, D., Beaujuge, P.M., van Elsbergen, V., Ribierre, J.C., Eds.; SPIE: New York, NY, USA, 2016. [CrossRef]
13. Rabe, T. *Materialien und Bauelementstrukturen für Organische Laserdioden*; Cuvillier Verlag: Goettingen, Germany, 2010.
14. Zaharov, V.V.; Farahi, R.H.; Snyder, P.J.; Davison, B.H.; Passian, A. Karhunen–Loève treatment to remove noise and facilitate data analysis in sensing, spectroscopy and other applications. *Analyst* **2014**, *139*, 5927–5935. [CrossRef] [PubMed]
15. Rabe, T.; Döring, S.; Hildebrandt, N.; Riedl, T.; Kowalsky, W.; Scherf, U. Optical Gain in Foerster Energy Transfer Based Organic Guest-Host-Systems. *MRS Proc.* **2009**, *1197*. [CrossRef]
16. Chiasera, A.; Dumeige, Y.; Féron, P.; Ferrari, M.; Jestin, Y.; Conti, G.N.; Pelli, S.; Soria, S.; Righini, G. Spherical whispering-gallery-mode microresonators. *Laser Photonics Rev.* **2010**, *4*, 457–482. [CrossRef]
17. Canazza, G.; Scotognella, F.; Lanzani, G.; Silvestri, S.D.; Zavelani-Rossi, M.; Comoretto, D. Lasing from all-polymer microcavities. *Laser Phys. Lett.* **2014**, *11*, 035804. [CrossRef]
18. Tsiminis, G.; Wang, Y.; Kanibolotsky, A.L.; Inigo, A.R.; Skabara, P.J.; Samuel, I.D.W.; Turnbull, G.A. Nanoimprinted Organic Semiconductor Laser Pumped by a Light-Emitting Diode. *Adv. Mater.* **2013**, *25*, 2826–2830. [CrossRef] [PubMed]
19. Samuel, I.D.W.; Turnbull, G.A. Organic Semiconductor Lasers. *Chem. Rev.* **2007**, *107*, 1272–1295. [CrossRef] [PubMed]

20. Bülow, T.; Gargouri, H.; Siebert, M.; Rudolph, R.; Johannes, H.H.; Kowalsky, W. Moisture barrier properties of thin organic-inorganic multilayers prepared by plasma-enhanced ALD and CVD in one reactor. *Nanoscale Res. Lett.* **2014**, *9*, 223. [CrossRef] [PubMed]
21. Ebeling, K.J. *Integrierte Optoelektronik*; Springer: Berlin/Heidelberg, Germany, 1992. [CrossRef]
22. Kogelnik, H.; Shank, C.V. Coupled-Wave Theory of Distributed Feedback Lasers. *J. Appl. Phys.* **1972**, *43*, 2327–2335. [CrossRef]
23. Lowery, A.; Novak, D. Performance comparison of gain-coupled and index-coupled DFB semiconductor lasers. *IEEE J. Quantum Electron.* **1994**, *30*, 2051–2063. [CrossRef]

Article

A High Peak Power and High Beam Quality Sub-Nanosecond Nd:YVO_4 Laser System at 1 kHz Repetition Rate without SRS Process

Yutao Huang [1,2,3], Hongbo Zhang [1,2,3], Xiaochao Yan [1,2], Zhijun Kang [1,2], Fuqiang Lian [1,2] and Zhongwei Fan [1,2,3,*]

1 Academy of Opto-Electronics, Chinese Academy of Sciences, Beijing 100094, China; yutaohuang@126.com (Y.H.); hbzhang999@126.com (H.Z.); yanxc123@126.com (X.Y.); kzjun1221@126.com (Z.K.); aoefiberlaser@126.com (F.L.)
2 National Engineering Research Center for DPSSL, Beijing 100094, China
3 University of Chinese Academy of Sciences, Beijing 100049, China
* Correspondence: fanzhongwei@aoe.ac.cn; Tel.: +86-10-8217-8609

Received: 6 November 2019; Accepted: 27 November 2019 ; Published: 2 December 2019

Abstract: We present a compact sub-nanosecond diode-end-pumped Nd:YVO_4 laser system running at 1 kHz. A maximum output energy of 65.4 mJ without significant stimulated Raman scattering (SRS) process was obtained with a pulse duration of 600 ps, corresponding to a pulse peak power of 109 MW. Laser pulses from this system had good beam quality, where $M^2 < 1.6$, and the excellent signal to noise ratio was more than 42 dB. By frequency doubling with an LBO crystal, 532 nm green light with an average power of 40.5 W and a power stability of 0.28% was achieved. The diode-end-pumped pump power limitation on a high peak power amplifier caused by the SRS process and thermal fracture in bulk Nd:YVO_4 crystal is also analyzed.

Keywords: sub-nanosecond laser; high peak power; Nd:YVO_4; stimulated Raman scattering (SRS); thermal fracture

---

1. Introduction

Compact diode-pumped solid state lasers with high-energy sub-nanosecond pulses at kHz repetition rates have a variety of applications, such as laser ranging [1,2], materials processing [3], nonlinear optical conversion [4,5] and Mid-IR optical parametric processes [6–10], etc.

The scaling of energy from sub-nanosecond oscillators is strongly limited by optical damage and thermal effect. Laser radiation with high beam quality, short pulse duration, high repetition rate, and high intensity can be obtained with the Master Oscillator Power Amplifier (MOPA) approach [11,12]. Both Yb^{3+}-doped lasers [13] and Nd^{3+}-doped lasers [14] can use the MOPA approach to achieve high energy sub-nanosecond pulses near 1 μm. Yb^{3+}-doped lasers have wide gain bandwidth, high Stokes efficiency, and high saturation fluence and achieve pulse energy with hundreds of mJ. In Yb^{3+}-doped lasers, cryogenic cooling is needed to acquire a four-level system, which will increase complexity of laser system [15,16]. Nd^{3+}-doped lasers with four-level systems operating in room temperature have higher gain and simpler structure, making them more attractive.

Several diode-pumped solid-state sub-nanosecond laser systems with different Nd^{3+}-doped materials geometry have been reported recently. A Nd^{3+}-doped amplifier in bounce geometry has higher small-signal gain to amplify sub-nanosecond oscillator with low energy and high repetition. A sub-nanosecond laser using a side-pumped Nd:YVO_4 bounce amplifier operating at 1–10 kHz was demonstrated with 577 ps duration and 545 μJ energy (1 MW peak power) in 2010 [17]. In a 2011 study, a sub-nanosecond Nd:YAG laser system with bounce geometry has been reported, the output energy of which was 0.8 mJ with peak power of 9.35 MW [18]. However, due to the small volume of the active medium, the pulse energy of a sub-nanosecond laser using bounce geometry is less than 1 mJ. In comparison, Nd^{3+}-doped lasers based on bulk crystal may acquire high pulse energy at the kHz level [19–21]. A sub-nanosecond single frequency MOPA laser system operating at 500 Hz was reported, which generated pulses with a width of 830 ps, energy of 13 mJ, peak power of 15.7 MW, and $M^2 < 1.4$ [22]. A Nd:YAG laser system based on the MOPA approach operating at 300 Hz and using selectable 350–600 ps nominally flat temporal pulse shape has been reported, which generate pulses with 127 mJ energy and maximum peak power of 400 MW [23], but beam quality M^2 were not measured. As is already known, beam quality always gets worse due to thermally induced birefringence in Nd:YAG crystal [21,24]. Laser systems based on end-pumped bulk Nd:YVO_4 crystal have better beam quality because of natural birefringence architecture, which can alleviate thermally induced birefringence [25–27]. At the same time, Nd:YVO_4 is a self-excited Raman scattering crystal [28–30]. Stimulated Raman scattering (SRS) is a third-order nonlinear optical effect, which must be taken into consideration in high peak power amplification. Through the SRS process in YVO_4 crystal, 1064 nm fundamental wave will be converted to Raman light, which consists of different Stokes waves, such as first Stokes wave λ = 1176 nm. Raman light brings in many disadvantages, for example, Raman light consumes amplified fundamental pulse energy and reduces the amplification efficiency. Morever, Raman light with high peak power may enter into the end-pumped laser diode and leads to damage of the laser diode. Because of this, the SRS process should be averted to improve amplification efficiency and keep optical components safety.

In this paper, we present a Sub-Nanosecond Nd:YVO_4 MOPA system consisting of a master and a series of amplifiers. By suppressing the SRS process during amplification, we obtain 65.4 mJ energy at 1064 nm with 600 ps and peak power above 100 MW at 1 kHz. The M^2 is less than 1.6 and the signal to noise ratio can be up to 42 dB. We also analyze the influence of SRS on working parameters in a high peak power Nd:YVO_4 amplifier. After frequency doubling, 40.5 mJ 532 nm green light is achieved with conversion efficiency of 61.8% and one hour power stability of 0.28% (RMS).

2. Experiment Setup

Figure 1 shows the layout of the laser system, which comprises one master oscillator, one four-pass high gain pre-amplifier, and three stages of main amplifier.

The master oscillator is a commercial industrial-grade sub-nanosecond Nd:YVO_4 Q-switched laser (Picolo AOT, InnoLas Laser, Munich, Germany), providing 1064 nm light, with 55 μJ, 450 ps at 1 kHz. Pulses from the Nd:YVO_4 Q-switched laser pass through an optical isolator composed of a PBS, an FR, and a HWP to avoid unwanted reflected light or even optical damage to the master oscillator. Before they are injected into the pre-amplifier, laser pulses are aligned and expanded to 2.0 mm with a beam expander (BE1), which suits the pump mode of the pre-amplifier. The pre-amplifier is an off-axis four-pass diode-end-pumped Nd:YVO_4 amplifier, where the Nd:YVO_4 crystal is $4 \times 4 \times 10$ mm^3, a-cut, 0.3%-doped, and has double-end-wedged cut at 2°. Both sides of the Nd:YVO_4 have been coated to transmission of 1064 nm and 808 nm. The crystal is end-pumped by a 500 W quasi-CW (QCW) fiber-coupled diode (E15F8S22-808.3-500Q-IS39, DILAS, Mainz, Germany), with pump cycle 100 μs at 808 nm. Light from the pump diode is coupled into an 800 μm core fiber, the output of which is focused into Nd:YVO_4 crystal with a spot diameter of 2.4 mm. A plane–concave lens (Lens1) is used for compensating thermal lens of Nd:YVO_4. To achieve compact architecture, a prism is used to reflect light and decrease off-axis angle to keep uniformity of amplified beam. A TFP, an FR, and a HWP are used as an isolator to output amplified pulses.

The pulses are further amplified in a main amplifier which composes of three stages, as illustrated in Figure 1. The main amplifier with single-pass form has high optical efficiency. In consideration of suppression of SRS effect and thermal fracture in gain medium, each amplifier stage has two identical sub-amplifiers. The gain medium in the main amplifier is 0.2 at.% doped Nd:YVO_4 crystals with the same length of 15 mm and double-end-wedged cut at 2°, but have different apertures: crystals at stage 1 are 5 × 5 mm^2, crystals at stage 2 are 6 × 6 mm^2, and crystals at stage 3 are 7 × 7 mm^2. Both sides of the Nd:YVO_4 have been coated to transmission of 1064 nm and 808 nm. Each sub-amplifier was pumped by 50 mJ, 100 µs (10% duty cycle) pulses provided by a 500W QCW fiber coupled diode (E15F8S22-808.3-500Q-IS39, DILAS, Mainz, Germany) operating at 808 nm. Coupling ratios of the pump beam at each amplifier stage are 1:3.7/1:5/1:5.5. Beam expanders (BE2–BE4) are used to mode matching of different stages for good beam quality. Focal length of plane–concave lens (Lens2–Lens4) used for compensating thermal lens of every stage's Nd:YVO_4 crystals are 500 mm, 800 mm, and 1000 mm respectively. A 9 × 9 × 20 mm^3 type I LBO (θ = 90°, ϕ = 10.8°) crystal is used for the extra-cavity frequency doubling.

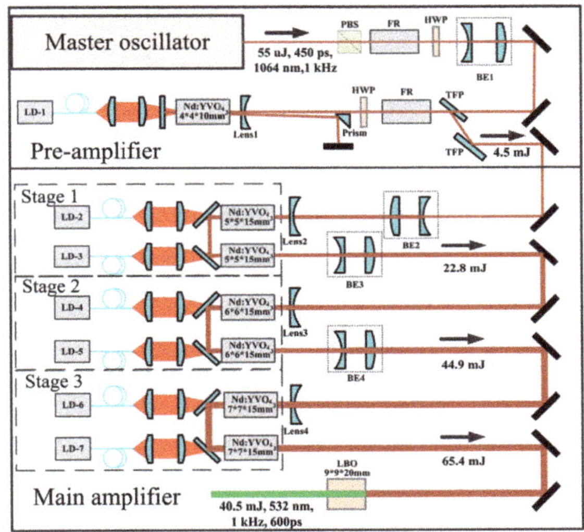

Figure 1. Schematic of the laser system. PBS: polarization beam splitter; FR: Faraday rotator; HWP: half-wave plate; BE: beam expander; TFP: thin-film plate; LD: laser diode.

3. Experimental Results and Discussions

3.1. Pre-Amplifier

A regenerative amplifier has also been used as the pre-amplifier. The regenerative amplifier with Nd:YAG as gain medium could suppress the ASE effect through loss of control by the pockels cell, and achieved higher energy of about 10 mJ. However, the time signal to noise ratio after regenerative amplifier was only 23 dB due to leakage pre-pulses from the pockels cell. The time signal to noise ratio after further amplification decreased to only 10 dB, and the pre-pulse energy was amplified to 7.7 mJ, which could consume energy stored in gain medium, as shown in Figure 2. A pulse picker after regenerative amplification based on electro-optic switches can eliminate pre-pulses and improve the time signal to noise ratio to 40 dB [25], but this will increase system complexity. In order to improve time signal to noise ratio and decrease system complexity, we designed an off-axis four-pass diode-end-pumped Nd:YVO_4 amplification architecture as the pre-amplifier, which could provide high gain with excellent time signal to noise ratio. With an input energy of 55 µJ and 28.6 mJ pump energy into the pre-amplifier stage, up to 4.5 mJ pulse energy at 1 kHz was acheived, and the gain of the pre-amplifier stage could reach 81. With further amplification, the time signal to noise ratio after main amplification was still excellent, which was measured by a photodetector (DET10A, Thorlabs), as shown in Figure 3. To increase the dynamic range of the time signal to noise ratio, the photodetector was overexposed to get the pre-pulse amplitude (2 mV), as shown in Figure 3a. Then the main pulse was reduced to 1/100 with a filter. The amplitude of the main pulse was 385 mV within the linear range of the photodetector. The time signal to noise ratio was better than 42 dB.

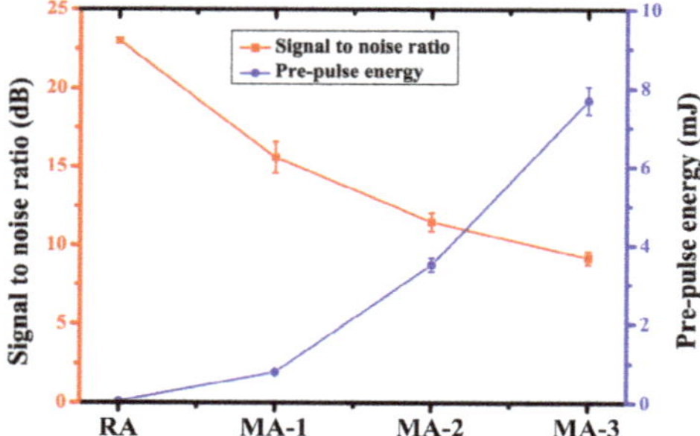

Figure 2. Signal to noise ratio and pre-pulse energy at different amplifiers when a regenerative amplifier was used as the pre-amplifier.

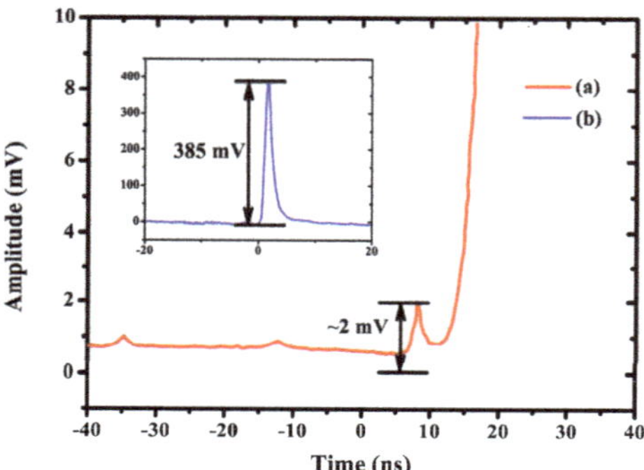

Figure 3. Time signal to noise ratio of maximum output energy: (**a**) Pre-pulse amplitude when the photodetector was overexposed; (**b**) main pulse amplitude when the photodetector was operated in the linear range.

3.2. Main Amplifier and Suppression of SRS effect

In the Nd:YVO_4 crystal of the main amplifier, the maximum peak power of the amplified pulse can reach about 100 MW, which can easily bring about stimulated Raman scattering (SRS). In the progress of SRS, the Stokes wave of Raman light has power gain $exp(G)$. The exponent G is given by

$$G = g_R IL \tag{1}$$

where g_R is the Raman gain, I is the amplified fundamental pulse intensity, and L is the length of the nonlinear Raman medium. The first Stokes wavelength beams will appear when G is 25–30 [31]. The Raman gain of g_R of YVO_4 crystal is 16.13 cm/GW [32]. So when

$$IL \leq 1.55\ \text{GW/cm} \tag{2}$$

the Raman gain in YVO_4 crystal will be below the threshold of SRS progress, and YVO_4 crystal cannot produce the Stokes wave. In our experiment, a 1064 nm laser with intensity of 2.65 GW/cm^2 was once injected into a Nd:YVO_4 crystal with a length of 1.5 cm, which brought about the first Stokes wave (λ = 1175.84 nm), as shown in Figure 4. The Raman shift of Nd:YVO_4 was 890 cm^{-1}, which was almost the same as that of YVO_4 crystal. After passing through Nd:YVO_4 crystal, the energy loss of fundamental 1064 nm laser reached to 20%. Therefore, the generation and accumulation of Raman light must be prevented in pulse amplification. However, this will affect the working parameters of the Nd:YVO_4 amplifier.

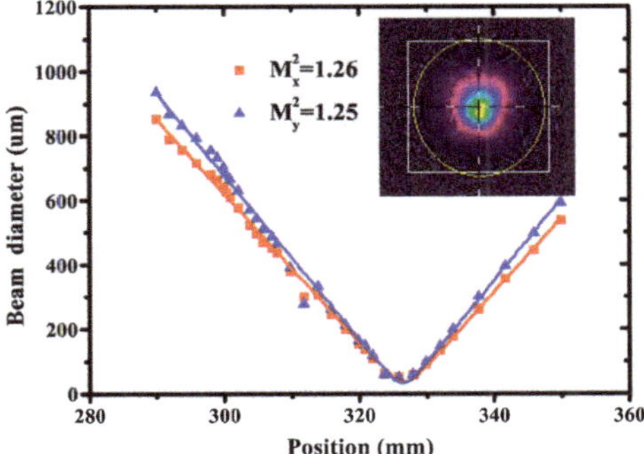

Figure 4. Obtained spectrum of first Stokes wave of Raman light as high intensity laser passing through the Nd:YVO_4 crystal.

The laser intensity continues to increase as the amplification based on the Nd:YVO_4 amplifier increases. To avoid the SRS process, Equation (2) must be satisfied in a high peak power amplifier.

$$\int_0^L I(l)dl \leq 1.55 \text{ GW/cm} \tag{3}$$

What is more, if

$$I_{max}L \leq 1.55 \text{ GW/cm} \tag{4}$$

There will be no significant Raman light. I_{max} is the maximum intensity in the amplification process. So the crystal length L in an amplifier based on Nd:YVO_4 is limited to the maximum amplified intensity for averting the SRS process. At the same time, crystal length L also affects the absorption efficiency of pump light of end-pumped Nd:YVO_4 crystal. To obtain enough absorption efficiency, absorption coefficient α and crystal length L in an end-pumped configuration should satisfy Equation (5) [33].

$$\alpha L = 4 \sim 5 \tag{5}$$

However, absorption coefficient α is an important factor of crystal thermal fracture. According to Reference [33,34], thermal fracture limits pump power for end-pumped Nd:YVO_4 crystal. During the QCW pump operation, the generally accepted way to avoid thermal fracture is pumping the crystal with an average power less than thermal fracture pump power P_{av}, which is less than the fracture pump power P_{cw} of CW pump operation. The maximum pump power at QCW operation is given by

$$P_{av} = \eta P_{cw} = \eta \frac{1}{\alpha} \frac{4\pi R}{\xi} \tag{6}$$

η is the correction factor, which depends on QCW pump duty cycle and repetition, considering a QCW pump duty cycle of 10% in Nd:YVO_4 crystal, $\eta \approx 0.6$ [34]. R is thermal shock parameter depending on the mechanical and thermal properties of the host material. ξ is the fractional thermal loading. When Nd:YVO_4 crystal is pumped with 808 nm, $\frac{4\pi R}{\xi}$ is about 250 W/cm [33]. Based on Equation (6), we can see that thermal fracture pump power P_{av} strongly depends on α. Lowering the dopant concentration can decrease the absorption coefficient and increase the fracture-limited pump power. But in high peak power amplifier, suppression of the SRS process and pump light absorption efficiency also limit the min absorption coefficient. With Equations (4)–(6), to avert the SRS process, when operating the QCW pump, the thermal fracture pump power P_{av} depends on the pulse intensity and can be given by:

$$P_{av} \simeq 100\frac{\eta}{I_{max}} \tag{7}$$

Figure 5 shows the relationship between the maximum pump power and the maximum laser intensity at QCW pump duty cycle of 10% and 1 kHz level for Nd:YVO_4 amplifier, which is described as Equation (7). It can be seen that, the higher amplified intensity is the lower the maximum pump power will be. When the amplified intensity is 1 GW/cm^2, the maximum pump power approaches 60 W. But when the amplified intensity increases to 4 GW/cm^2, the maximum pump power decreases to 15 W. At high peak power amplifier, the diode-end-pumped power for bulk Nd:YVO_4 is limited to suppression of the SRS process and thermal fracture of bulk Nd:YVO_4 crystal. Therefore, the amplification ability of an end-pumped Nd:YVO_4 amplifier is limited to the laser intensity.

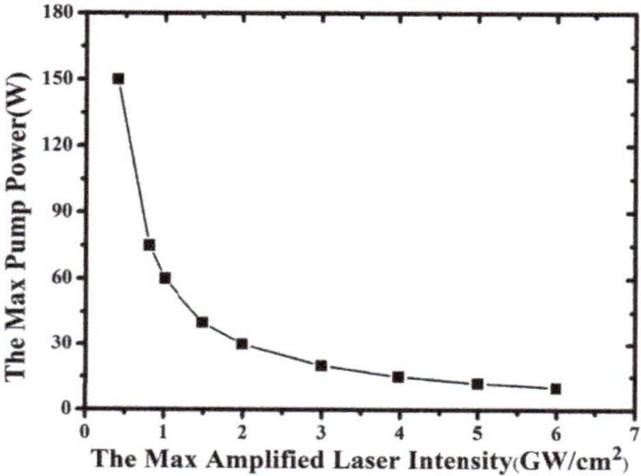

Figure 5. The relationship between the maximum pump power and the maximum laser intensity in an end-pumped Nd:YVO_4 amplifier at a QCW pump duty cycle of 10%.

In the main amplifier, the absorption efficiency of Nd:YVO_4 (0.2%, 1.5 cm) with absorption coefficient was 2.5 cm^{-1} and could reach 95%. Based on Equation (6), the calculated maximum average pump power was about 60 W at 1 kHz with QCW pump duty cycle of 10%, so each sub-amplifier pumped by 50 mJ pulses can work safely. Table 1 shows parameters of the picosecond amplification stages. With an input energy of 4.5 mJ and a total of 300 mJ pump energy into the main amplifier, we achieved maximum output energy of 65.4 mJ, corresponding to an optical efficiency of 20.3%. The spot size in different stages were gradually increased to ensure the laser intensity was below 1 GW/cm^2 and reduce the Raman gain. At different amplified stages, $I_{max} * L_{crystal}$ in each stage had the maximum value of 1.31 and below the SRS threshold of 1.55. In addition, catadioptric mirrors between the two amplifiers are anti-reflective coated to reduce the accumulation of Raman light. Via these measures, we achieved tens of mJ energy sub-ns pulse through a Nd:YVO_4 amplifier without the SRS process, as shown in Figure 6. There is no Raman light at a wavelength of 1176 nm (first Stokes wave) and wavelength 1313 nm (second Stokes wave) in the maximum laser output.

Table 1. Parameters of the sub-ns amplification stages.

Stage	E_{pump} [mJ]	E_{out} [mJ]	D_{out} [mm]	η_{o-o} [%]	I_{max} [GW/cm^2]	$L_{crystal}$ [cm]	$I_{max} * L_{crystal}$ [GW/cm]
Seed	-	0.055	-	-	-	-	-
Pre-amplifier	28.6	4.5	2.0	16.7	0.32	1.0	0.32
Main amplifier 1	100	22.8	2.6	18.3	0.72	1.5	1.08
Main amplifier 2	100	44.9	3.4	22.1	0.82	1.5	1.23
Main amplifier 3	100	65.4	4.0	20.5	0.87	1.5	1.31

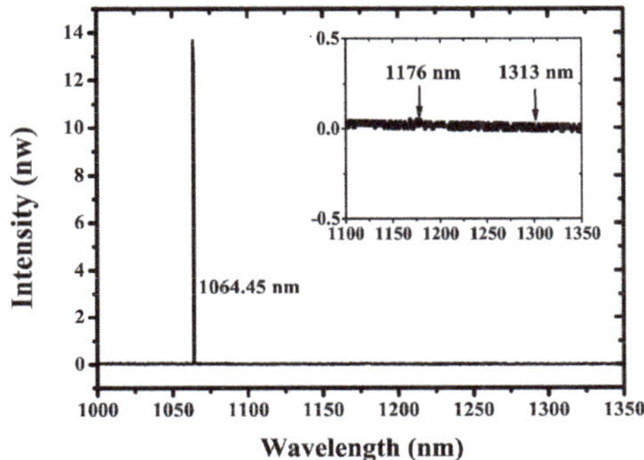

Figure 6. Obtained spectrum of the maximum laser output.

Figure 7 shows the spectra of the oscillator and amplifying systems. This sub-nanosecond oscillator had several longitudinal modea. The main longitudinal mode got more amplification, and, after pulse amplification, spectrum bandwidth became narrower. Narrower spectrum caused by gain narrowing would increase pulse width. A change of pulse width after amplification was shown in Figure 8. The pulse from the oscillator and amplifying systems was measured by a 8 GHz analog oscilloscope and an InGaAs photodiode (UPD-70-IR2-P, ALPHALAS). The pulse width of the oscillator is 450 ps. The pulse width of amplifying systems was 600 ps, corresponding to 109 MW pulse peak power.

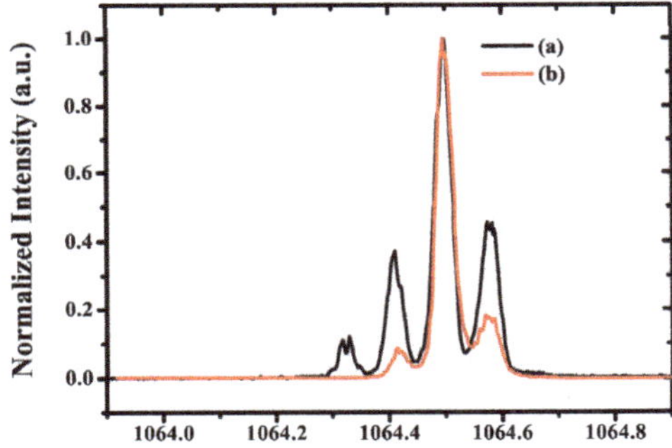

Figure 7. Pulse spectrum of: (**a**) the oscillator (**b**) via the amplifying system.

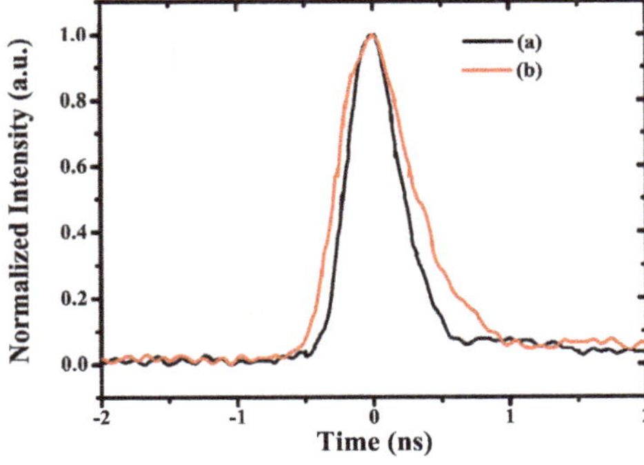

Figure 8. Pulse temporal width of: (**a**) the oscillator (**b**) via the amplifying system.

The beam images of this amplifying system are shown in Figure 9. The beam images were measured with a commercial CCD setting at a position which was about 1 m away from each amplifier. The laser beam of pre-amplifier had good Gaussian distribution. With the increment of pulse energy in the main amplifier, beam center intensity gradually increased due to self-focusing, which could lead to little beam deterioration, as illustrated in Figure 10. The output beam of the oscillator was near diffraction limitation with $M_x^2 \times M_y^2 = 1.28 \times 1.20$. The M^2 for the horizontal and vertical axes via the amplifying system were 1.50 and 1.57, respectively. The results show the amplifier system still kept good beam quality using diode-end-pumped Nd:YVO_4 amplifier architecture when running at 1 kHz.

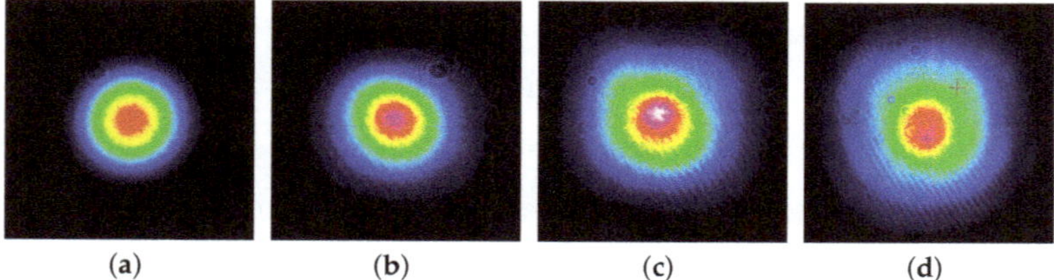

Figure 9. Beam images of each stage. (**a**) pre-amplifier; (**b**) stage 1 in main amplifier; (**c**) stage 2 in main amplifier; (**d**) stage 3 in main amplifier.

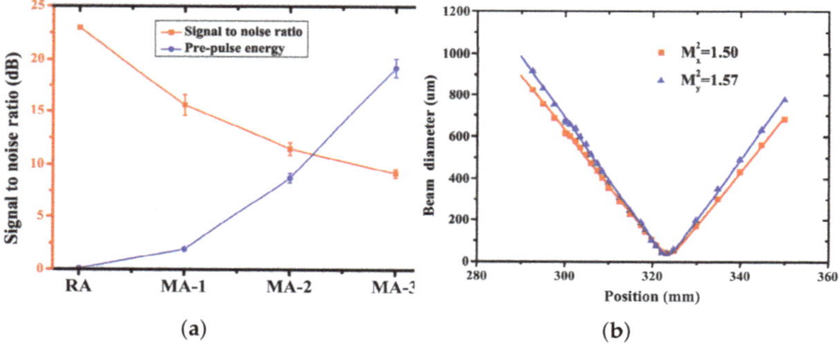

Figure 10. Beam quality factor M^2 of (**a**) the oscillator (**b**) via the amplifying system.

A 9 × 9 × 20 mm^3 type I LBO (θ = 90°, ϕ = 10.8°) crystal was used for extra-cavity frequency doubling. Maximum pulse 532 nm green light energy of 40.5 mJ was obtained at the maximum pump energy. The conversion efficiency was 61.8% and corresponding green light power was 40.5 W. The one-hour power stability has thus been tested, and 0.28% RMS was achieved, as shown in Figure 11. Figure 12 shows beam quality of green light were 1.26 and 1.25 for the horizontal and vertical axes. The beam distribution is also showed as the color image insets in Figure 12. The beam quality of green light was better than that of fundamental light, which is because the nonlinear process during frequency doubling can filter out the stray light with low peak power.

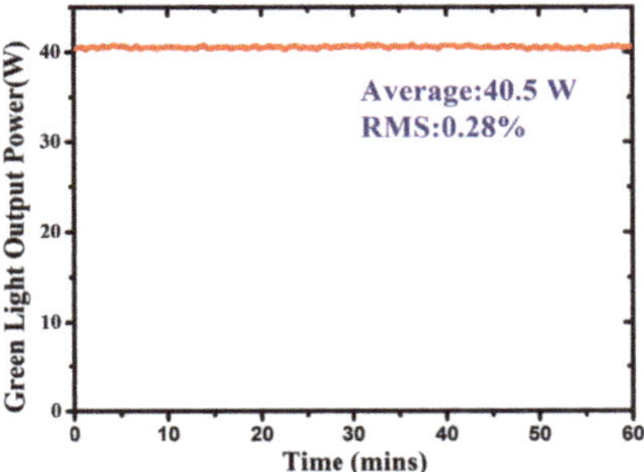

Figure 11. Energy stability of amplified pulses from the whole amplifier.

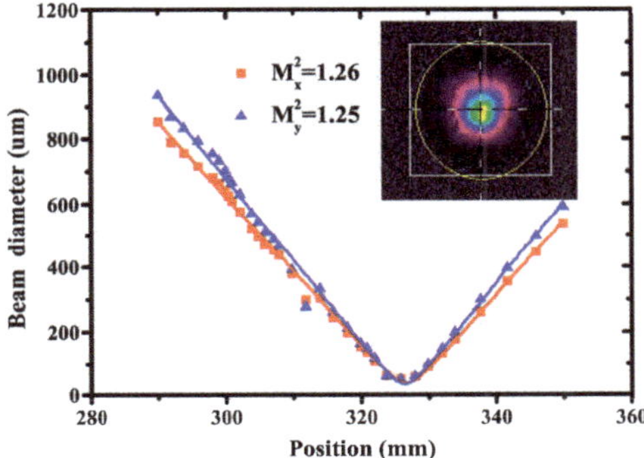

Figure 12. Beam quality factor M^2 of green light. The insets show the beam distribution of green light.

3.3. Conclusions

In conclusion, we have demonstrated a compact and efficient sub-nanosecond diode-end-pumped Nd:YVO_4 laser system. An off-axis four-pass diode-end-pumped Nd:YVO_4 amplification architecture was designed as a pre-amplifier. The main amplifier is composed of three stages, which include two single pass end-pumped Nd:YVO_4 sub-amplifiers. A maximum output energy of 65.4 mJ at 1 kHz without the stimulated Raman scattering (SRS) process was obtained with pulse duration of 600 ps, corresponding to 109 MW peak power. Good beam quality was obtained with $M^2 < 1.6$, and the excellent signal to noise ratio was more than 42 dB. Using LBO crystal, 40.5 W of 532 nm green light and a 0.28% power stability (RMS) were achieved. We also analyzed the influence of the SRS process on working parameters of diode-end-pumped bulk Nd:YVO_4 amplifier. At the high peak power amplifier, the limitation of diode-end-pumped pump power of bulk Nd:YVO_4 crystal due to suppression of the SRS process and thermal fracture. Our work offers an approach to estimate the maximum amplified ability of end-pumped Nd:YVO_4 crystal for high peak power laser amplification. This end-pumped Nd:YVO_4 sub-nanosecond laser system with high energy and good beam quality has simple and compact structure, which makes this laser an attractive light source for remote laser ranging.

Author Contributions: We confirm that all authors contributed substantially to the reported work. Y.H. was the main originator of this study and conceived most of the experiments; H.Z.; X.Y. performed the experiments and analyzed the data; Z.K. and F.L. participated in the research design and laser system adjustment; Z.F. supervised the project and provided the facilities. All the authors discussed and interpreted the results. All the authors read the final manuscript.

Acknowledgments: The work was supported by the project of National a thousand program from Academy of opto-electronics, Chinese academy of sciences (Y80B22A13Y) and the Development of High Power Nanosecond Laser Precision & Detecting Instrument Foundation (Grant No.ZDYZ2013-2).

Conflicts of Interest: The authors declare no conflict of interest.

Abbreviations

The following abbreviations are used in this manuscript:

MDPI	Multidisciplinary Digital Publishing Institute
DOAJ	Directory of open access journals
TLA	Three letter acronym
LD	linear dichroism

References

1. Ostermeyer, M.; Kappe, P.; Menzel, R.; Wulfmeyer, V. Diode-pumped Nd: YAG master oscillator power amplifier with high pulse energy, excellent beam quality, and frequency-stabilized master oscillator as a basis for a next-generation lidar system. *Appl. Opt.* **2005**, *44*, 582–590. [CrossRef] [PubMed]
2. Zhai, D.; Li, Y.; Xu, R.; Fu, H.; Zhang, H.; Li, Z.; Xiong, Y. Design and Realization of Single Telescope Transmitting and Twin Receiving Laser Ranging System at Yunnan Observatories. *Astron. Res. Technol.* **2017**, *14*, 310–316.
3. O'Neill, W.; Li, K. High-quality micromachining of silicon at 1064 nm using a high-brightness MOPA-based 20-W Yb fiber laser. *IEEE J. Sel. Top. Quantum Electron.* **2009**, *15*, 462–470. [CrossRef]
4. Deyra, L.; Martial, I.; Balembois, F.; Diderjean, J.; Georges, P. Megawatt peak power, 1 kHz, 266 nm sub nanosecond laser source based on single-crystal fiber amplifier. *Appl. Phys. B* **2013**, *111*, 573–576. [CrossRef]
5. Major, A.; Sukhoy, K.; Zhao, H.; Lima, I. Green sub-nanosecond microchip laser based on BiBO crystals. *Laser Phys.* **2011**, *21*, 57–60. [CrossRef]
6. Petrov, V.; Marchev, G.; Schunemann, P.G.; Tyazhev, A.; Zawilski, K.T.; Pollak, T.M. Sub-nanosecond, 1-kHz, temperature-tuned, non-critical mid-IR OPO based on CdSiP 2 crystal pumped at 1064 nm. In Proceedings of the Conference on Lasers and Electro-Optics, San Jose, CA, USA, 16–21 May 2010.
7. Chuchumishev, D.; Gaydardzhiev, A.; Fiebig, T.; Buchvarov, I. 0.7 mJ, 0.5 kHz Mid-IR Tunable PPSLT Based OPO Pumped at 1064 nm. In Proceedings of the Advanced Solid-State Photonics, San Diego, CA, USA, 29 January–1 February 2012.
8. Marchev, G.; Petrov, V.; Tyazhev, A.; Pasiskevicius, V.; Thilmann, N.; Laurell, F.; Buchvarov, I. Sub-nanosecond, 1-kHz, low-threshold, non-critical OPO based on periodically-poled KTP crystal pumped at 1064 nm. In *Nonlinear Frequency Generation and Conversion: Materials, Devices, and Applications XI*; International Society for Optics and Photonics: San Francisco, CA, USA, 15 February 2012; Volume 8240, p. 82400D.
9. Chuchumishev, D.; Gaydardzhiev, A.; Richter, C.; Buchvarov, I. 5 mJ, sub-nanosecond PPSLT OPA at 0.5 kHz, tunable in the water absorption band at 3 microns. In Proceedings of the IEEE 2013 Conferenceon and International Quantum Electronics Conference Lasers and Electro-Optics Europe (CLEO EUROPE/IQEC), Munich, Germany, 12–16 May 2013.
10. Chuchumishev, D.; Gaydardzhiev, A.; Fiebig, T.; Buchvarov, I. Subnanosecond, mid-IR, 0.5 kHz periodically poled stoichiometric LiTaO 3 optical parametric oscillator with over 1 W average power. *Opt. Lett.* **2013**, *38*, 3347–3349. [CrossRef]
11. Kyusho, Y.; Arai, M.; Mukaihara, K.; Yamane, T.; Hotta, K.; Kuwano, Y.; Saito, S. High-energy subnanosecond compact laser system with diode-pumped, Q-switched Nd: YVO_4 laser. In Proceedings of the Advanced Solid State Lasers, San Francisco, CA, USA, 31 January 1996.
12. Gaydardzhiev, A.; Trifonov, A.; Fiebig, T.; Buchvarov, I. High-power diode pumped Nd: YAG master oscillator power amplifier system. In *International Conference on Ultrafast and Nonlinear Optics 2009*; International Society for Optics and Photonics: Bourgas, Bulgaria, 2009; Volume 7501, p. 750105.
13. Yu, A.; Krainak, M.; Betin, A.; Hendry, D.; Hendry, B.; Sotelo, C. Highly Efficient Yb: YAG Master Oscillator Power Amplifier Laser Transmitter for Lidar Applications. In Proceedings of the Conference on Lasers and Electro-Optics 2012, San Jose, CA, USA, 6–11 May 2012.
14. Druon, F.; Balembois, F.; Georges, P.; Brun, A. High-repetition-rate 300-ps pulsed ultraviolet source with a passively Q-switched microchip laser and a multipass amplifier. *Opt. Lett.* **1999**, *24*, 499–501. [CrossRef]
15. Höiminger, C.; Zhang, G.; Moser, M.; Keller, U.; Johannsen, I.; Giesen, A. Diode-pumped thin disc Yb:YAG regenerative amplifier. In Proceedings of the Advanced Solid State Lasers, Coeur d'Alene, ID, USA, 2 February 1998.

16. Kaksis, E.; Andriukaitis, G.; Flöry, T.; Pugžlys, A.; Baltuška, A. 30-mJ 200-fs cw-Pumped Yb:CaF2 Regenerative Amplifier. In Proceedings of the Conference on Lasers and Electro-Optics, San Jose, CA, USA, 5–10 June 2016.
17. Agnesi, A.; Dallocchio, P.; Pirzio, F.; Reali, G. Sub-nanosecond single-frequency 10-kHz diode-pumped MOPA laser. *Appl. Phys. B* **2010**, *98*, 737–741. [CrossRef]
18. Jelínek, M.; Kubeček, V.; Čech, M.; Hiršl, P. 0.8 mJ quasi-continuously pumped sub-nanosecond highly doped Nd: YAG oscillator-amplifier laser system in bounce geometry. *Laser Phys. Lett.* **2011**, *8*, 205. [CrossRef]
19. Michailovas, K.; Smilgevičius, V.; Michailovas, A. High average power effective pump source at 1 kHz repetition rate for OPCPA system. *Lith. J. Phys.* **2014**, *54*. [CrossRef]
20. Liu, J.; Wang, W.; Wang, Z.; Lv, Z.; Zhang, Z.; Wei, Z. Diode-pumped high energy and high average power all-solid-state picosecond amplifier systems. *Appl. Sci.* **2015**, *5*, 1590–1602. [CrossRef]
21. Oreshkov, B.; Chuchumishev, D.; Iliev, H.; Trifonov, A.; Fiebig, T.; Richter, C.P.; Buchvarov, I. 52-mJ, kHz-Nd: YAG laser with diffraction limited output. In Proceedings of the IEEE 2014 Conference on Lasers and Electro-Optics (CLEO), San Jose, CA, USA, 8–13 June 2014; pp. 1–2.
22. Chuchumishev, D.; Gaydardzhiev, A.; Trifonov, A.; Buchvarov, I.C. 13-mJ, single frequency, sub-nanosecond Nd: YAG laser at kHz repetition rate with near diffraction limited beam quality. In Proceedings of the CLEO: Applications and Technology, San Jose, CA, USA, 6–11 May 2012.
23. Honig, J.; Halpin, J.; Browning, D.; Crane, J.; Hackel, R.; Henesian, M.; Peterson, J.; Ravizza, D.; Wennberg, T.; Rieger, H.; et al. Diode-pumped Nd: YAG laser with 38 W average power and user-selectable, flat-in-time subnanosecond pulses. *Appl. Opt.* **2007**, *46*, 3269–3275. [CrossRef] [PubMed]
24. Dong, J.; Liu, X.S.; Peng, C.; Liu, Y.Q.; Wang, Z.Y. High Power Diode-Side-Pumped Q-Switched Nd: YAG Solid-State Laser with a Thermoelectric Cooler. *Appl. Sci.* **2015**, *5*, 1837–1845. [CrossRef]
25. Bai, Z.A.; Fan, Z.W.; Bai, Z.X.; Lian, F.Q.; Kang, Z.J.; Lin, W.R. Optical fiber pumped high repetition rate and high power Nd: YVO_4 picosecond regenerative amplifier. *Appl. Sci.* **2015**, *5*, 359–366. [CrossRef]
26. Shen, Y.; Zhang, W.; Gong, M.; Meng, Y.; Wang, Y.; Fu, X. Four-Dimensional Thermal Analysis of 888 nm Pumped Nd: YVO_4 Dual-Rod Acousto-Optic Q-Switched Laser. *Appl. Sci.* **2017**, *7*, 470. [CrossRef]
27. Bai, Z.; Bai, Z.; Kang, Z.; Lian, F.; Lin, W.; Fan, Z. Non-Pulse-Leakage 100-kHz Level, High Beam Quality Industrial Grade Nd: YVO_4 Picosecond Amplifier. *Appl. Sci.* **2017**, *7*, 615. [CrossRef]
28. Li, R.; Bauer, R.; Lubeigt, W. Continuous-Wave Nd:YVO_4 self-Raman lasers operating at 1109 nm, 1158 nm and 1231 nm. *Opt. Express* **2013**, *21*, 17745–17750. [CrossRef]
29. Huang, G.; Yu, Y.; Xie, X.; Zhang, Y.; Du, C. Diode-pumped simultaneously Q-switched and mode-locked YVO_4/Nd: YVO 4/YVO_4 crystal self-Raman first-Stokes laser. *Opt. Express* **2013**, *21*, 19723–19731. [CrossRef]
30. Li, R.; Griffith, M.; Laycock, L.; Lubeigt, W. Controllable continuous-wave Nd:YVO_4 self-Raman lasers using intracavity adaptive optics. *Opt. Lett.* **2014**, *39*, 4762–4765. [CrossRef]
31. Kaminskii, A.A.; Ueda, K.i.; Eichler, H.J.; Kuwano, Y.; Kouta, H.; Bagaev, S.N.; Chyba, T.H.; Barnes, J.C.; Gad, G.M.; Murai, T.; et al. Tetragonal vanadates YVO_4 and $GdVO_4$–new efficient χ (3)-materials for Raman lasers. *Opt. Commun.* **2001**, *194*, 201–206. [CrossRef]
32. Zong, N.; Zhang, X.; Li, C.; Cui, D.; Xu, Z.; Zhang, H.; Wang, J. Stimulated Raman scattering of picosecond pulses in a YVO_4 crystal. *Laser Phys.* **2008**, *18*, 1544–1545. [CrossRef]
33. Chen, Y.F. Design criteria for concentration optimization in scaling diode end-pumped lasers to high powers: influence of thermal fracture. *IEEE J. Quantum Electron.* **1999**, *35*, 234–239. [CrossRef]
34. Bernhardi, E.; Forbes, A.; Bollig, C.; Esser, M.D. Estimation of thermal fracture limits in quasi-continuous-wave end-pumped lasers through a time-dependent analytical model. *Opt. Express* **2008**, *16*, 11115–11123. [CrossRef] [PubMed]

MDPI
St. Alban-Anlage 66
4052 Basel
Switzerland
Tel. +41 61 683 77 34
Fax +41 61 302 89 18
www.mdpi.com

Applied Sciences Editorial Office
E-mail: applsci@mdpi.com
www.mdpi.com/journal/applsci

www.ingramcontent.com/pod-product-compliance
Lightning Source LLC
LaVergne TN
LVHW071033170726
843515LV00006B/1277

* 9 7 8 3 0 3 6 5 0 8 6 6 5 *